T0256503

SPECIAL PAPERS IN PALAEONTOLOGY NO. 73

CONODONT BIOLOGY AND PHYLOGENY: INTERPRETING THE FOSSIL RECORD

EDITED BY

MARK A. PURNELL

and PHILIP C. J. DONOGHUE

TECHNICAL EDITING BY

DAVID J. BATTEN

with 3 plates, 9 tables and 79 text-figures

THE PALAEONTOLOGICAL ASSOCIATION

LONDON

March 2005

CONTENTS

[Special Papers in Palaeontology, 73, 2005, pp. 5]

PREFACE

The collection of papers in this volume arose directly from a symposium entitled Bias and Completeness in the Conodont Fossil Record held at ECOS VIII (Eighth International Conodont Symposium held in Europe) in Toulouse during 2002. But the idea to bring together conodont workers to discuss the nature and quality of the conodont fossil record had a longer period of gestation and followed from numerous discussions during the excellent field excursions of the 1998 ECOS VII in Italy, some debate on the con-nexus conodont listserver (http://www.conodont.net), and a conversation in a hire-car between Flagstaff and Phoenix. We would like to thank: the organisers of these meetings, especially Marie-France Perret-Mirouse, organiser of ECOS VIII; The Palaeontological Association, whose financial support allowed us to invite people who would otherwise not have been able to participate in the symposium; and all those who contributed to the stimulating and sometimes heated exchange of ideas regarding the conodont fossil record in the years leading up to the publication of this volume.

The symposium and the discussion that arose from it were well received by the assembled audience in Toulouse, and we are grateful to all those who accepted our invitations and contributed talks on the day (Dick Aldridge; Howard Armstrong; Chris Barnes; Jim Barrick and Peep Männik; Tom Broadhead and John Repetski; Jerzy Dzik; Lennart Jeppsson; Oliver Lehnert, Jim Miller and Steve Leslie; Glen Merrill; Mike Orchard; Peter Roopnarine, Mike Murphy and N. Buening; Immo Schülke; Peter von Bitter). As will be clear from a perusal of the contents of this volume, many of these contributors also agreed to the more onerous task of committing their work to paper; what is less obvious is that we also relied on their willingness to act as reviewers. Without these efforts this volume could not have been published.

MARK A. PURNELL
Department of Geology, University of Leicester, University Road, Leicester LE1 7RH, UK

and PHILIP C. J. DONOGHUE
Department of Earth Sciences, University of Bristol, Wills Memorial Building, Queens Road, Bristol BS8 1RJ, UK

[Special Papers in Palaeontology, 73, 2005, pp. 7–25]

BETWEEN DEATH AND DATA: BIASES IN INTERPRETATION OF THE FOSSIL RECORD OF CONODONTS

by MARK A. PURNELL* *and* PHILIP C. J. DONOGHUE†

*Department of Geology, University of Leicester, University Road, Leicester LE1 7RH, UK; e-mail: map2@le.ac.uk
†Department of Earth Sciences, University of Bristol, Wills Memorial Building, Queens Road, Bristol BS8 1RJ, UK; e-mail: phil.donoghue@bristol.ac.uk

Abstract: The fossil record of conodonts may be among the best of any group of organisms, but it is biased nonetheless. Pre- and syndepositional biases, including predation and scavenging of carcasses, current activity, reworking and bioturbation, cause loss, redistribution and breakage of elements. These biases may be exacerbated by the way in which rocks are collected and treated in the laboratory to extract elements. As is the case for all fossils, intervals for which there is no rock record cause inevitable gaps in the stratigraphic distribution of conodonts, and unpreserved environments lead to further impoverishment of the recorded spatial and temporal distributions of taxa. On the other hand, because they are resistant to abrasion and can withstand considerable metamorphism conodonts can preserve evidence of otherwise lost sequences or environments through reworking.

We have conducted a preliminary investigation into how the various forms of gross collecting bias arising from period to period variation in intensity of research effort and in preserved outcrop area have affected the conodont fossil record. At the period level, we are unable to reject the hypothesis that sampling, in terms of research effort, is biased. We have also found evidence of a relationship between outcrop area and standing generic diversity. Analysis of epoch/stage-level data for the Ordovician–Devonian interval suggests that there is generally no correspondence between research effort and generic diversity, and more research is required to determine whether this indicates that sampling of the conodont record has reached a level of maturity where few genera remain to be discovered. One area of long-standing interest in potential biases and the conodont record concerns the pattern of recovery of different components of the skeleton through time. We have found no evidence that the increasing abundance of P elements relative to S and M elements in later parts of the conodont record reflects evolutionary changes in the composition of the apparatus.

Ignoring the biases and incompleteness of the conodont fossil record will inevitably lead to unnecessary errors and misleading or erroneous conclusions. Taking biases into account has the potential to enhance our understanding of conodonts and their application to geological and biological questions of broad interest.

Key words: completeness, gaps, microfossil, preservation, taphonomy, vertebrate skeletons.

OUR purpose with this contribution is to introduce and provide an overview of an issue that underlies all palaeontological study and provides the common theme of this collection of papers: how we interpret the fossil record. To what extent do perceived changes in morphology, skeletal composition, abundance and diversity through time reflect changes in biology and evolutionary history and how has this primary signal been biased by post-mortem processes? How do biases affect the ways in which we use the record for evolutionary, biological and biostratigraphic purposes?

Conodonts provide a particularly interesting window through which to view the sometimes uneasy relationship between interpretations of the fossil record and hypotheses of bias. The quality of the conodont fossil record is generally held to be among the best of any group of organisms (Foote and Sepkoski 1999; Sweet and Donoghue 2001), and because of their near ubiquity and ease of recovery from marine strata of Late Cambrian to latest Triassic age conodonts have attained an almost unrivalled reputation for biostratigraphic utility. This in turn has fuelled a widespread tacit assumption that because conodont biostratigraphy works, biases in their fossil record cannot be significant (Donoghue 2001*a*,*b*; Wickström and Donoghue, 2005). The record must be 'close enough' to the original signal.

Yet few would argue that post-mortem factors have not played some role in shaping what we see, and it must therefore be true that if the record as we perceive it reflects both biological-evolutionary patterns and post-mortem biases, failure to take both into account will decrease the reliability and accuracy of any interpretations. Every fossil sample lies somewhere in a spectrum that ranges from complete preservation to complete loss, and the papers in this volume explore the fertile ground of interactions between bias and biology.

Biases, for the purpose of this paper, are taken as factors that distort or selectively filter the patterns of spatial and temporal distribution of fossils, as revealed through analysis of collections, causing them to deviate from a perfect record of 'true' biological and evolutionary history. This is more than taphonomy, as we include other biasing factors such as sampling, collecting and processing methods, and consider how assumptions and methods, especially phylogenetic methods, can bias interpretations.

Any simple classification will inevitably underemphasize the complex interactions and feedbacks that occur between biases, but in order to provide a framework for discussion and to be consistent with the overall structure of this collection of papers we consider biases primarily in terms of when they exert their influence, as summarized in Text-figure 1. It is important to note that not all conodonts are equally susceptible to different biases, with various aspects of conodont biology and evolution making some species, or some types of elements within the apparatus more likely to be lost. Similarly, different elements or species may be more susceptible to bias at different stages in the transition from death to data, and biases at one stage can make elements more or less susceptible to the effects of bias during subsequent stages. We have attempted to highlight these factors in Text-figure 1 and in the discussion below. We also present a more detailed discussion of the potential effects of biology and bias on the relative abundance of different components of the conodont skeleton in collections of isolated elements.

PRE- AND SYNDEPOSTIONAL NON-PRESERVATION AND SELECTIVE LOSS

Predation and scavenging

Numerous examples of elements or apparatuses preserved within predators and scavengers (Scott 1969, 1973; Melton and Scott 1973; Nicoll 1977; Conway Morris 1990; Purnell 1993; Purnell and Donoghue 1998) or in coprolites (Higgins 1981) demonstrate that conodonts were food for other animals. It is possible that many elements in conodont collections have passed through the guts of other animals, but there has been no systematic study of how this may have affected what is preserved. Given the well-known effects on the enamel of gnathostome teeth of passage through a gut (Fisher 1981), it is likely that conodont elements, composed primarily of enamel-like tissues (Donoghue 1998, 2001*c*), could be partially or completely dissolved in the process of digestion by some conodont eaters. Fragmentation is also possible. Species with small elements, or the more gracile

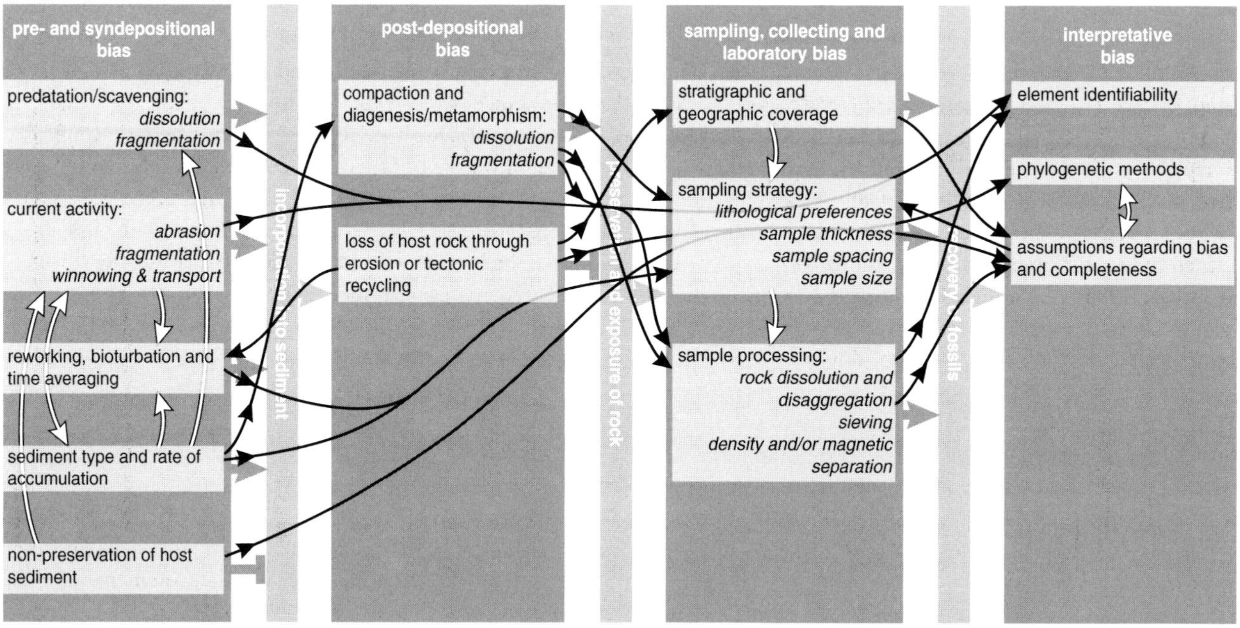

TEXT-FIG. 1. Biases that act to distort recovery of conodont elements. The diagram summarizes when different biases exert their influence and indicates how they interact.

elements in an apparatus are more likely to be lost or fragmented during digestion. Compaction of a coprolitic mass may also result in higher levels of fragmentation because of the close juxtaposition of elements. On the other hand, incorporation of elements into a coprolite may enhance their chances and quality of preservation if it is mineralized or lithified before significant sedimentary compaction.

Current activity

That conodont elements were subject to post-mortem current sorting has long been recognized (e.g. Ellison 1968; von Bitter 1972). More recent experimental work confirmed that the susceptibility of elements to current entrainment, transport and sorting is correlated with their hydrodynamic properties, which in turn are correlated with size and shape (Broadhead *et al.* 1990; McGoff 1991). Studies by Broadhead and Driese (1994) indicated that elements carried in aqueous suspension with carbonate grains are relatively resistant to abrasion and are unlikely to be destroyed, even after prolonged transport. Simulated aeolian transport with quartz grains, however, resulted in significant abrasion of elements. This work also suggests that current activity does not cause significant breakage of elements. Current sorting is likely to amplify the effects of any differential fragmentation of elements resulting from other predepositional factors, such as predation, leading to increased levels of bias.

Reworking, bioturbation and time-averaging

The effects of bioturbation on element fragmentation are unknown. Its potential for producing time-averaged faunas, however, is beyond doubt. Of particular concern is the fact that bioturbation may have been most intense where it is least evident; the lack of any clear burrowing may indicate that a bed and its conodont elements have been completely homogenized by bioturbation (Droser and Bottjer 1986), possibly resulting in the amalgamation of depositional events (and conodont populations) spanning many thousands of years. Reworking and winnowing may also lead to time-averaging and differential size bias, and because they are relatively resistant elements may be reworked following erosion of their host rock (see below). Reworking and bioturbation may produce conodont faunas of mixed age and environmental affinities, and, given the evidence of time-averaging in macrofossil groups (see Behrensmeyer *et al.* 2000 for a review), it is highly likely that most conodont faunas are significantly time-averaged, possibly representing tens of thousands or even hundreds of thousands of years. This has obvious implications for temporal and spatial resolution of interpretations that draw directly on stratigraphic ordering of fossils (see Barrick and Männik 2005; Dzik 2005; Roopnarine 2005). Reworking is also likely to result in significant element size bias.

Transport of elements, either all the elements of a species or just the more easily entrained components of the skeleton, may result in their removal to different depositional settings. This can ultimately result in their complete loss from the record if those environments are less likely to be collected (lithological sampling bias), are more difficult or impossible to process effectively, or are more likely to be subject to tectonic recycling. Quiet, offshore, deep-water environments are particularly susceptible to these biases.

Sediment type and rate of accumulation

Many of the effects of sedimentation on conodont faunas are mediated by other potential biases. Rapid sedimentation, for example, will tend to remove elements more quickly from predators, scavengers and burrowers, reducing the bias arising from these factors. However, high rates of sediment input will result in fewer elements per unit rock volume that, depending on the downstream effects of compaction and sample size, may result in lower element recovery, which in turn can have a significant impact on interpretation (see Jeppsson 2005). Slow net rates of sedimentation will increase the potential for reworking and time-averaging. Interactions between sediment type, compaction and diagenesis also have a significant effect on fragmentation. Shales, some of which may be low-density, soupy sediment at the time of deposition (see Purnell and Donoghue 1998 for a discussion of black shale density and conodont taphonomy), are subject to higher levels of compaction, and thus higher levels of element fragmentation (von Bitter and Purnell 2005). Carbonate sediments, especially framework-supported lithologies such as grainstones, or other sediments liable to rapid cementation, will be less compacted and elements consequently less fragmented (although the high depositional energy of grainstones will result in winnowed and sorted faunas; e.g. Krumhardt *et al.* 1996).

Non-preservation of host sediment

The effects of current activity range from light disturbance of elements within natural assemblages through various degrees of winnowing, transport and hydrodynamic sorting, to complete loss or non-deposition of elements and host sediment. Whether or not conodont elements themselves are removed to a different depositional setting, destroyed or remain as part of a lag deposit will depend on the specific hydrodynamic regime. The net results will vary from time-averaging to loss of record; the downstream effects on interpretation are discussed above.

POST-DEPOSITIONAL BIAS

The various processes that a sedimentary unit undergoes during its incorporation into the stratigraphic record will significantly affect the conodont elements it contains. Compaction, for example, will result in element fragmentation and possible downstream loss (see below), whereas early cementation will reduce fragmentation. Elements are more resistant to the effects of pressure solution than carbonate grains, and this can result in penetration of calcareous fossils by conodont elements. Conodont elements can survive hydrothermal alteration, contact metamorphism and regional metamorphism up to greenschist facies and more (Rejebian *et al.* 1987), but the biases introduced by declining element identifiably increase as elements become more tectonically deformed, recrystallized or covered with mineral encrustations (e.g. Kovács and Árkai 1987; Rejebian *et al.* 1987). Cement mineralogy also exerts a bias that is linked to processing and collection. Rocks cemented with quartz or other minerals that are insoluble in buffered acetic or formic acids are less likely to be collected by conodont workers, leading to significant lithological collecting bias (see below). If collected, such rocks are likely to be processed using more aggressive chemical or mechanical techniques that will tend to increase element loss through dissolution, fragmentation or decreased identifiably (Jeppsson 2005; for illustrations of conodonts recovered using hydrofluoric acid, see Barrick 1987; Orchard 1987).

Loss of rock through erosion or tectonic recycling varies according to tectonic and depositional setting, and sequence architecture. The longest surviving strata are found on stable cratonic areas, continental rift margins and aulacogens (Behrensmeyer *et al.* 2000), but sequences from these areas may be far from complete, containing numerous depositional hiatuses and erosional unconformities. Barrick and Männik (2005) and Lehnert *et al.* (2005) discuss the implications of these factors for analyses of conodont biostratigraphy and evolution.

In most cases loss of a sedimentary unit will result in loss of the elements it contained, but this is not always true. For example, conodont faunas from redeposited clasts or olistoliths have been used to reconstruct otherwise unpreserved inner shelf palaeoenvironments of the Ordovician Cow Head Group of Newfoundland (Pohler *et al.* 1987; Pohler and James 1989), shallow-marine carbonate and flysch sequences from a cryptic Ordovician arc terrane in northern Britain (Armstrong *et al.* 2000), and a lost Devonian carbonate shelf reconstructed on the basis of polymictic clasts in the Viséan of the Holy Cross Mountains (Belka *et al.* 1996). Lehnert *et al.* (2005) discuss more examples.

Element fragmentation

Several pre-, syn- and post-depositional processes (Text-fig. 1) will result in element fragmentation, as will certain collecting and processing methods (see Jeppsson 2005 for discussion). The downstream affects of fragmentation, particularly after sieving or decanting to separate elements from sediment, are potentially huge, with those elements most susceptible to fragmentation being completely lost, or rendered unidentifiable. At the interpretation stage, this can result in the effective loss (through non-recovery) of species with small or gracile elements (Jeppsson 2005) and in itself, without current sorting, is sufficient to bias the relative abundance of element types recovered (von Bitter and Purnell 2005).

Controlling for bias in the sedimentary record

Although the sedimentary rocks within which conodont elements are entombed were accumulated episodically in response to changes in sea level resulting from a combination of eustatic and local effects, the apparent completeness of a particular sedimentary record is relative and contingent upon the time span and the resolution of the time intervals required (Strauss and Sadler 1989; Sadler and Strauss 1990). The coarser the temporal resolution required the more complete a section will be perceived to be over a given time span. Thus, if a sedimentary section has accumulated over a few million years, it will provide a much better record with respect to 100-ky intervals than to 10- or 1-ky intervals. However, sequences deposited over shorter periods of time are generally more complete because the longer the time span the more likely the sedimentary record is to include significant gaps (Sadler 1981; Schindel 1980).

It is possible to overcome the limitations of individual sections through compilation of numerous stratigraphic sections that represent the time span of interest, using the method of graphic correlation for example (Shaw 1964; Sweet 2005). If gaps are distributed randomly throughout the component sections then it is likely that individual sections will compensate for one another and the completeness of the composite section will increase as more sections are included. Valentine *et al.* (1991) calculated the increasing probabilities of completeness for composite sections by dividing the average sediment accumulation rate for the time span of interest by the average rate for the resolution interval (rates were based on comparable modern marine environments; Sadler 1981). Thus, for a 1-my resolution interval, a 30-my time span and average accumulation rates for carbonate sediments, the probability that any given interval is represented by some sediment at one site (at the very least) rises from 0·33 in one section to 0·98 for ten independent sections, with probability increasing still further if either the resolution interval or the number of independent sections is increased. However, this improvement in completeness only applies if the gaps within the component sections are randomly distributed within and therefore between the sections.

Despite the veracity of the global record at the given time span and resolution interval, Valentine *et al.* (1991) also showed that the precision of correlation between component sections is much lower. Given that the probability of any 1-my interval being represented by at least some sediment over a time span of 30-my in a marine carbonate sediment setting is 0·33, the likelihood that the same interval is represented in further independent sections drops to less than 0·11 for two sections and $1{\cdot}53 \times 10^{-5}$ for ten sections. As they argue, this observation should severely limit the application and resolution of biostratigraphic correlation, but this technique is universally applied and 1-my resolution is often achieved and exceeded, particularly with respect to graphically correlated global composite sections (Sweet 2005).

The reason is that the gaps in sedimentary sections are non-independent; they are controlled by regional and eustatic fluctuations in sea level. One consequence is that there are gaps in the record that may never be filled because there are intervals of geological time for which there is no stratigraphic record over large geographical areas (Valentine 2004, p. 159). The conodont record is further impoverished because there are also intervals of geological time for which there is no sedimentary record of the environments in which conodonts lived and died.

So although global composite sections derived from graphic correlation provide the best means of recovering the available record, they will nevertheless include undetected gaps. This has implications for those scientists who attempt to derive the phylogenetic relationships of organisms using stratigraphic range data because we have no means of directly recovering the absolute and sometimes even the relative stratigraphic ranges of taxa (Wickström and Donoghue 2005). Cladistics, on the other hand, because it eschews stratigraphic data in the formulation of phylogenetic hypotheses, can reveal the existence of gaps through post hoc calibration of cladograms to the stratigraphic ranges of their component taxa (Norell 1992; Benton 1995*b*; Donoghue 2001*a*). The logic of this is simple: because sister taxa diverge from their most recent common ancestor at the same time, they should exhibit coincident first appearances in the stratigraphic record. If one member (species A) of a pair appears before the other (species B), the interval between the first recorded appearance of species A and that of B gives a minimum estimate of the unpreserved range of species B. Such gaps are termed 'ghost lineages' (Norell 1992) but ghost ranges can be accounted for not only by a gap in the range of a known taxon, but also by the existence of hitherto undescribed taxa that are more closely related to one of the pair of sister taxa. Wickström and Donoghue (2005) show that it is possible to discriminate between these two possibilities by employing confidence intervals to constrain the expected range of known taxa. Donoghue *et al.* (2003) showed that even existing hypotheses of conodont intrarelationships, which are heavily based on stratigraphic data, indicate that the conodont fossil record is far from complete.

SAMPLING, COLLECTING AND LABORATORY BIAS

Although the potential biases discussed above act and interact in different ways at different times over millions of years, a number of them are only realized as actual biases after a sample has been collected and processed. For these biases, particularly those that increase element fragmentation, different sampling, collecting and laboratory processing can exaggerate or reduce their impact on the collection of conodont elements that results. Much of this is intuitively obvious and has been touched upon in previous discussion: collecting small samples of rock deposited under high rates of sedimentation, for example, will yield few elements; sieving will result in loss of all elements and fragments below the minimum screen size used. Perhaps the dominant factor at this stage, however, is the lithological sampling bias imposed by the preferred method of recovering conodont elements from their host rock; since the 1940s dilute acids have been used to extract elements by dissolution of carbonates (Ellison and Graves 1941). We are aware of no rigorously collected data concerning lithological selection, but estimates based on a straw poll of conodont workers who subscribe to the con-nexus listserver (http://www.conodont.net) indicate that percentages for sample lithologies processed by different laboratories average out at *c.* 80 per cent limestones, *c.* 10–15 per cent dolomites, *c.* 5 per cent shales, and 2 or 3 per cent 'other lithologies'. There is some variation from one laboratory to another, and with different geographical sampling areas, but the data confirm that the vast majority of recent conodont collections are derived from carbonates, with many conodont workers collecting other lithologies only when they have no alternative, or where stratigraphic/ecological coverage demands. This will obviously impose certain biases on conodont collections in general, especially when coupled with the fact that the volume and area of deposition and preservation of these different lithologies varies through the Phanerozoic (Bluth and Kump 1991). It is also worth noting that lithological sampling bias has changed with time. The work of Hass, active between the 1930s and 1960s, and Huddle, active in the 1930s to mid 1940s, then again in the 1960s and 1970s, for example, included a much higher proportion of shale samples, with the majority of Hass's samples being obtained from shales, either by examination of split surfaces or by disaggregation of rock by boiling or slaking. As acid processing became more routine, shale sampling percentages declined because of the increase in opportunity to process carbonate rocks, so that Huddle's overall black shale percentages probably were more like 50 per cent (J. E. Repetski, pers. comm. 2004).

The biases inherent in sampling, collection and laboratory processing are touched on by a number of authors in this volume, but are dealt with most comprehensively by Jeppsson (2005), who also provides suggestions for monitoring, evaluating and minimizing their biasing effects.

Temporal and spatial collecting bias

Stratigraphic collecting bias and research effort through time. In addition to these intrinsic and operational biases, the conodont fossil record is also subject to cultural biases. If we are to read anything into diversity curves through the stratigraphic range of conodonts we must assume that there has been a consistent degree of research effort expended throughout, or at least that the record has been randomly sampled. However, this does not square easily with the fact that most conodont workers specialize on particular stratigraphic intervals, with factors other than the desire for uniform stratigraphic coverage involved in determining their specialism.

To test for sampling biases in the conodont fossil record we searched the ISI® Science Citation Index (SCI) database (1945–July 2004) using the search string <conodont* and [period name]> for each geological period (including Mississippian and Pennsylvanian in totals for the Carboniferous). Because conodonts only occur in the later parts of the Cambrian, we have excluded this period from our analysis. The SCI by no means captures all publications presenting or using conodont data but we have assumed that its sampling of the palaeontological literature does not vary from period to period, and our data should thus provide a representative sample of the number of publications containing conodont data for each of the periods within the stratigraphic range of conodonts. Such publication counts provide a relatively crude measure of research effort, but they do allow us to explore the possibility that the record is biased by uneven sampling through time. The results of this search, presented in Text-figure 2A, demonstrate that the number of publications per period varies considerably, with the Ordovician and Devonian having particularly high research productivity. However, even if research productivity were uniform throughout we should not expect equal productivity on a period-by-period basis because of the varying duration of Phanerozoic periods. Rather we should expect longer periods to exhibit the highest research productivity (in terms of total publications), with the number of publications correlated with period duration. The result of a Spearman's Rank Correlation test ($r_s = 0{\cdot}257$, $P = 0{\cdot}623$), however, indicates that they are not (we have used the timescale of Gradstein *et al.* 2004 throughout). [The non-parametric Spearman's Rank Correlation is used because n is small, and we have no evidence that our data are normally distributed; analyses were carried out using PAST (Hammer *et al.* 2001).] Plotting publications onto duration (Text-fig. 2B) provides graphic confirmation of this lack of correlation. If sampling is uniform through time, then standardizing research effort per unit time (number of publications divided by period duration) should produce a correlation close to zero, but once again these data do not conform to this expectation ($r_s = -0{\cdot}429$, $P = 0{\cdot}396$). This negative correlation is non-significant, but given the limitations of our dataset it is high enough to suggest that there may be a weak negative relationship that better data might reveal. If the correlation were real, it would mean either that longer periods have higher research effort (publications my^{-1}), or that shorter periods have lower effort. Our limited data suggest the latter, but clearly we cannot with any confidence reject the hypothesis that sampling of the conodont record at period level is non-uniform.

However, if we look deeper within the stratigraphic hierarchy for the periods that show the highest absolute and/or standardized research productivity (Ordovician, Silurian, Devonian), a different pattern emerges. The same search strategy was conducted using epoch (for the Ordovician and Silurian) and stage names (for the Devonian) in the search string. We conducted searches for stratigraphic units of unequal rank because some stage names have been in widespread international use for decades while others have been conceived only relatively recently, and we chose to use those stratigraphic names that have been in circulation longest for each period. The use of stratigraphic units of differing rank should not affect the analysis because if the record has been evenly sampled research productivity per unit time should be independent of rank (see below). More importantly, however, in terms of duration there is no clear distinction between the Ordovician–Silurian epochs and Devonian stages. Two of the three shortest intervals are epochs

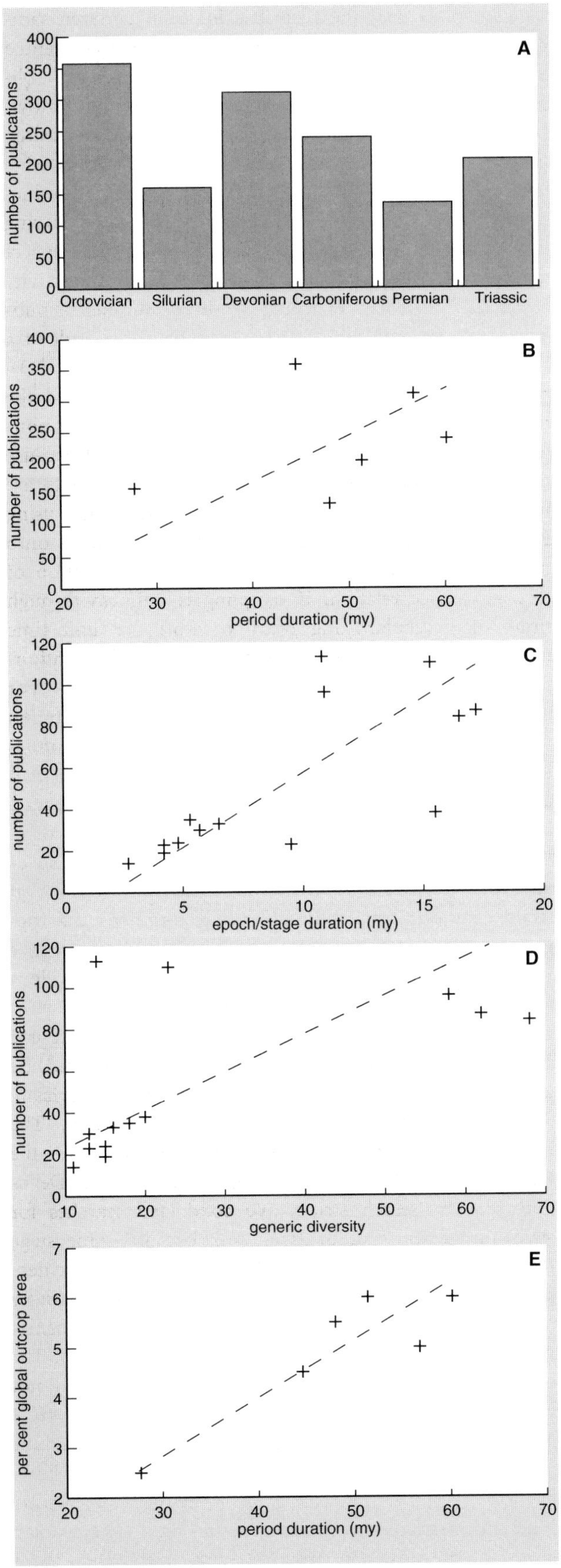

TEXT-FIG. 2. Histogram (A) and bivariate plots of data (B–E) pertaining to analysis of bias in research effort expended on conodonts through geological periods and epochs/stages. Dashed lines indicate the results of RMA regression (calculated with PAST; Hammer *et al.* 2001); they are intended solely as a guide to general trends in the data and cannot be assumed to be significant (for discussion of correlations between variables and significance, see text).

(Přídolí and Ludlow) and the fourth longest interval is a stage. Furthermore, although its reliability is open to question because of the nature of our data, the results of a Welch test (an unequal variance t statistic) suggest that mean stage length (8·11) is not significantly different from mean epoch length (10·33; $t = 0{\cdot}791$, $P = 0{\cdot}447$).

The same caveats apply to these data as those for periods, and once again our analysis should be viewed as exploratory in nature. Nevertheless, the stage/epoch data are better than the period data in that n is higher, and analysis reveals a highly significant correlation between raw research productivity and epoch/stage duration ($r_s = 0{\cdot}798$, $P = 0{\cdot}0006$). Epoch/stage duration and standardized publication data are not correlated ($r_s = 0{\cdot}004$, $P = 0{\cdot}988$). Both these results are what we would expect from uniform sampling through this stratigraphic interval, but there are clearly intervals where publication effort differs markedly from the norm (Text-fig. 2C), and this observation is compatible with a sampling bias of uneven research effort. These intervals cannot simply be ignored, but alternative explanations remain which must be rejected before it can be concluded that sampling of the conodont fossil record is significantly biased.

It is possible that those stratigraphic units exhibiting unusually high and low levels of research productivity are those that exhibit highest and lowest levels of diversity. This relationship has been observed in analyses of the Phanerozoic fossil record as a whole (Raup 1976) and can be interpreted as indicating either that high diversity stimulates high palaeontological interest, or that high levels of activity produce more taxa (Raup 1977; Sheehan 1977); that is, the pattern of diversity is either real, as advocated by Valentine (1969, 1973), or it is an artefact of sampling effort (Raup 1972, 1976). We are not about to try and resolve this ongoing debate (e.g. Benton 1995*a*, 2003; Peters and Foote 2001, 2002; Smith 2001; Smith *et al.* 2001). Rather, our purpose is to determine whether any of those Ordovician–Devonian epochs/stages identified as exhibiting markedly high or low levels of research productivity coincide with intervals of markedly high or low conodont generic diversity. Raup's (1976) database is clearly better than ours in that it is more heavily researched and, being based on the Zoological Record, attempts to recover data from all literature, not just from those journals sampled by the SCI. However, Raup's database records dates of description of new taxa, rather than research effort *per se*. It is research effort that we are interested in, and in this respect our data may be a better reflection of the maturity with which the record has been sampled and so better serve our objectives.

As might be expected, we have found a significant correlation between conodont generic diversity (diversity data modified from Aldridge 1988) and absolute research productivity ($r_s = 0{\cdot}69$, $P = 0{\cdot}006$) (data shown graphically in Text-fig. 2D; we also obtain a significant correlation if diversity data are detrended using methods similar to Smith 2001). We have tested the hypothesis that one is biasing the other in two ways. Firstly, if inequalities in research productivity are biasing diversity, then standardized productivity (publications my^{-1}) not just raw productivity should be correlated with diversity, but it is not ($r_s = 0{\cdot}062$, $P = 0{\cdot}833$; detrended diversity data are also uncorrelated). Although this does not itself support the hypothesis that diversity is biasing research effort, it does not directly test whether intervals with publication effort differing markedly from the norm have unusually high or low diversity, and this leads to our second test. If we take the fifth and ninety-fifth percentiles in the productivity data as cut-off points, this identifies the Frasnian and the Přídolí as intervals of abnormally high and low absolute research productivity, respectively, but this is biased by the short duration of the Přídolí (we have already established a strong correlation between research effort and stage/epoch duration). The extremes of the data for standardized research productivity (my^{-1}) are the Frasnian (high) and the Emsian (low; it is also worth noting that the Llandovery, 2·45 publications my^{-1} is very close to the fifth percentile at 2·42 publications my^{-1}). The generic diversity for the Frasnian and Emsian stages is comparable (13 and 14, respectively), and Llandovery diversity is higher (20), providing no support for a relationship between abnormal research productivity and diversity. A similar pattern emerges if we apply the same percentile-based test to the diversity data. The Early Ordovician (unusually high diversity) and the Přídolí (unusually low) both have research productivity of just over 5 publications my^{-1} (in terms of publications my^{-1}, mean productivity = 5·61, median = 5·14). Thus none of our tests provides support for the hypothesis that conodont research effort, including intervals where it is unusually high or low, is biased by conodont generic diversity.

Another possibility is that peaks and troughs in research effort are an artefact of poor correlation between available rock area or volume for sampling and the temporal duration of stratigraphic units. Although the period-level data are limited, this hypothesis can be tested by looking at the relationship between global outcrop area (period-level data from Blatt and Jones 1975) and period duration. The result (Text-fig. 2E) demonstrates some scatter of these data, and a general trend is evident; the correlation between area and duration is relatively high ($r_s = 0{\cdot}783$) but with this limited dataset we are unable to reject the null hypothesis of no correlation ($P = 0{\cdot}066$). There are also no significant correlations between outcrop area and raw or standardized research productivity ($r_s = -0{\cdot}116$, $P = 0{\cdot}827$; $r_s = -0{\cdot}754$, $P = 0{\cdot}083$, respectively) and between genera and area ($r_s = -0{\cdot}464$, $P = 0{\cdot}354$). However, outcrop area decreases with increasing geological age, and the correlation between area and time is significant ($r_s = -0{\cdot}841$, $P = 0{\cdot}04$), thus detrended outcrop data (residuals from a least squares regression of area onto geological time) may provide a better means of investigating the potential biasing effects of outcrop area. Neither raw nor standardized research effort are correlated with detrended area ($r_s = 0{\cdot}6$, $P = 0{\cdot}208$; $r_s = 0{\cdot}14$, $P = 0{\cdot}787$, respectively), but the correlation between detrended diversity data and detrended outcrop area is highly significant ($r_s = 0{\cdot}9443$, $P = 0{\cdot}004$). In summary, at period level we can find no evidence to support the hypothesis that decoupling of the availability of rock for sampling from the duration of stratigraphic units might provide an explanation for intervals of extreme research productivity. Our data and analysis are limited, but the relationship between detrended area and detrended diversity data suggests either a sampling bias in recorded conodont diversity, or a relationship between habitable marine settings and conodont diversity. Without a detailed analysis of variation in facies distribution and preservation through time we are unable to test which of these alternative explanations is closer to reality.

We are unaware of any data for outcrop area by stage/epoch for the Palaeozoic, but we have tested the hypothesis using formation counts (data from Peters and Foote 2002), and find that generic diversity through the Ordovician–Devonian interval is not correlated with number of formations ($r_s = 0{\cdot}304$, $P = 0{\cdot}290$). The meaning of this result, however, is open to question given the doubts raised by Crampton *et al.* (2003) concerning the validity of formation counts as a proxy for outcrop area in analyses of this kind.

Thus, it appears from the analysis of data for all periods through which conodonts range (excluding the Cambrian) that our sampling of the conodont fossil record is neither uniform nor random, and that outcrop area may have biased our sampling of diversity. Looking at epoch/stage data for the Ordovician–Devonian interval, however, our analysis is consistent with even sampling throughout this part of the record. This is noteworthy. The lack of relationship between research productivity and conodont generic diversity is counterintuitive in that it suggests that more intense research effort has not resulted in more genera. This could mean that our publication counts provide an inadequate proxy for research effort, and although we have no reason to believe that this is the case, further research into this possibility may be fruitful. Alternatively, the sampling of the Ordovician–Devonian conodont record could have reached a level of maturity where there are few new genera left to discover (as noted above). This would be consistent with other studies that have concluded that the fossil records of taxa with robust skeletons are often well sampled (for recent reviews, see Foote and Sepkoski 1999; Forey *et al.* 2004).

As we have explicitly pointed out, our assumptions and methods of data acquisition mean that our investigation is relatively crude, intended only as a first exploratory step in analysing the potential gross temporal biases in the sampling of the conodont record. But our preliminary results are encouraging. At the stage/epoch level for the Ordovician–Devonian interval we have been unable to detect significant temporal biases in the conodont fossil record, suggesting that the recovered diversity pattern contains a significant biological signal. Given the possible area-related sampling bias for period-level data, however, there is clearly a need for more work, in particular investigating the relationship between diversity and outcrop area at stage/epoch level. Similarly, work using collector curves or other methods (Paul 1982; Wickström and Donoghue 2005) is required to investigate the relationship between research productivity and diversity in more detail.

One factor that we have not investigated and which may have significantly biased our recovered record is the varying availability for sampling of rocks that represent those particular environments within which conodonts lived and died, or those lithofacies from which conodont elements can be recovered by standard laboratory methods (see above). It is possible that intervals of particularly high or low apparent diversity have been influenced by this facies bias. However, it is also likely that cultural biases impinge upon the collection of data. In either instance, the raw data for conodont diversity cannot be taken at face value as a literal record and studied in an uncritical manner; biases must be taken into account. This is of particular concern in the case of the Frasnian (with anomalously high research productivity) and the Llandovery (low research productivity) because the diversity data for both intervals are integral to understanding two of the five greatest mass extinction events in Earth history, the recovery phase from end-Ordovician extinction event(s) in the instance of the Llandovery, and the pre-event conditions for the Frasnian–Famennian extinctions. These potential biases cannot be ignored when considering the nature and magnitude of these events and the intrinsic and extrinsic mechanisms that have been implicated.

Spatial (geographical) collecting biases. Whether our sampling of the conodont fossil record has been even with respect to stratigraphy is only one dimension of the problem of sampling biases; the other is biases with respect to geography, both modern and ancient. It is widely appreciated that there is a geographical bias in the sampling of the fossil record as a whole that is concentrated on north-west Europe, the USA and Russia (Raup 1976). This is because palaeontological science began in these regions and so they have engaged the greater number of palaeontologists for the greatest period of time. Thus, geographical regions that are relatively new to modern palaeontology are those areas in which the most surprising discoveries are being made; it is not by virtue of these regions being special in any way, whether evolutionarily or preservational, it is just that they have been sparsely sampled to date and, by analogy to a collector curve (e.g. Paul 1982; Wickström and Donoghue 2005), we remain on the steep, initial part of the curve and can expect more novel discoveries for some time to come. Hence the remarkable discoveries of fossilized remains of all groups and from all time intervals that have been made in China in recent years (Gee 2001).

The same appears to hold true for the conodont fossil record. The census of conodont-related research publications compiled by Ellison (1988) demonstrates that, although sampling of the North American record began within a couple of decades of their discovery in Europe, it took almost 90 years for the search to spread to other continents, with the first discoveries in Asia in the early 1960s (e.g. China; Jin 1960), and in Antarctica in the 1980s. While some regions have made rapid progress in remedying this situation in the years that have elapsed since those first records, the vast majority remain sparsely sampled, in terms of both time and space (South America, eastern Europe, Asia, Antarctica). This uneven and non-random sampling of the record within the spatial dimension indicates that the few composite standards that exist are considerably less than global and that our perception of the conodont fossil record is biased as a result, both in terms of palaeogeography and in terms of the total stratigraphic ranges of individual taxa that can be observed directly.

BIAS AND BIOLOGY: ELEMENT RELATIVE ABUNDANCES AND SECULAR TRENDS IN APPARATUS ARCHITECTURE

The majority of conodont studies have mapped patterns of stratigraphic distribution, diversity, ecology or phylogeny directly from the fossil record. Until recently, little attention has been paid to the potential biases affecting the record, with one notable exception: the relative abundance of the different types of isolated elements that make up most conodont collections. This is no coincidence. Almost all aspects of modern conodont palaeontology, including systematics, taxonomy, palaeoecology and palaeobiology, rely to some extent on an understanding of conodonts as skeletal apparatuses, not just as isolated elements. But almost all taxa are found as disarticulated, isolated elements, and the reconstruction of apparatuses is thus a fundamental, underpinning activity of conodont research. A variety of methods of reconstruction have been employed, but all except reconstructions drawn directly from articulated skeletons, rely on co-occurrences of elements as data. Ideally, then, the fossil record should furnish information concerning all elements of a conodont skeleton and their relative abundance in the apparatus. Any biases that may have acted to alter the likelihood of preservation and recovery of one component of the skeleton compared with another are of paramount importance to those engaged in reconstruction.

Although study of conodonts was initiated in the mid-nineteenth century, questions concerning the meaning and significance of the relative abundance of element types only came to the fore in the mid 1960s, since when major efforts have been expended in developing a biologically meaningful taxonomy of conodonts rather that simply assigning a different name to every different part of the skeleton. This has involved the development and refining of techniques for the recognition of groups of elements that originally comprised the skeleton of a single taxon (for a review, see Sweet 1988), and it soon became clear as part of this process that the numbers of different element types observed in the comparatively rare fossils that preserve complete skeletons (natural assemblages) do not match those found in collections of isolated elements. Although interpretations of the number of elements in natural assemblages have varied a little over the years, the basic message has not: the vast majority represent taxa assigned to the Ozarkodinida and contain 15 elements, 2 M, 9 S, and 4 P elements (of which two are P_1, or platform elements) (for a discussion of natural assemblages, apparatus architecture, homology and element notation, see Purnell and Donoghue 1997, 1998; Purnell *et al.* 2000); collections of isolated elements, on the other hand, contain far more P_1 elements than would be expected. In some cases the ratio of P_1 to S + M elements exceeds 25:1, 139 times more P_1 elements than would be expected based on the 0·18:1 ratio of the elements in natural assemblages (for further discussion and examples see von Bitter and Purnell 2005).

Various evolutionary and biological hypotheses have been proposed to explain this. For example, Carls (1977), following Schmidt and Müller (1964) to some extent, concluded that post-mortem sorting was unlikely in many cases and that P_1 elements are overrepresented in collections because they were shed and replaced more frequently than other parts of the apparatus (see also Krejsa *et al.* 1990; Armstrong 2005 for discussion). Merrill and Powell (1980) also favoured a developmental explanation but suggested that the composition of the apparatus varied through life: juveniles bore only ramiform elements (= S and M elements), later stages bore ramiform (S and M) and platform elements (= P_1), adult and gerontic animals bore only platform elements, their ramiform elements having been resorbed. They proposed this as a general model of skeletal development, and suggested (p. 1073) that 'the "normal" condition, especially in some environments, increasingly became one of platform-only apparatuses later in the Paleozoic and Mesozoic' thereby implying an hypothesis that changes in element ratios through time reflected heterochronic shifts in apparatus development. Sweet (1985, 1988) developed this idea further, presenting data to support long-term evolutionary trends in apparatus reduction. He suggested (1988, p. 144) that under conditions of limited phosphate availability S and M elements did not mineralize, the apparent patterns in the record thus being explained by most Ordovician and Silurian species (and younger species represented by natural assemblages) having inhabited environments in which phosphate was not a limiting factor, whereas species represented largely or entirely by P elements may have been adapted to phosphate-poor conditions.

Several of these authors, and others such as Ellison (1968) and von Bitter (1972), did consider the possibility that to some extent the apparent overrepresentation of P_1 elements reflects post-mortem processes such as current sorting, but only later was the potential role of element hydrodynamics investigated (Broadhead *et al.* 1990; McGoff 1991). As noted above, these authors were able to demonstrate that element size and shape are correlated with their susceptibility to current entrainment, transport and sorting.

That post-mortem factors influence the elemental composition of conodont collections is now beyond doubt, but hydrodynamic sorting as the source of bias in element relative abundance data is difficult to reconcile with the hypothesis, well known among conodont workers, that there is a long-term trend in P element overrepresentation (Sweet 1985, 1988). This apparent pattern continues to provide compelling support for the view that a biological signal of S and M element loss through time lurks beneath the taphonomic noise (Merrill 2002). Even when more recent evidence that apparatus composition did not vary through ontogeny and that elements were not shed is taken into account (Purnell 1994; Donoghue and Purnell 1999; Armstrong 2005), the pattern cannot simply be ignored.

Text-figure 3A is based on Sweet's well-known plot (Sweet 1985, fig. 9; 1988, fig. 6.7) showing ratios between ramiform (S and M) and pectiniform (P) elements in Ordovician–Triassic conodonts. This is probably the clearest published evidence to support the hypothesis of long-term unidirectional trends in apparatus composition, but does the hypothesis stand up to analysis based on what we now know of conodonts? We would like to consider three questions that have a direct bearing on this problem: (1) What is the evidence for variation in the composition of the conodont skeleton through time? (2) Can the trends be explained by changes in diversity of taxa bearing apparatuses with different pectiniform–ramiform ratios? (3) How do alternative hypotheses of element homology, especially among taxa assigned to Prioniodinida, affect the pattern?

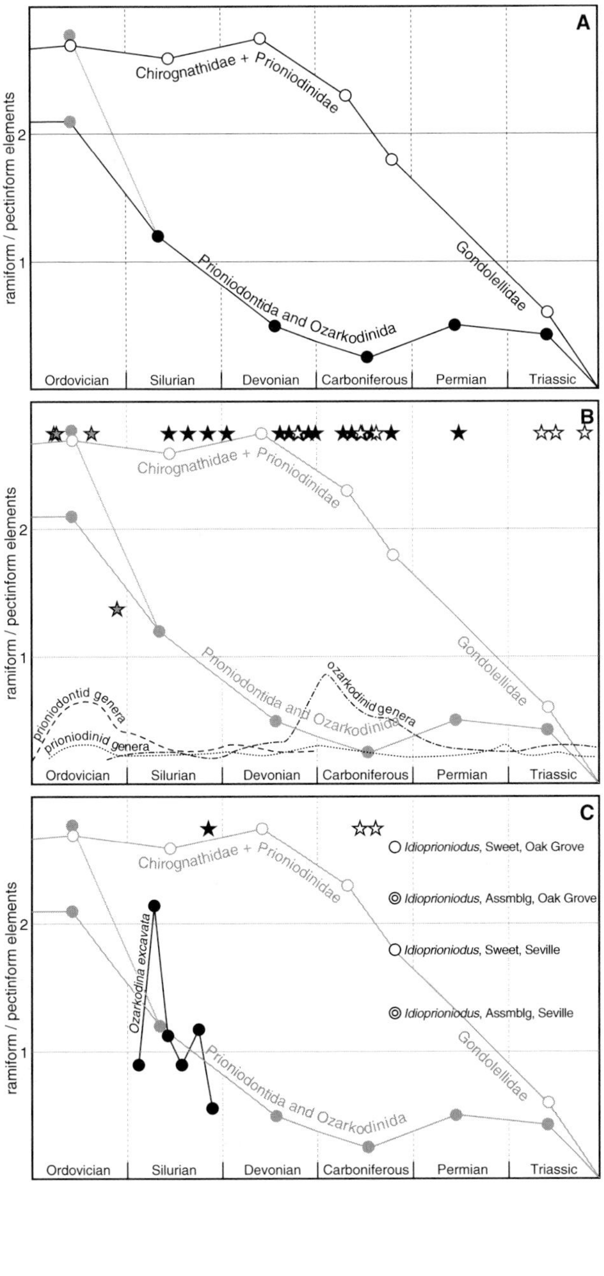

One of Sweet's (1985) original purposes with his plot (and his preceding fig. 8) was to highlight changes in apparatus composition through time. At that time the degree of stability in apparatus composition was unclear, and Sweet discussed three different types of apparatuses characterized as bimembrate (two element types), tri- and quadrimembrate, and quinque- to septimembrate (see also Barnes *et al.* 1979). Apparatuses bearing only P elements have also been advocated by some authors, mainly because ramiform–pectiniform ratios are low to the point that ramiforms are considered to be absent from collections. If conodont taxa with a variety of different apparatus structures existed, each with a different ratio of ramiform to pectiniform elements, then changes in the relative abundance of taxa could cause the overall pattern of ramiform–pectiniform ratios to change through time. The hypothesis certainly has the potential to explain the pattern, but what is the evidence for variation in skeletal composition among Prioniodontids, Ozarkodinids and Prioniodinids (the taxa upon which the plot is based)? Pectiniform–ramiform ratios for natural assemblages are shown in Text-figure 3B as stars. With one exception they plot at 2·75 because current evidence indicates that the structure of the apparatus in these conodonts was remarkably stable. Natural assemblages are now known for more than 20 taxa, ranging from the Ordovician through to the Triassic, and including both primitive and derived members of Prioniodontida, Ozarkodinida and Prioniodinida. Except for one species, they all bear 2 M,

TEXT-FIG. 3. Plot of conodont pectiniform to ramiform ratios through time. A, ratios calculated from collections of isolated elements: grey circles, taxa assigned by Sweet (1988) to Prioniodontida; black, taxa assigned by Sweet to Ozarkodinida; white, taxa assigned by Sweet to Prioniodinida (Gondolellidae in Triassic). See Appendix for sources of data (modified from Sweet 1988, fig. 6.7). B, element ratios in natural assemblages: grey stars, Prioniodontida; black stars, Ozarkodinida; white stars, Prioniodinida. Lower broken lines show relative approximate generic diversity of Prioniodontida, Ozarkodinida and Prioniodinida. See Appendix for sources of data. C, noise in different collections of the same taxon and arising from alternative hypotheses of element homology. Black circles and lines, ratios for *Ozarkodina excavata* calculated in six different collections (see von Bitter and Purnell 2005 for details; points plotted in order, but stratigraphic positions are approximate); black star shows true ratio in natural assemblage. White circles, calculations of element ratios for *Idioprioniodus* (Prioniodinida) based on different data sets (Oak Grove and Seville units of Merrill and King 1971) and alternative hypotheses of element homology (Sweet 1988, and Natural Assemblages; see text for details). White stars shows true ratio in natural assemblages.

9 S and 4 P elements (for more details, see Purnell and Donoghue 1998; Purnell *et al.* 2000). The argument that these taxa, and others, bore a complete apparatus only in certain environments and it is only from these environments that we recover complete natural assemblages (Sweet 1988) can never be fully refuted as it relies on negative evidence, but each discovery of an additional taxon with the 15-element skeletal plan makes the hypothesis weaker. Furthermore, natural assemblages have been found in depositional settings ranging from near abyssal depths to the shallow shelf, from cherts, shales and carbonates; the hypothesis is not strong.

To date, only one taxon provides clear evidence of a different skeletal structure: *Promissum pulchrum*, the giant prioniodontid from the Late Ordovician of South Africa, has 2 M, 9 S and 8 P elements (Aldridge *et al.* 1995). It is likely that some other taxa currently assigned to Prioniodontida also bore more than 4 P elements and some species have been reconstructed as such (e.g. *Pranognathus*: Männik and Aldridge 1989; *Pterospathodus*: Männik 1998; *Coryssognathus*: Miller and Aldridge 1993), but the discovery of Prioniodontids with 15 elements [*Phragmodus*: Repetski *et al.* 1998, and *Oepikodus*: Stewart and Nicoll 2003; *Paracordylodus* (Tolmacheva and Purnell 2002) is probably also a prioniodontid] indicates that the 19-element apparatus may be limited to a few taxa, possibly a single clade within Prioniodontida (*sensu* Aldridge and Smith 1993). Even if the 19-element apparatus were widespread among prioniodontids it could not explain the apparent pattern of pectiniform–ramiform ratios. The ratio for the *Promissum* apparatus is 1:1·375 (or 1:1·833 if P_4 elements are not included among P counts), so periods of high relative generic diversity of prioniodontids (shown as dashed curve on Text-fig. 3B) should correspond to periods of lower pectiniform–ramiform ratios. The opposite is true.

The argument that the pattern of element ratios is influenced by the existence of P-only apparatuses which, if they became more abundant through time would lead to reduced pectiniform–ramiform ratios, is similarly problematic in that, because it relies on negative evidence, complete refutation is impossible. However, other than pectiniform–ramiform ratios there is little if any direct support for the hypothesis. Many taxa for which S and M elements are uncommon seem to have had a standard apparatus, including taxa that have long been held to contain only P elements, such as *Eoplacognathus* (Löfgren and Zhang 2003; see Jeppsson 2005 for further discussion) and *Mestognathus* (Purnell and von Bitter 1993) and contrary to much of the literature, recent work utilizing low-diversity collections from relatively offshore and/or low-energy settings and complete articulated skeletons has demonstrated that even Triassic gondolellids, most of which (following Sweet 1970; Kozur and Mostler 1971) have been reconstructed as having P_1 elements only, had a standard 15-element apparatus (Rieber 1980; Orchard and Rieber 1999; Orchard 2005).

Hypotheses of element homology also have a direct bearing on the problem. Identifying homologous elements in some taxa, almost all members of Ozarkodinida for example, is relatively straightforward. There is a relatively high degree of conservatism in the morphology of homologous elements, different elements in the apparatus exhibit clear morphological differentiation, natural assemblages are comparatively common and apparatus structure is stable over long periods of time. Consequently there is little difference of opinion regarding element homology in ozarkodinids. For other taxa, members of Prioniodinida especially, this is not the case. Until comparatively recently few natural assemblages of prioniodinids were known, and they had failed to yield clear evidence of skeletal architecture. Furthermore, because there is less morphological differentiation within the apparatus it is much more difficult to identify homologous elements in collections of isolated elements. Obviously, this could have a significant impact on calculations of pectiniform–ramiform ratios (or more correctly in this context, ratios of P to S + M elements), and this is illustrated in Text-figure 3C with alternative calculations for *Idioprioniodus*.

Idioprioniodus provides a good example of the problem of homology: sound hypotheses of its apparatus composition were established decades ago (Merrill and Merrill 1974), and these hypotheses have now been confirmed in natural assemblages (Purnell and von Bitter 1996, 2002), but hypotheses of element homology have been far from stable. The significance of this for calculation of ramiform–pectinform ratios is shown by the difference between the points labelled 'Sweet, Seville' and 'Assmblg, Seville', and those labelled 'Sweet, Oak Grove' and 'Assmblg, Oak Grove' in Text-figure 3C (based on published data for two collections of isolated elements, one from the Oak Grove Member and the other from the Seville Member from the Pennsylvanian of Illinois; Merrill and King 1971). 'Sweet, Seville' and 'Sweet, Oak' are calculations of ratios based on the hypothesis of homology advocated by Sweet (1988); 'Assmblg, Seville' and 'Assmblg, Oak' are calculations based on the same data, but using hypotheses of element homology derived from natural assemblages. In both cases, Sweet's hypothesis significantly overestimated the pectiniform–ramiform ratios in these collections, but places one directly on his curve. The true value for the apparatus is 2·75. Hypotheses of homology can vary between conodont workers, and workers tend to concentrate on one or two geological periods more than others. Differences and uncertainties in hypotheses of homology probably add significantly to the noise in the data for prioniodinid ratios, but it is also possible that they could contribute towards the trends in element ratios through time.

Differentiating noise from signal in these data is clearly a major problem. This is highlighted by the calculations of pectiniform–ramiform ratios based on data from six Silurian collections containing *Ozarkodina excavata* (Aldridge 1972; Jeppsson 1974; Klapper and Murphy 1974; Rexroad *et al.* 1978; Helfrich 1980; Simpson and Talent 1995). A more sophisticated analysis of element ratios in this species is presented by von Bitter and Purnell (2005), but the plot shows clearly how widely ratios fluctuate in a single species through a single period. The most parsimonious explanation of this is that the signal is dominated by noise (i.e. post-mortem biases) and that collections of isolated elements rarely preserve elements in their original relative abundance.

Perhaps the most obvious test of the pattern of pectiniform–ramiform element ratios would be to repeat Sweet's analysis based on randomly selected published records of element abundance. This is beyond the scope of the present study. Such an analysis may or may not reproduce Sweet's pattern, but given the evidence presented here against there being a long-term trend in apparatus composition within complex conodonts the question should perhaps be rephrased: is there an alternative mechanism capable of explaining trends in calculated ratios in collections of discrete elements? This applies equally to secular trends as outlined by Sweet, and similar apparent ecological, spatial and taxonomic trends in relative ramiform retention (Merrill *et al.* 1991). We propose that there is such a mechanism, and that apparent relative ramiform retention reflects interaction between the biological and post-mortem variables as follows: conodont taxa vary in the relative robustness and relative size of their S, M and P elements (those taxa that have been subject to analysis have similar rates of apparatus growth, but the size differential of S and M elements relative to P elements varies: Purnell 1994; Tolmacheva and Purnell 2002). Taxa that have small and/or fragile S and M elements relative to P elements are less likely to have their S and M elements preserved in collections that contain P elements because S and M elements will be more susceptible to winnowing, current entrainment and removal, and fragmentation (see 'Element fragmentation' above). Such taxa are more likely to have P, M and S elements preserved together in quieter deposition settings, and in lithologies where factors leading to element fragmentation are reduced. It may seem obvious, but it is worth noting here that relative sizes of P, M and S elements in the apparatus of a conodont species can be determined only by measurement of elements in natural assemblages. It is most unlikely that the relative sizes of skeletal components in collections of isolated elements are unaffected by post-mortem processes. If our model is correct, then spatial and secular trends in apparent pectiniform–ramiform ratios within clades should correspond to variation in relative size and robustness of S and M elements relative to P elements.

CONCLUDING DISCUSSION: HOW LESS CAN BE MORE

A complete record of conodont diversity through time was never preserved and can never be recovered. The record we have reflects the interplay of diverse biases, ranging from the differential hydrodynamic effects upon the various element morphologies that constitute apparatuses, through environmental and stratigraphic biases in the sedimentary record, to cultural, operational and interpretational biases imposed by our retrieval and study of rocks and the conodonts they contain. Very few conodont studies have taken these biases into account and even fewer have attempted to control for their influence upon their data. As a result, conclusions drawn from these data are subject to unconstrained artefact to an unknown degree. For instance, studies of conodont evolutionary history through the Phanerozoic (e.g. Aldridge 1988; Sweet 1988) have implicitly assumed that diversity patterns read from the fossil record preserve a strong biological signal. Our preliminary investigation of potential biases provides only limited support for this view, and highlights the need to take secular variation in research effort and outcrop area into account before apparent diversity patterns can be interpreted primarily as the results of evolution and extinction. Our analysis, however, is exploratory in nature, and more work is required before the full effects of these gross biases on sampling *sensu lato* can be evaluated with confidence.

The data upon which most models of conodont palaeoecology and biofacies are based are also likely to be biased to some degree. Post-mortem hydrodynamic sorting (Broadhead *et al.* 1990; McGoff 1991) and differential redistribution of elements have affected the final facies distributions of different elements and species in different ways. The degree to which this has influenced apparent associations of elements and species, and perceived environmental ranges, is largely unknown, but, as noted by McGoff (1991), samples containing elements that reflect a narrow range of Reynolds numbers and drag coefficients are very likely to be the product of post-mortem sorting. This applies equally to samples dominated by a single taxon and those containing a more mixed fauna. In either case the sample cannot be used in biofacies analysis (McGoff 1991). A rigorous study of the prevalence of post-mortem sorting would be worthwhile, and has the potential to add significantly to the usefulness of conodonts in palaeoenvironmental analysis. It may be possible, for example, to establish conodont taphofacies (cf. Speyer and Brett 1986) based on analysis of the size and shape of elements in a sample, attributes that are known to correlate with their relative hydrodynamic properties (Broadhead *et al.* 1990; McGoff 1991).

Hydrodynamic biases also have a bearing on hypotheses that attribute variation in the recovery of pectiniform and ramiform elements from strata of different ages to evolutionary trends in apparatus composition or to changes in differential shedding or resorption of P, S and M elements. We have found no evidence to support these hypotheses, and interpret differences in recovered element abundances as the result, largely, of the interplay between hydrodynamic factors and relative size and fragility of elements within the skeleton. Finally, it is clear from a consideration of biases and the incomplete nature of the stratigraphic record that attempts to reconstruct phylogeny through the assembly of species-level lineages can never lead to a composite tree for all conodonts.

Notwithstanding its imperfections, the conodont fossil record remains among the best of any group of organisms, with clear utility across a range of geological and biological contexts. Attempts to control for biases can be problematic and for some research agenda may prove impossible, but acknowledging the biases that affect the record can actually enrich understanding of it (cf. Behrensmeyer and Kidwell 1985). This, we hope, is clear from our discussions of hydrodynamic sorting, palaeoecology and pectiniform–ramiform ratios, but there are also stratigraphic implications that should not be ignored. Recognizing that the ranges of taxa vary from section to section and from region to region provides an opportunity for closer investigation of the relationship between the spatial and temporal distribution of conodont taxa at a variety of scales. How, for example, have distributions varied in response to extrinsic environmental events? A prerequisite for such studies is the application of more rigorous, quantitative methods of biostratigraphy (e.g. Armstrong 1999). The cause of graphic correlation has been championed on the basis of the conodont fossil record (Shaw 1969) but, with a few exceptions (see Sweet 2005 and references therein), conodont workers have yet to capitalize on the potential and benefits of this technique in resolving rates of sedimentation, detecting otherwise imperceptible hiatuses in sections, and for producing a high-resolution composite chronostratigraphic timescale of global relevance. Confidence intervals, too, provide a basis for assessing the significance of first and last appearances in local sections and in constraining the veracity of apparent bioevents (Paul 1982; Strauss and Sadler 1989; Marshall 1990); protocols for calculating confidence intervals compatible with standard micropalaeontological sampling strategies have been available for some time (Weiss and Marshall 1999).

A range of variables have together conspired to produce what we recover as taxon ranges, element distributions and diversity data. Taking these variables into account will improve our understanding of patterns of conodont palaeobiology, palaeoecology, palaeogeography and phylogeny and their controls while at the same time enhancing the utility of conodonts as geological tools, and highlighting new avenues for research. Above all, the clear message of this collection of papers (Purnell and Donoghue 2005) is that to pay no regard to post-mortem processes and the nature of the conodont fossil record risks overlooking factors that have significantly biased our primary data. Ignorance is not bliss.

Acknowledgements. Thanks to: Marie-France Perret Mirouse, for allowing us to hijack a day of she scientific proceedings of ECOS VIII; all those who contributed talks and participated in the lively discussion during the symposium on bias and completeness, especially those who have produced written contributions to this volume; Walt Sweet for unpublished information and discussion of the data used in constructing his whatzits graph; Dick Aldridge for his review and helpful comments on the manuscript; those Panderers who provided estimated percentages of sample lithologies they have processed for conodonts. The funding of NERC (GT5/98/4/ES and NER/J/S/2002/00673 to MAP, GT5/99/ES/2 to PCJD) and the Royal Society (PCJD) is gratefully acknowledged. We also thank The Palaeontological Association for their financial support of the symposium.

REFERENCES

ALDRIDGE, R. J. 1972. Llandovery conodonts from the Welsh Borderland. *Bulletin of the British Museum (Natural History), Geology*, **22**, 125–231.

—— 1988. Extinction and survival in the Conodonta. 231–256. *In* LARWOOD, G. P. (ed.). *Extinction and survival in the fossil record.* Systematics Association, Special Volume, **34**. Clarendon press, Oxford, 376 pp.

—— BRIGGS, D. E. G., SMITH, M. P., CLARKSON, E. N. K. and CLARK, N. D. L. 1993. The anatomy of conodonts. *Philosophical Transactions of the Royal Society of London, Series B*, **340**, 405–421.

—— PURNELL, M. A., GABBOTT, S. E. and THERON, J. N. 1995. The apparatus architecture and function of *Promissum pulchrum* Kovács-Endrödy (Conodonta, Upper Ordovician), and the prioniodontid plan. *Philosophical Transactions of the Royal Society of London, Series B*, **347**, 275–291.

—— and SMITH, M. P. 1993. Conodonta. 563–572. *In* BENTON, M. J. (ed.). *The fossil record 2.* Chapman & Hall, London, 864 pp.

ARMSTRONG, H. A. 1999. Quantitative biostratigraphy. 181–226. *In* HARPER, D. A. T. (ed.). *Numerical palaeobiology: computer-based modelling and analysis of fossils and their distribution.* Wiley, Chichester, 478 pp.

—— 2005. Modes of growth in the euconodont oral skeleton: implications for bias and completeness in the fossil record.

27–38. *In* PURNELL, M. A. and DONOGHUE, P. C. J. (eds). *Conodont biology and phylogeny: interpreting the fossil record.* Special Papers in Palaeontology, **73**, 218 pp.
—— OWEN, A. W. and CLARKSON, E. N. K. 2000. Ordovician limestone clasts in the Lower Old Red Sandstone, Pentland Hills, southern Midland Valley terrane. *Scottish Journal of Geology*, **36**, 33–37.
BARNES, C. R., KENNEDY, D. J., MCCRACKEN, A. D., NOWLAN, G. S. and TARRANT, G. A. 1979. The structure and evolution of Ordovician conodont apparatuses. *Lethaia*, **12**, 125–151.
BARRICK, J. E. 1987. Conodont biostratigraphy of the Caballos Novaculite (Early Devonian – Early Mississippian), northwestern Marathon Uplift, West Texas. 120–135. *In* AUSTIN, R. L. (ed.). *Conodonts, investigative techniques and applications.* British Micropalaeontological Society Series. Ellis Horwood, Chichester, 422 pp.
—— and MÄNNIK, P. 2005. Silurian conodont biostratigraphy and palaeobiology in stratigraphic sequences. 103–116. *In* PURNELL, M. A. and DONOGHUE, P. C. J. (eds). *Conodont biology and phylogeny: interpreting the fossil record.* Special Papers in Palaeontology, **73**, 218 pp.
BEHRENSMEYER, A. K. and KIDWELL, S. M. 1985. Taphonomy's contributions to paleobiology. *Paleobiology*, **11**, 105–119.
—— —— and GASTALDO, R. A. 2000. Taphonomy and paleobiology. *Paleobiology*, **26**, 103–147.
BELKA, Z., SKOMPSKI, S. and SOBON-PODGORSKA, J. 1996. Reconstruction of a lost carbonate platform on the shelf of Fennosarmatia: evidence from Viséan polymictic debrites, Holy Cross Mountains, Poland. 315–329. *In* STROGEN, P., SOMERVILLE, I. D. and JONES, G. L. (eds). *Recent advances in Lower Carboniferous geology.* Geological Society, Special Publication, 464 pp.
BENTON, M. J. 1995*a*. Diversification and extinction in the history of life. *Science*, **268**, 52–58.
—— 1995*b*. Testing the time axis of phylogenies. *Philosophical Transactions of the Royal Society of London, Series B*, **349**, 5–10.
—— 2003. The quality of the fossil record. 66–90. *In* DONOGHUE, P. C. J. and SMITH, M. P. (eds). *Telling the evolutionary time: molecular clocks and the fossil record.* CRC Press, London, 296 pp.
BERGSTRÖM, S. M. and SWEET, W. C. 1966. Conodonts from the Lexington Limestone (Middle Ordovician) of Kentucky and its lateral equivalents in Ohio and Indiana. *Bulletins of American Paleontology*, **50**, 271–441.
BLATT, H. and JONES, R. L. 1975. Proportions of exposed igneous, metamorphic, and sedimentary rocks. *Geological Society of America, Bulletin*, **86**, 1085–1088.
BLUTH, G. J. S. and KUMP, L. R. 1991. Phanerozoic paleogeology. *American Journal of Science*, **291**, 284–308.
BROADHEAD, T. W. and DRIESE, S. G. 1994. Experimental and natural abrasion of conodonts in marine and eolian environments. *Palaios*, **9**, 564–560.
—— —— and HARVEY, J. L. 1990. Gravitational settling of conodont elements; implications for paleoecologic interpretations of conodont assemblages. *Geology*, **18**, 850–853.
CARLS, P. 1977. Could conodonts be lost and replaced? – Numerical relations among disjunct conodont elements of certain Polygnathidae (late Silurian – Lower Devonian, Europe). *Neues Jahrbuch für Geologie und Paläontologie, Abhandlungen*, **155**, 18–64.
CONWAY MORRIS, S. 1990. *Typhloesus wellsi* (Melton and Scott, 1973), a bizarre metazoan from the Carboniferous of Montana, U.S.A. *Philosophical Transactions of the Royal Society of London, Series B*, **327**, 545–624.
CRAMPTON, J. S., BEU, A. G., COOPER, R. A., JONES, C. M., MARSHALL, B. and MAXWELL, P. A. 2003. Estimating the rock volume bias in paleobiodiversity studies. *Science*, **301**, 358–360.
DONOGHUE, P. C. J. 1998. Growth and patterning in the conodont skeleton. *Philosophical Transactions of the Royal Society of London, Series B*, **353**, 633–666.
—— 2001*a*. Conodonts meet cladistics: recovering relationships and assessing the completeness of the conodont fossil record. *Palaeontology*, **44**, 65–93.
—— 2001*b*. Corrigendum. Conodonts meet cladistics: recovering relationships and assessing the completeness of the conodont fossil record (vol. 44, p. 65, 2001). *Palaeontology*, **44**, 1237.
—— 2001*c*. Microstructural variation in conodont enamel is a functional adaptation. *Proceedings of the Royal Society of London, Series B*, **268**, 1691–1698.
—— and PURNELL, M. A. 1999. Growth, function, and the conodont fossil record. *Geology*, **27**, 251–254.
—— SMITH, M. P. and SANSOM, I. J. 2003. The origin and early evolution of chordates: molecular clocks and the fossil record. 190–223. *In* DONOGHUE, P. C. J. and SMITH, M. P. (eds). *Telling the evolutionary time: molecular clocks and the fossil record.* CRC Press, London, 296 pp.
DROSER, M. L. and BOTTJER, D. J. 1986. A semiquantitative field classification of ichnofabric. *Journal of Sedimentary Petrology*, **56**, 558–559.
DZIK, J. 1991. Evolution of the oral apparatuses in the conodont chordates. *Acta Palaeontologica Polonica*, **36**, 265–323.
—— 2005. The chronophyletic approach: stratophenetics facing an incomplete fossil record. 159–183. *In* PURNELL, M. A. and DONOGHUE, P. C. J. (eds). *Conodont biology and phylogeny: interpreting the fossil record.* Special Papers in Palaeontology, **73**, 218 pp.
ELLISON, S. P. Jr 1968. Conodont census studies as evidence of sorting. *Abstracts and Program of the Annual Meeting of the Geological Society of America, North-Central Section, Iowa City, Iowa (May 8–11, 1968)*, p. 42.
—— 1988. Conodont bibliography to January 1, 1987. *Courier Forschungsinstitut Senckenberg*, **103**, 1–245.
—— and GRAVES, R. W. Jr 1941. Lower Pennsylvanian (Dimple limestone) conodonts of the Marathon region, Texas. *University of Missouri School of Mines and Metallurgy, Technical Series, Bulletin*, **14**, 1–21.
FISHER, D. W. 1981. Crocodilian scatology, microverterbrate concentrations, and enamel-less teeth. *Paleobiology*, **7**, 262–275.
FOOTE, M. and SEPKOSKI, J. J. 1999. Absolute measures of the completeness of the fossil record. *Nature*, **398**, 415–417.

FOREY, P. L., FORTEY, R. A., KENRICK, P. and SMITH, A. B. 2004. Taxonomy and fossils: a critical appraisal. *Philosophical Transactions of the Royal Society of London, Series B*, **359**, 639–653.

GABLE, K. M. 1973. Conodonts and stratigraphy of the Olentangy Shale (Middle and Upper Devonian), central and south-central Ohio. Unpublished MSc thesis, The Ohio State University.

GEE, H. (ed.) 2001. *Rise of the dragon: readings from Nature on the Chinese fossil record.* University of Chicago Press, London, 256 pp.

GRADSTEIN, F. M., COOPER, R. A., SADLER, P. M., HINNOV, L. A., SMITH, A. G., OGG, J. G., VILLENEUVE, M., McARTHUR, M., HOWARTH, R. J., AGTERBERG, F. P., ROBB, L. J., KNOLL, A. H., PLUMB, K. A., SHIELDS, G. A., STRAUSS, H., VEIZER, J., BLEEKER, W., SHERGOLD, J. H., MELCHIN, M. J., HOUSE, M. R., DAVYDOV, V., WARDLAW, B. R., LUTERBACHER, H. P., ALI, J. R., BRINKHUIS, H., HOOKER, J. J., MONECHI, S., POWELL, J., RÖHL, U., SANFILIPPO, A., SCHMITZ, B., LOURENS, L., HILGEN, F., SHACKLETON, N. J., LASKAR, J., WILSON, D., GIBBARD, P. and VAN KOLFSCHOTEN, T. 2004. *A geologic time scale 2004.* Cambridge University Press, Cambridge, 384 pp.

HAMMER, Ø., HARPER, D. A. T. and RYAN, P. D. 2001. PAST: Paleontological Statistics Software Package for Education and Data Analysis. *Palaeontologia Electronica*, **4**, 9 pp. http://palaeo-electronica.org/2001_1/past/issue1_01.htm.

HELFRICH, C. T. 1980. Late Llandovery–early Wenlock conodonts from the upper part of the Rose Hill and the basal part of the Mifflintown formations, Virginia, West Virginia, and Maryland. *Journal of Paleontology*, **54**, 557–569.

HIEKE, W. 1967. Feinstratigraphie und Palaeogeographie des Trochitenkalkes zwischen Leinetal-Graben und Rhoen. *Geologica et Palaeontologica*, **1**, 57–86.

HIGGINS, A. C. 1981. Coprolitic conodont assemblages from the lower Westphalian of north Staffordshire. *Palaeontology*, **24**, 437–441.

JEPPSSON, L. 1974. Aspects of late Silurian conodonts. *Fossils and Strata*, **6**, 1–54.

—— 2005. Biases in the recovery and interpretation of micropalaeontological data. 57–71. *In* PURNELL, M. A. and DONOGHUE, P. C. J. (eds). *Conodont biology and phylogeny: interpreting the fossil record.* Special Papers in Palaeontology, **73**, 218 pp.

JIN YU-GAN (CHING YU-KAN). 1960. Conodont fossils from the Kufeng Suite (Formation) of Longtan, Nanking. *Acta Palaeontologica Sinica*, **8**, 242–248. [In Chinese, English abstract].

KLAPPER, G. and MURPHY, M. A. 1974. Silurian–lower Devonian conodont sequence in the Roberts Mountains Formation of central Nevada. *University of California, Publications in Geological Sciences*, **111**, 1–62.

KOHUT, J. J. 1969. Determination, statistical analysis and interpretation of recurrent conodont groups in Middle and Upper Ordovician strata of the Cincinnati Region (Ohio, Kentucky, and Indiana). *Journal of Paleontology*, **43**, 392–412.

KOVÁCS, S. and ÁRKAI, P. 1987. Conodont alteration in metamorphosed limestones from northern Hungary, and its relationship to carbonate texture, illite crystallinity and vitrinite reflectance. 209–229. *In* AUSTIN, R. L. (ed.). *Conodonts, investigative techniques and applications.* British Micropalaeontological Society Series. Ellis Horwood, Chichester, 422 pp.

KOZUR, H. and MOSTLER, H. 1971. Probleme der Conodontenforschung in der Trias. *Geologische und Paläontologische Mitteilungen, Innsbruck*, **1**, 1–19.

KREJSA, R. J., BRINGAS, P. and SLAVKIN, H. C. 1990. A neontological interpretation of conodont elements based on agnathan cyclostome tooth structure, function, and development. *Lethaia*, **23**, 359–378.

KRUMHARDT, A. P., HARRIS, A. G. and WATTS, K. F. 1996. Lithostratigraphy, microlithofacies, and conodont biostratigraphy and biofacies of the Wahoo Limestone (Carboniferous), eastern Sadlerochit Mountains, northeast Brooks Range, Alaska. *United States Geological Survey, Professional Paper*, **1568**, 1–70.

LANGE, F. G. 1968. Conodonten-gruppenfunde aus Kalken des tieferen Oberdevon. *Geologica et Palaeontologica*, **2**, 37–57.

LEHNERT, O., MILLER, J. F., LESLIE, S. A., REPETSKI, J. E. and ETHINGTON, R. L. 2005. Cambro-Ordovician sea-level fluctuations and sequence boundaries: the missing record and the evolution of new taxa. 117–134. *In* PURNELL, M. A. and DONOGHUE, P. C. J. (eds). *Conodont biology and phylogeny: interpreting the fossil record.* Special Papers in Palaeontology, **73**, 218 pp.

LÖFGREN, A. and ZHANG, J. H. 2003. Element association and morphology in some middle Ordovician platform-equipped conodonts. *Journal of Paleontology*, **77**, 721–737.

MÄNNIK, P. 1998. Evolution and taxonomy of the Silurian conodont *Pterospathodus. Palaeontology*, **41**, 1001–1050.

—— and ALDRIDGE, R. J. 1989. Evolution, taxonomy and relationships of the Silurian conodont *Pterospathodus. Palaeontology*, **32**, 893–906.

MARSHALL, C. R. 1990. Confidence-intervals on stratigraphic ranges. *Paleobiology*, **16**, 1–10.

MASHKOVA, T. V. 1972. *Ozarkodina steinhornensis* (Ziegler) apparatus, its conodonts and biozone. *Geologica et Palaeontologica*, **1**, 81–90.

MASTANDREA, A., IETTO, F., NERI, C. and RUSSO, F. 1997. Conodont biostratigraphy of the Late Triassic sequence of monte Cocuzzo (Catena Costiera, Calabria, Italy). *Rivista Italiana di Paleontologia e Stratigrafia*, **103**, 173–181.

—— NERI, C., IETTO, F., RUSSO, F., KOZUR, M. U. and MOCK, F. E. 1999. *Miskella ultima* Kozur and Mock, 1991: first evidence of Late Rhaetian conodonts in Calabria (southern Italy). 497–506. *In* SERPAGLI, E. (ed.). *Studies on conodonts. Proceedings of the Seventh European Conodont Symposium.* Bollettino della Società Paleontologica Italiana, **37**, 557 pp.

McGOFF, H. J. 1991. The hydrodynamics of conodont elements. *Lethaia*, **24**, 235–247.

MELTON, W. and SCOTT, H. W. 1973. Conodont-bearing animals from the Bear Gulch Limestone, Montana. 31–65. *In* RHODES, F. H. T. (ed.). *Conodont paleozoology.* Geological Society of America, Special Paper, **141**, 296 pp.

MERRILL, G. K. 2002. Ramiform retention in time and ecospace. *Strata, Série 1, Abstracts of the Eighth International Conodont Symposium held in Europe*, **12**, 45.

—— and KING, C. W. 1971. Platform conodonts from the lowest Pennsylvanian rocks of northwestern Illinois. *Journal of Paleontology*, **45**, 645–664.

—— and MERRILL, S. M. 1974. Pennsylvanian nonplatform conodonts; IIa, The dimorphic apparatus of *Idioprioniodus*. *Geologica et Palaeontologica*, **8**, 119–130.

—— and POWELL, R. J. 1980. Paleobiology of juvenile (nepionic?) conodonts from the Drum Limestone (Pennsylvanian, Missourian, Kansas City area) and its bearing on apparatus ontogeny. *Journal of Paleontology*, **54**, 1058–1074.

—— VON BITTER, P. H. and GRAYSON, R. C., Jr 1991. The generic concept in conodont paleontology; growth, changes and developments in the last two decades. *Courier Forschungsinstitut Senckenberg*, **118**, 397–408.

MILLER, C. G. and ALDRIDGE, R. J. 1993. The taxonomy and apparatus structure of the Silurian distomodontid conodont *Coryssognathus* Link & Druce, 1972. *Journal of Micropalaeontology*, **12**, 241–255.

NICOLL, R. S. 1977. Conodont apparatuses in an Upper Devonian palaeoniscoid fish from the Canning Basin, Western Australia. *Bureau of Mineral Resources, Journal of Australian Geology and Geophysics*, **2**, 217–228.

—— 1985. Multielement composition of the conodont species *Polygnathus xylus xylus* Stauffer, 1940 and *Ozarkodina brevis* (Bischoff and Ziegler, 1957) from the Upper Devonian of the Canning Basin, Western Australia. *Bureau of Mineral Resources, Journal of Australian Geology and Geophysics*, **9**, 133–147.

—— and REXROAD, C. B. 1987. Re-examination of Silurian conodont clusters from northern Indiana. 49–61. *In* ALDRIDGE, R. J. (ed.). *Palaeobiology of conodonts*. Ellis Horwood, Chichester, 180 pp.

NORBY, R. D. 1976. Conodont apparatuses from Chesterian (Mississippian) strata of Montana and Illinois. Unpublished PhD thesis, University of Illinois at Urbana-Champaign.

NORELL, M. A. 1992. Taxic origin and temporal diversity: the effect of phylogeny. 89–118. *In* NOVACEK, M. J. and WHEELER, Q. D. (eds). *Extinction and phylogeny*. Columbia University Press, New York, 253 pp.

ORCHARD, M. J. 1987. Conodonts from western Canadian chert: their nature, distribution and stratigraphic application. 94–119. *In* AUSTIN, R. L. (ed.). *Conodonts, investigative techniques and applications*. British Micropalaeontological Society Series. Ellis Horwood, Chichester, 442 pp.

—— 2005. Multielement conodont apparatuses of Triassic Gondolelloidea. 73–101. *In* PURNELL, M. A. and DONOGHUE, P. C. J. (eds). *Conodont biology and phylogeny: interpreting the fossil record*. Special Papers in Palaeontology, **73**, 218 pp.

—— and RIEBER, H. 1999. Multielement *Neogondolella* (Conodonta, Upper Permian–Middle Triassic). 475–488. *In* SERPAGLI, E. (ed.). *Studies on conodonts. Proceedings of the Seventh European Conodont Symposium*. Bollettino della Società Paleontologica Italiana, **37**, 557 pp.

PAUL, C. R. C. 1982. The adequacy of the fossil record. 75–117. *In* JOYSEY, K. A. and FRIDAY, A. E. (eds). *Problems of phylogenetic reconstruction*. Systematics Association, Special Volume, **21**. Academic Press, London, 442 pp.

PETERS, S. E. and FOOTE, M. 2001. Biodiversity in the Phanerozoic: a reinterpretation. *Paleobiology*, **27**, 583–601.

—— —— 2002. Determinants of extinction in the fossil record. *Nature*, **416**, 420–424.

POHLER, S. M. L., BARNES, C. R. and JAMES, N. P. 1987. Reconstructing a lost realm: conodonts from megaconglomerates of the Ordovician Cow Head Group, western Newfoundland. 341–362. *In* AUSTIN, R. L. (ed.). *Conodonts, investigative techniques and applications*. British Micropalaeontological Society Series. Ellis Horwood, Chichester, 442 pp.

POHLER, S. M. L. and JAMES, N. P. 1989. Reconstruction of a Lower/Middle Ordovician carbonate shelf margin; Cow Head Group, western Newfoundland. *Facies*, **21**, 189–262.

PURNELL, M. A. 1993. The *Kladognathus* apparatus (Conodonta, Carboniferous): homologies with ozarkodinids and the prioniodinid Bauplan. *Journal of Paleontology*, **67**, 875–882.

—— 1994. Skeletal ontogeny and feeding mechanisms in conodonts. *Lethaia*, **27**, 129–138.

—— and DONOGHUE, P. C. J. 1997. Architecture and functional morphology of the skeletal apparatus of ozarkodinid conodonts. *Philosophical Transactions of the Royal Society of London, Series B*, **352**, 1545–1564.

—— —— 1998. Skeletal architecture, homologies and taphonomy of ozarkodinid conodonts. *Palaeontology*, **41**, 57–102.

—— —— (eds) 2005. Conodont biology and phylogeny: interpreting the fossil record. *Special Papers in Palaeontology*, **73**, 218 pp.

—— —— and ALDRIDGE, R. J. 2000. Orientation and anatomical notation in conodonts. *Journal of Paleontology*, **74**, 113–122.

—— and VON BITTER, P. H. 1993. The apparatus and affinities of *Mestognathus* Bischoff (Conodonta, Lower Carboniferous). *Geological Society of America, Abstracts with Programs*, **25** (3), 73.

—— —— 1996. Bedding-plane assemblages of *Idioprioniodus*, element locations, and the Bauplan of prioniodinid conodonts. 48. *In* DZIK, J. (ed.). *Sixth European conodont symposium, abstracts*. Instytut Paleobiologii PAN, Warszawa, 70 pp.

—— —— 2002. Natural assemblages of *Idioprioniodus* (Conodonta, Vertebrata) and the first three-dimensional skeletal model of a prioniodinid conodont. *Strata, Série 1, Abstracts of the Eighth International Conodont Symposium held in Europe*, **12**, 100.

RAMOVŠ, A. 1977. Skelettapparat von *Pseudofurnishius murcianus* (Conodontophorida) in der Mitteltrias Sloweniens (NW Jugoslawien). *Neues Jahrbuch für Geologie und Paläontologie, Abhandlungen*, **153**, 361–399.

—— 1978. Mitteltriassische Conodonten-clusters in Slovenien, NW Jugoslawien. *Paläontologische Zeitschrift*, **52**, 129–137.

RAMSEY, N. J. 1969. Upper Emsian–upper Givetian conodonts from the Columbus and Delaware limestones (middle Devonian) and lower Olentangy shale (upper Devonian) of central Ohio. Unpublished MS thesis, Ohio State University.

RAUP, D. M. 1972. Taxonomic diversity during the Phanerozoic. *Science*, **177**, 1065–1071.

—— 1976. Species diversity in the Phanerozoic: a tabulation. *Paleobiology*, **2**, 279–288.

—— 1977. Species diversity in the Phanerozoic: systematists follow the fossils. *Paleobiology*, **3**, 328–329.

REJEBIAN, V. A., HARRIS, A. G. and HUEBNER, J. S. 1987. Conodont color and textural alternation: an index to regional metamorphism, contact metamorphism, and hydrothermal alteration. *Geological Society of America, Bulletin*, **99**, 471–479.

REPETSKI, J., PURNELL, M. A. and BARRETT, S. F. 1998. The apparatus architecture of *Phragmodus*. 91–92. *In* BAGNOLI, G. (ed.). *Seventh International Conodont Symposium held in Europe (ECOS VII), Abstracts*. Tipografia Compositori, Bologna, 136 pp.

REXROAD, C. B., NOLAND, A. V. and POLLOCK, C. A. 1978. Conodonts from the Louisville Limestone and the Wabash Formation (Silurian) in Clark County, Indiana and Jefferson County, Kentucky. *Indiana Geological Survey, Special Report*, **16**, 1–15.

RIEBER, H. 1980. Ein Conodonten-Cluster aus der Grenzbitumenzone (Mittlere Trias) des Monte San Giorgio (Kt. Tessin/Schweiz). *Annalen, Naturhistorische Museum Wien*, **83**, 265–274.

RITTER, S. M. and BAESEMANN, J. F. 1991. Early Permian conodont assemblages from the Wolfcamp Shale, Midland Basin, West Texas. *Journal of Paleontology*, **65**, 670–677.

ROOPNARINE, P. D. 2005. The likelihood of stratophenetic-based hypotheses of genealogical succession. 143–157. *In* PURNELL, M. A. and DONOGHUE, P. C. J. (eds). *Conodont biology and phylogeny: interpreting the fossil record*. Special Papers in Palaeontology, **73**, 218 pp.

SADLER, P. M. 1981. Sediment accumulation rates and the completeness of stratigraphic sections. *Journal of Geology*, **89**, 569–584.

—— and STRAUSS, D. J. 1990. Estimation of completeness of stratigraphical sections using empirical-data and theoretical-models. *Journal of the Geological Society*, **147**, 471–485.

SCHINDEL, D. E. 1980. Microstratigraphic sampling and the limits of paleontologic resolution. *Paleobiology*, **6**, 408–426.

SCHMIDT, H. and MÜLLER, K. J. 1964. Weitere Funde von Conodonten-Gruppen aus dem oberen Karbon des Sauerlandes. *Paläontologische Zeitschrift*, **38**, 105–135.

SCOTT, H. W. 1969. Discoveries bearing on the nature of the conodont animal. *Micropaleontology*, **15**, 420–426.

—— 1973. New Conodontochordata from the Bear Gulch Limestone (Namurian, Montana). *Publications of the Museum, Michigan State University, Paleontological Series*, **1**, 81–100.

SHAW, A. B. 1964. *Time in stratigraphy*. McGraw-Hill, New York, 365 pp.

—— 1969. Adam and Eve, paleontology, and the non-objective sciences. *Journal of Paleontology*, **43**, 1085–1098.

SHEEHAN, P. M. 1977. Species diversity in the Phanerozoic: a reflection of labor by systematists? *Paleobiology*, **3**, 325–328.

SIMPSON, A. J. and TALENT, J. A. 1995. Silurian conodonts from the headwaters of the Indi (upper Murray) and Buchan rivers, southeastern Australia, and their implications. *Courier Forschungsinstitut Senckenberg*, **182**, 79–215.

SMITH, A. B. 2001. Large-scale heterogeneity of the fossil record: implications for Phanerozoic biodiversity studies. *Philosophical Transactions of the Royal Society of London, Series B, Biological Sciences*, **356**, 351–367.

—— GALE, A. S. and MONKS, N. E. A. 2001. Sea-level change and rock-record bias in the Cretaceous: a problem for extinction and biodiversity studies. *Paleobiology*, **27**, 241–253.

SPEYER, S. E. and BRETT, C. E. 1986. Trilobite taphonomy and Middle Devonian taphofacies. *Palaios*, **1**, 312–327.

STEWART, I. and NICOLL, R. S. 2003. Multielement apparatus structure of the Early Ordovician conodont *Oepikodus evae* Lindström from Australia and Sweden. *Courier Forschungsinstitut Senckenberg*, **245**, 361–387.

STRAUSS, D. and SADLER, P. M. 1989. Stochastic-models for the completeness of stratigraphic sections. *Mathematical Geology*, **21**, 37–59.

SWEET, W. C. 1970. Uppermost Permian and Lower Triassic conodonts of the Salt Range and Trans-Indus ranges, West Pakistan. 207–275. *In* KUMMEL, B. and TEICHERT, C. (eds). *Stratigraphic boundary problems: Permian and Triassic of West Pakistan*. University of Kansas, Department of Geology, Special Publication **4**, 474 pp.

—— 1985. Conodonts: those fascinating little whatzits. *Journal of Paleontology*, **59**, 485–494.

—— 1988. *The Conodonta: morphology, taxonomy, paleoecology, and evolutionary history of a long-extinct animal phylum*. Oxford Monographs on Geology and Geophysics, **10**. Clarendon Press, Oxford, 212 pp.

—— 2005. Graphical refinement of the conodont data base: examples and a plea. 135–141. *In* PURNELL, M. A. and DONOGHUE, P. C. J. (eds). *Conodont biology and phylogeny: interpreting the fossil record*. Special Papers in Palaeontology, **73**, 218 pp.

—— and DONOGHUE, P. C. J. 2001. Conodonts: past, present, future. *Journal of Paleontology*, **75**, 1174–1184.

TOLMACHEVA, T. Yu and PURNELL, M. A. 2002. Apparatus composition, growth, and survivorship of the Lower Ordovician conodont *Paracordylodus gracilis* Lindström, 1955. *Palaeontology*, **45**, 209–228.

VALENTINE, J. W. 1969. Patterns of taxonomic and ecological structure of the shelf benthos during Phanerozoic time. *Palaeontology*, **12**, 684–709.

—— 1973. *Evolutionary paleoecology of the marine biosphere*. Prentice Hall, Englewood Cliffs, 512 pp.

—— 2004. *On the origin of phyla*. University of Chicago Press, Chicago, 614 pp.

—— AWRAMIK, S. M., SIGNOR, P. W. and SADLER, P. M. 1991. The biological explosion at the Precambrian-Cambrian boundary. *Evolutionary Biology*, **25**, 279–356.

VON BITTER, P. H. 1972. Environmental control of conodont distribution in the Shawnee Group (Upper Pennsylvanian) of eastern Kansas. *University of Kansas Paleontological Contributions*, **59**, 1–105, 16 pls.

—— and PURNELL, M. A. 2004. Articulated skeletons of *Ctenognathodus* (Conodonta, Vertebrata) from the Eramosa Member of the Guelph Formation (Silurian) at Hepworth, Ontario, Canada. *Joint Annual Meeting of Geological Association of Canada and Mineralogical Association of Canada, St. Catherines, Ontario, Abstracts Volume*, **29**, 82.

—— —— 2005. An experimental investigation of post-depositional taphonomic bias in conodonts. 39–56. *In* PURNELL, M. A. and DONOGHUE, P. C. J. (eds). *Conodont biology and phylogeny: interpreting the fossil record.* Special Papers in Palaeontology, **73**, 218 pp.

WEISS, R. E. and MARSHALL, C. R. 1999. The uncertainty in the true end point of a fossil's stratigraphic range when stratigraphic sections are sampled discretely. *Mathematical Geology*, **31**, 435–453.

WICKSTRÖM, L. M. and DONOGHUE, P. C. J. 2005. Cladograms, phylogenies and the veracity of the conodont fossil record. 185–218. *In* PURNELL, M. A. and DONOGHUE, P. C. J. (eds). *Conodont biology and phylogeny: interpreting the fossil record.* Special Papers in Palaeontology, **73**, 218 pp.

APPENDIX

Data for Text-figure 3

Except as noted in figure caption, isolated element ratio data are taken from the following sources (Sweet, pers. comm. 2002): Ordovician, Prioniodontida: *Phragmodus undatus,* based on whole-collection numbers for the 4-component elements in the large dataset assembled by Kohut (1969), the higher point is based on counts in Bergström and Sweet (1966). Silurian: unpublished material from the Bainbridge Formation (since donated to Hans Peter Schönlaub). Devonian: based on element counts of Gable (1973) and Ramsey (1969). Carboniferous: unpublished Pennsylvanian collections of Sweet. Triassic, Ozarkodinida: based on data for *Hindeodus typicalis* in Sweet (1970; P elements = *Anchignathodus typicalis* and LA element of *Ellisonia teicherti*, S and M elements = remainder of *E. teicherti*). Using the same dataset, calculation of ratios for *Neospathodus cristagalli* (reconstruction of Purnell *et al.* in prep.) gives the same result (P elements = *Neospathodus cristagalli* and *Xaniognathus deflectus*, S and M elements = *Ellisonia gradata*). Triassic, Gondolellidae: based on counts by Sweet in most complete samples in the collections of Hieke (1967). Natural assemblage data are based on the following sources: Prioniodontida – O, *Paracordylodus gracilis* (Tolmacheva and Purnell 2002), *Phragmodus* (Repetski *et al.* 1998), *Oepikodus* (Stewart and Nicoll 2003), *Promissum pulchrum* (Aldridge *et al.* 1995); Ozarkodinida – S, *Ozarkodina steinhornensis* (Nicoll and Rexroad 1987), *Ozarkodina excavata* (von Bitter and Purnell 2005), *Ctenognathodus muchisoni* (von Bitter and Purnell 2004); D, *Criteriognathus* (Mashkova 1972; Purnell and Donoghue 1998), *Palmatolepis* (Lange 1968; Donoghue 2001*a*), *Bispathodus aculeatus* (Purnell and Donoghue 1998), *Polygnathus xylus xylus* (Nicoll 1985), *Polygnathus nodocostatus* (Dzik 1991; illustrated as *Hemilistrona*); C, *Clydagnathus windsorensis* (Aldridge *et al.* 1993; Purnell and Donoghue 1998), *Lochriea commutata* and *Gnathodus bilineatus* (Schmidt and Müller 1964; Norby 1976), *Cavusgnathus unicornis* (pers. obs.), *Adetognathus unicornis* (Purnell and Donoghue 1998); P, *Sweetognathus* (Ritter and Baesemann 1991); Prioniodinida – D, *Hibbardella angulata* (Nicoll 1977), *Prioniodina*? (pers. obs., unpublished cluster from the Gogo Formation); C, *Kladognathus* (Purnell 1993), *Idioprioniodus* (Purnell and von Bitter 1996); Ꝛ, *Neogondolella* (Rieber 1980; Orchard and Rieber 1999), *Pseudofurnishius* (Ramovš 1977, 1978), *Misikella* (Mastandrea *et al.* 1997, 1999).

[Special Papers in Palaeontology, 73, 2005, pp. 27–38]

MODES OF GROWTH IN THE EUCONODONT ORAL SKELETON: IMPLICATIONS FOR BIAS AND COMPLETENESS IN THE FOSSIL RECORD

by HOWARD A. ARMSTRONG

Department of Earth Sciences, University of Durham, Durham DH1 3LE, UK; e-mail: h.a.armstrong@durham.ac.uk

Abstract: Measuring abundance, one measure of an organism's success, is vital for describing the ecological context and community characteristics of fossil species and evaluating biases related to preservation and sampling. All euconodont 'death assemblages' will be biased by life history; short-lived individuals or those that moulted will be overrepresented compared with long-lived individuals. If a true measure of abundance is to be obtained we have to account for this bias. By comparison with living vertebrates it has been hypothesized that the euconodont oral skeleton, the first part of the vertebrate skeleton to mineralize, was deciduous; either elements or the whole apparatus could have been shed. This hypothesis has received little scrutiny and could have a direct bearing on the biological and palaeoecological inferences that can be made from species-abundance data. Natural assemblages support retention of elements whilst histological analysis indicates polycyclic growth comprising alternating periods of function and overgrowth that probably continued throughout life. This study challenges the idea of whole apparatus shedding by utilizing biometric data from time-averaged collections of *Idiognathodus* natural assemblages. Scatter plots of element linear dimensions exhibit clustering into size cohorts that can be used to elucidate population structure. Published time-specific and dynamic survivorship curves for the Silurian ostracod *Beyrichia jonesi* are used as a model for the population structure of a moulting organism. These curves are similar, having roughly the same average slope and identical mean rates of mortality per age class. The same survivorship curves for the *Idiognathodus* collection have different average slopes and mean mortality rates. It is therefore concluded that the apparatus was retained throughout life. At least in this taxon, and in the absence of preservational or sampling biases, element-abundance data are a true measure of a species' success. However, quantitative palaeoecology of euconodonts will not become fully rigorous until the age of euconodonts can be determined.

Key words: euconodonts, growth, survivorship, biometrics.

MEASURING abundance and hence the success of an organism in a fossil assemblage is not the same as measuring abundance in living communities. Organisms with short and long lifespans are treated the same and samples are almost always time-averaged (Vermeij and Herbert 2004). In addition to preservational and sampling biases, all euconodont 'death assemblages' are biased by life history; short-lived individuals or those that shed elements or the apparatus will be overrepresented in samples compared with long-lived individuals. This will have a direct bearing on the completeness (the proportion of taxa sampled), bias (relative proportions of each taxon sampled) and hence the fidelity of biological and palaeoecological inferences that can be made from euconodont species-abundance data. If euconodonts were shown to have shed either their elements or their apparatuses, palaeoautecology analyses would have to be restricted to presence-absence data and the absence of a size–age relationship would mean element abundances would reflect shedding rates rather than the number of individuals (Donoghue and Purnell 1999). The palaeosynecology of euconodonts would be non-viable unless it were possible to determine shedding rates of particular species and this remained constant through time. If a true measure of abundance is to be obtained we have to reject the shedding hypothesis.

COMPETING MODELS OF GROWTH

The oropharyngeal cavity of euconodonts bore an array of endoskeletal elements composed of dentine and enamel, homologous with these tissues in vertebrates (see Donoghue and Sansom 2002 for a review). Elements grew appositionally and the patterning of euconodont elements (Donoghue 1998) and their food gathering and processing function (Purnell 1993, 1995) are widely accepted. But few studies have attempted to document euconodont life histories.

Polycyclic growth model

The crowns of euconodonts are composed entirely of enamel that exhibit internal wear discontinuities and overgrowth patterns, developed over multiple growth cycles (Bengston 1976, 1983; Jeppsson 1979; Purnell 1995; Donoghue 1998, 2001; Donoghue and Purnell 1999). This contrasts with the deciduous teeth in other vertebrates that have relatively thin enamel crowns that develop rapidly, a pattern consistent with a short period of function before being replaced (Zhang *et al.* 1997; Donoghue 2001).

Müller and Nogami (1971) first illustrated and described polycyclic growth in platform euconodonts, each growth cycle beginning with a relatively thick lamella, with subsequent lamellae gradually decreasing in thickness to a point where the final lamella no longer covered the oral surface. Subsequent work has shown euconodont crown enamel typically consists of individual lamellae grouped into 'sets' of 8–10 (Müller 1972; Zhang *et al.* 1997; Donoghue 1998; Donoghue and Purnell 1999; Armstrong and Smith 2001).

Zhang *et al.* (1997) documented this pattern in the platform and ramiform elements in *Parapachycladina*, a taxon currently assigned to the order Prioniodinida. They concluded that the duration of growth was not discernible, but if the growth increments are daily and the cycles annual, then most of the annual cycle was spent functioning.

Armstrong and Smith (2001) distinguished lamellae and sets of lamellae as minor and major increments and concluded that minor increments are equivalent to the cross striations in hominoid enamel. They demonstrated that polycyclic growth also occurred in the coniform protopanderodontids *Protopanderodus varicostatus* (Sweet and Bergström, 1962) and *Drepanodus robustus* (Hadding, 1913). Both of these species bore a complex multidomain apparatus similar to that of *Panderodus* (Armstrong 1997, 2000). Elements of *D. robustus* contained five major increments whilst those of *P. varicostatus* contained nine (Armstrong and Smith 2001).

In *P. varicostatus* minor increments typically have a minimum thickness of *c.* 1 μm, and by analogy with other vertebrates, were likely to have been deposited in a day. Few minor increments were, however, 1 μm thick; more typically they were up to 7 μm, therefore reflecting multiple growth episodes lasting up to a week. Major increments represent periods of continuous growth over a month. Intervening periods of function have unknown durations. The growth of *P. varicostatus* was further characterized by two distinct phases: the production of a conical, asymmetrical 'proto-element', followed in a second phase by the development of the curved and twisted geometry of the adult element.

By analogy with fish, Armstrong and Smith (2001) concluded that element growth in *P. varicostatus* was initiated at the beginning of the juvenile stage; growth continued in the juvenile for approximately 1 month and was followed by a change in mode of growth at the onset of sexual maturity. Growth of the elements was therefore indeterminate and continued throughout life. Donoghue (1998) argued that in order to rationalize polycyclic growth it was necessary to reject the shedding hypothesis.

Shedding models

The hypothesis that euconodont elements were shed has developed from two sources: (1) collections of ozarkodinid elements (Carls 1977), particularly Pennsylvanian taxa such as *Idiognathodus* that show apparent platform element overrepresentation (for further discussion, see Purnell and Donoghue 2005; von Bitter and Purnell 2005); (2) a comparison of the growth of extant cyclostomes and euconodonts. Extant cyclostomes, the hagfish and lampreys, frequently shed their keratinous teeth after intense feeding activity and it has been speculated that shedding the teeth was a precursor to a period of growth in the animal (Kresja *et al.* 1990; Kresja 1990). If a euconodont–hagfish tooth element homology can be assumed and euconodonts were related to fossil or extant cyclostomes, then euconodont crowns could represent shed tooth coverings. Subsequent work on euconodont histology and the phylogenetic relationships of early vertebrates has shown both of these hypotheses to be incorrect (e.g. Sansom *et al.* 1992; Donoghue 1998; Donoghue and Sansom 2002).

The most clearly argued case for shedding based on apparent platform overrepresentation was presented by Carls (1977). Through a careful evaluation of the relative frequency of occurrence of different element types in collections of ozarkodinid taxa from the Silurian and Lower Devonian of Europe he argued that rates of shedding varied according to position in the apparatus. P elements, he concluded, were shed more frequently and were thus overrepresented in collections of isolated elements. Merrill and Powell (1980) also provided detailed evidence of P element overrepresentation derived from a comparison of element size-frequency distributions from the Lower and Upper Members of the Pennsylvanian Drum Limestone and an interbedded shale parting. They used complete elements to recalculate the dimensions of incomplete elements, using linear scaling functions for S element length (process height $\times$ 34) and P_1 length (platform length $\times$ 1·67). Platform elements from the shale parting were smaller that those from the adjacent limestone beds and were considered juvenile forms. P_1/S ratios, based on an assumption of apparatuses containing seven S elements rather than nine, indicated minimal platform overrepresentation for the shale fauna whilst limestone faunas were typically platform overrepresented. While P element shedding may provide an explanation for this pattern, Merrill and Powell concluded that juvenile forms had a fully developed apparatus whilst the adults commonly lost S elements through resorption, implying that either the mode of feeding must have changed or the adults did not feed.

DEVELOPING TESTS FOR COMPETING GROWTH HYPOTHESES

Element shedding

Natural assemblages preserve the oral skeleton as it was *in vivo*, allow for an analysis of growth excluding the vagaries of taphonomic sorting of individual elements, and provide a baseline condition from which to test the fidelity of discrete element collections. The element-shedding hypothesis can only be tested in taxa where the apparatus can be confirmed from natural assemblages and predicts that only partial natural assemblages of adults will be found. Purnell (1994) and Purnell and Donoghue (1998) noted that the vast majority of natural assemblages of *Idiognathodus* and *Gnathodus*, despite showing a range of sizes, contain no obviously smaller element and do not show systematic loss of the S elements.

Whole apparatus shedding

Demonstrating the retention of elements within the apparatus does not, however, preclude the possibility that the whole apparatus was shed. This hypothesis is more difficult to falsify (Donoghue and Purnell 1999); a test based on survivorship analysis forms the focus of this contribution.

Survivorship curves reveal the pattern of death which led to the formation of the fossil assemblage. In essence these curves show the relative survival of individuals as a function of time; plotted on semi-logarithmic axes, equal slopes equate to equal rates of death. The age structure of a fossil assemblage will vary according to whether it is composed of dead individuals or shed skeletons, and under what conditions the assemblage formed, either through the steady accumulation of dead animals or through mass mortality. A mass mortality event (assuming no moults are preserved) will provide a census of the age structure of a population at a moment in time (time-specific survivorship). Similarly, an assemblage that only represents the accumulation of shed skeletons, or moults, preserves a record of animals that were alive and is equivalent to a census population; it can thus be used directly to construct a time-specific survivorship curve. If, on the other hand, the fossils in the assemblage represent dead individuals, then the assemblage provides a record of death, analogous to following a cohort of individuals throughout life and measuring mortality rates; a dynamic survivorship analysis would be more appropriate. The survivorship curve based on this approach would be similar, but not identical, to the time-specific analysis. The classic example of dynamic and time-specific survivorship analysis in fossils is that of Kurtén (1964; also used by Raup and Stanley 1978 and Dodd and Stanton 1981). Kurtén analysed an assemblage comprising the moulted valves of the Silurian ostracod *Beyrichia jonesi* from Gotland and found that the dynamic and time-specific survivorship plots were very similar, having roughly the same average slope, and identical mean rates of mortality per age class (32%). This provides a test of the apparatus-shedding hypothesis: if the euconodont apparatus was shed, then dynamic and time-specific survivorship curves should be similar, having the same average slope and identical mean mortality rates.

In applying the dynamic survivorship method in the analysis of euconodont population structure it is important to note that it does not matter whether the preserved skeletons are the remains of animals spawned during a single time period or over several time periods if growth rate and survivorship are nearly constant within the population through time. For the purposes of this study it is assumed that element size is correlated with age: euconodonts grew by the appositional addition of enamel lamellae and there is clear evidence that larger euconodonts contained larger elements (Purnell 1994). Single element types are used herein as proxies for the size of the euconodont apparatus.

MATERIAL AND METHODS

All linear size data were provided by Dr M. A. Purnell (referred to in Purnell 1994) and obtained from orientated specimens to an error of 10 μm using an occular graticule on a Wild binocular microscope (Text-fig. 1A; Appendix: Linear Size Data and cluster analysis data). S element dimensions of *Idiognathodus* come from specimens within natural assemblages and include the collections of DuBois (1943), Rhodes (1952) and Avcin (1974) held by the Illinois Geological Survey, and new material currently in the Micropalaeontological Collections of the University of Leicester. Most are from the black shale at the top of the Modesto Formation at Bailey Falls (see Rhodes 1952); a few assemblages are from the Francis Creek Shale Member, Wolf Covered Bridge (Oak Grove locality 4 of Merrill and King 1971).

It is assumed that within a single time-averaged sample no morphological change has occurred over the duration of the sample (Bush *et al.* 2002). Craig and Hallam (1963) recognized that in natural populations the resulting size-frequency distribution depended upon growth rate and survivorship. Normally distributed size-frequency distributions were found in organisms with exponential or logistic growth and constantly increasing mortality rate, and can be taken to indicate an absence of taphonomic sorting. The size-frequency distribution was tested for skew and kurtosis using protocols in Microsoft Excel, and samples show only a slight positive skew and small negative kurtosis (Text-fig. 1B), indicating an absence of significant taphonomic sorting.

Complete Linkage cluster analysis, conducted using the cluster analysis protocol in MINITAB version 14, allows the discrimination of size cohorts along growth trajectories [a feature first recognized by Jeppsson (1976) in discrete element collections of *Ozarkodina confluens* from Gotland]. This procedure uses a hierarchical agglomerative method in which the largest distance between one cluster and its nearest surrounding cluster is used to build

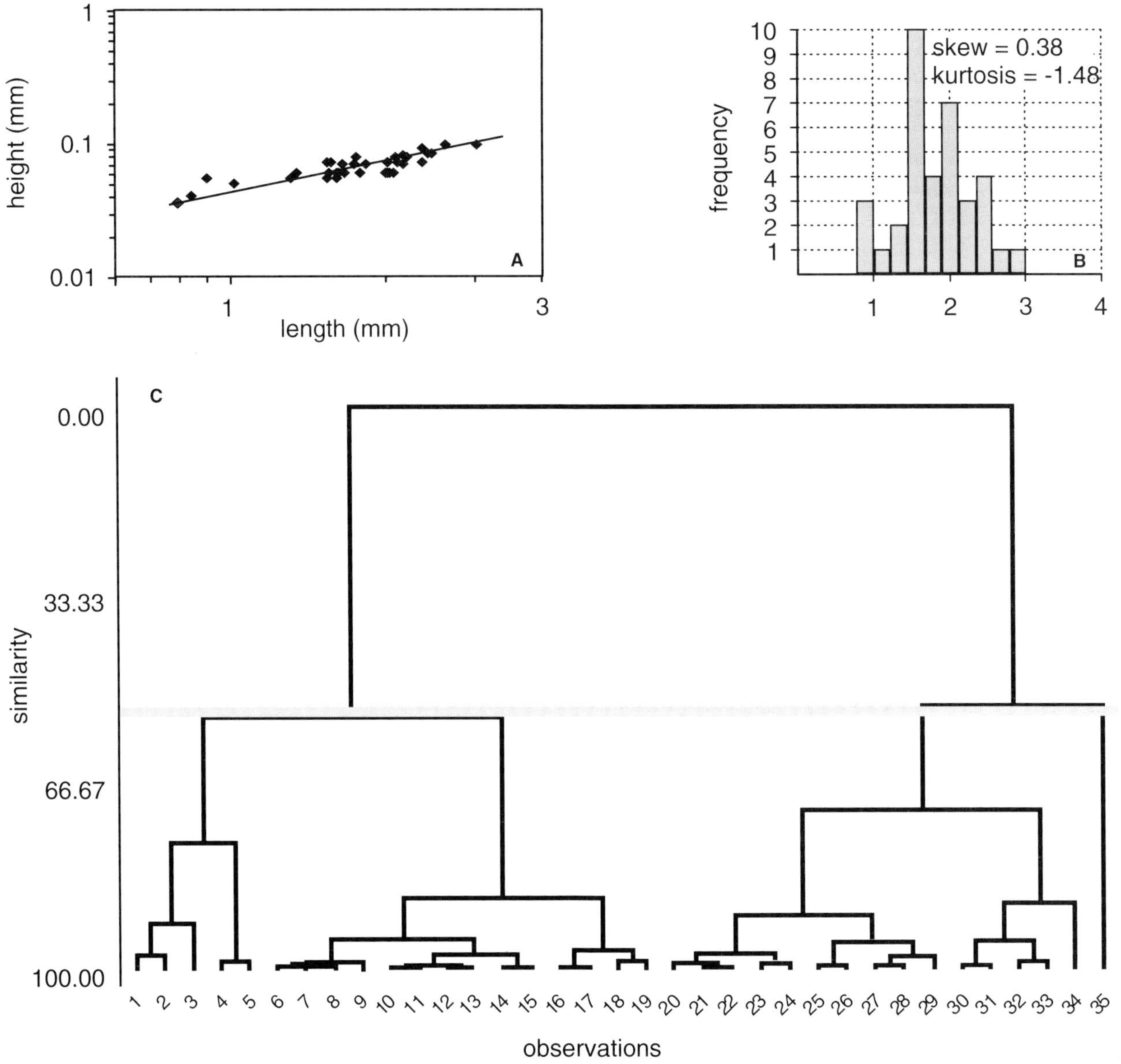

TEXT-FIG. 1. *Idiognathodus* biometric data. A, scatter plots of S element linear dimensions. B, size-frequency histogram for S element linear dimensions; histogram was produced in PAST (Hammer *et al.* 2001) with bins automatically selected to ten. C, complete linkage cluster analysis for *Idiognathodus* for S element linear dimensions; the dendrogram is cut (grey) at three clusters based on the results in Text-figure 2 (see text for explanation).

the cluster hierarchy (for an explanation of the advantages of this technique over other hierarchical methods, see Everitt 1993). It begins with all observations being separate, each forming its own cluster. In the first step, the two nearest observations in each group are joined and the distance recorded. In the next step, either a third observation joins the first two and the furthest distance is recorded or two other observations join together in a different cluster. The distance between the clusters thus formed is defined as that of the most distant pair of individuals, one from each cluster. This process will continue until all clusters are joined into a single cluster. The dendrogram for complete linkage cluster analysis represents the maximum cluster size at each partition. A single cluster is not useful for classification purposes; a decision has to be made as to how many groups are logical for the data. Subjective and statistical methods are available to inform this decision. The dendrogram provides a visual representation of the hierarchical cluster analysis but it is difficult to obtain direct quantitative information from the dendrogram (Text-fig. 1C). The final grouping is defined by the level of similarity at any step, i.e. the per cent of the minimum distance at that step relative to the maximum interobservation distance in the data. Re-running the clustering procedure using the specified similarity level gives a set number of groups for cutting the dendrogram. The resulting clusters in the final partition can be compared *a posteriori* to the data to see if the grouping seems logical.

However, the nature of hierarchical cluster analysis means that it is possible to generate dendrograms, which provide clustering solutions within any type of distribution, even within random distributions. It is therefore essential to obtain an independent estimate of the number of clusters expected in the measured dataset. Normalizing the results of the measured-data cluster analysis to those of a random distribution allows a measure of how different the cluster analysis results are from a random distribution (Eneva *et al.* 1992; Jeram *et al.* 1996; Jerram and Cheadle 2000; Text-fig. 2).

The information calculated from a complete linkage cluster analysis of a point distribution can be represented by two components: the maximum cluster size (or diameter) at each partition and the cluster frequency (number of clusters at each partition). The number of clusters needs to be expressed as a frequency to allow samples of different sizes to be compared and is given by

$$n_c/(N-1) \times 100\%$$

where n_c is the number of clusters and N is the population size. Cluster frequency is considered to be 100 per cent at the bottom of the hierarchy (lowest similarity), when the distribution consists of a single cluster of two points; the cluster 'seed' and the rest of the distribution is

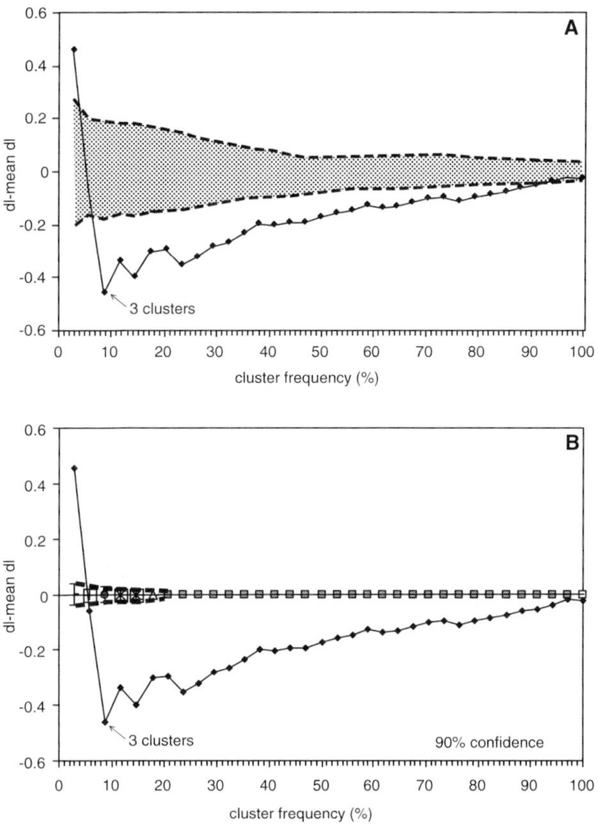

TEXT-FIG. 2. Jerram and Cheadle diagrams. A, cluster frequency vs. residual distance minus mean residual distance (dl – mean dl) for complete linkage cluster analysis of ten random distributions of points with the same size range and population size as the measured data, normalized to the mean distance for these distributions; stippled area between dashed lines defines the envelope of these data; solid line is the residual distance of the *Idiognathodus* data, normalized to the mean distance of the random distributions. B, error bars define 90 per cent of the variation within the random distributions; the solid line defines the residual distance minus mean residual distance (dl – mean dl) of the *Idiognathodus* data, normalized to the mean distance of the random distributions.

made up of single member clusters (i.e. individual points). At the top of the hierarchy all of the population is contained in one cluster and the cluster frequency is $1/(N-1) \times 100\%$ (Jerram and Cheadle 2000).

Complete linkage hierarchical cluster analysis is performed on the measured data and ten populations of random numbers (generated within MINITAB version 14 using the size range of the original data). Normalizing the measured data to the data from the random distribution of points is performed by subtracting the mean distance recorded at each cluster number for the random cluster data from the distance recorded at the same cluster number for the measured sample to give a residual distance. It is useful to quantify the variation that can be expected between the results of the complete linkage cluster analysis of different random distributions. This gives error bars to the random distribution and helps to identify true cluster variations expressed in terms of confidence limits. Jeram *et al.* (1996) showed that for ten populations of randomly distributed points, as the population size becomes larger the variation among the different random populations becomes less; also that the largest variation in the distribution occurs at low cluster frequencies, where the largest distance is recorded in the cluster hierarchy.

The residual distance of the measured distribution is then plotted on the residual distance–cluster frequency plot, and the resulting curve measures, in terms of residual distance, how much the hierarchical cluster analysis for the measured sample differs from a random distribution of points. Where significant differences from a random distribution occur, residual distance in the measured data will plot outside the residual distance envelope of the random distributions. Significant steps in the hierarchy are reflected as troughs in the residual distance of the measured sample line (Text-fig. 2). Cluster number is determined from the cluster frequency at these points.

Dynamic and time-specific survivorship curves are plotted (Text-fig. 3) following the methodology of Kurtén (1953, 1964). The data required for these calculations are recorded in life history tables (Text-fig. 3). The mortality rate (q_z) is the ratio of the number dying in an age interval (d_z) to the survivors at the start of the interval (l_z). The data in the l_z column are the cumulative form of those in the d_z column: the first l_z value is the sum of all the d_z values, the second l_z value is the sum of the second through to last d_z values, and so on. Because of this relationship, l_z can always be calculated from d_z and vice versa. If we assume that the fossil assemblage is the result of a mass mortality or of moulting alone the frequency of the specimens in the age clusters are l_z data. If, on the other hand, we assume that the fossil assemblage formed from the gradual accumulation of individuals dying naturally then the frequency data are most appropriately considered as d_z data and l_z is calculated from them by the cumulative process (Raup and Stanley 1978). The survivorship curve is a plot of the cluster/age cohort vs. log(l_z) in the life table. The average slopes of the survivorship curves are described by an exponential function calculated in Microsoft Excel (Text-fig. 3).

Time specific	d_z	l_z	q_z
a	19	35	0·54
b	15	16	0·94
c	1	1	

Dynamic	d_z	l_z	q_z
a	4	19	0·21
b	14	15	0·93
c	-	1	

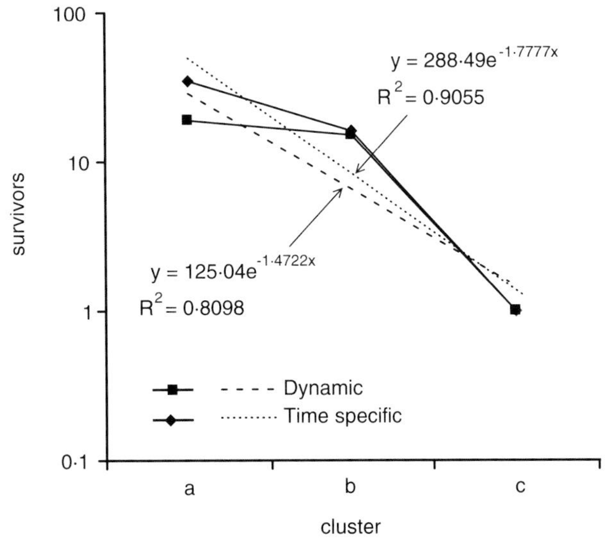

TEXT-FIG. 3. Life history tables and survivorship curves (solid); broken lines are trend lines described by exponentials; R^2 is the correlation coefficient.

RESULTS

Platform overrepresentation in Idiognathodus

The growth trajectory of the *Idiognathodus* apparatus determined from cross-plotting linear dimension data of P_1 and S elements in natural assemblages is identical to that calculated by Purnell (1993, 1994). The apparatuses in the natural assemblages have a mean S length/P_1 length ratio of 1·4.

Merrill and Powell (1980) used linear scaling from complete elements to obtain sizes for incomplete specimens. Recalculating element sizes in the Drum Limestone data using the regressions of element dimensions calculated using the natural assemblage data indicates the bias imparted by linear scaling: in all cases Merrill and Powell underestimated P_1 length and overestimated S length (Text-fig. 4). Recalculated S length/P_1 length ratios are 2·54, 3·2 and 2·08 in the Upper Limestone, shale parting and Lower Limestone samples, respectively. In terms of S and P_1 element relative dimensions, although the Upper and Lower Limestone specimens are larger than those in the assemblages the samples fall on the growth trajectory calculated from the natural assemblage data (Text-fig. 4).

Population structure as a test of whole apparatus shedding

The scatter plot of *Idiognathodus* S element linear dimensions has a growth trajectory that is described by a

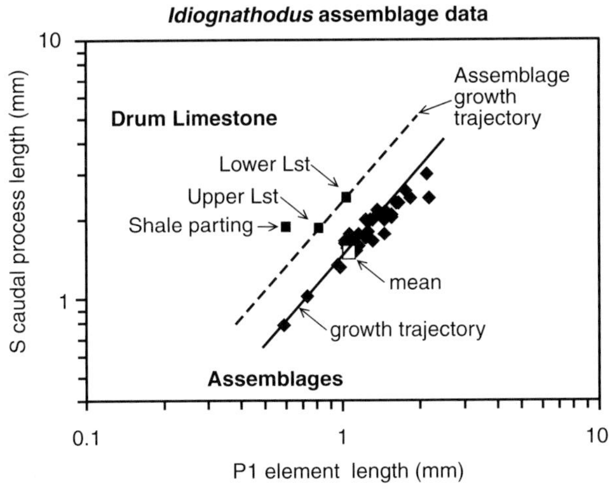

	Mean P length[1]	P length calculated[2]	Mean S length[1]	S length calculated[3]	S height[1]
Upper Limestone	0·65	0·82	2·48	2·07	0·073
Shale parting	0·47	0·61	2·38	1·95	0·07
Lower Limestone	0·92	1·13	2·72	2·36	0·08

[1] Data from Merrill and Powell (1972)
[2] Calculated using the RMA method from Purnell (1994)
[3] Calculated using the growth trajectory of S length to S height for assemblage data and the S height data in Merrill and Powell (1972)

TEXT-FIG. 4. Apparatus growth trajectories for collections of *Idiognathodus* based on P_1 and S element linear dimensions (see also Purnell 1994); assemblage data provided by Dr M. A. Purnell (see text); Drum Limestone data are from Merrill and Powell (1980); the Drum Limestone faunas were collected from the Lower and Upper Members of the limestone and an interbedded shale parting.

power function $y = 0{\cdot}044x^{0{\cdot}697}$ ($r = 0{\cdot}8622$). The data are divided into three clusters (Text-figs 1A, C, 2). Dynamic and time-specific survivorship curves are separate for the S element data and indicate constantly increasing death rate with increasing age (Text-fig. 3). The time-specific curve represents a mean mortality rate of 0·74 and slope described by the exponential $y = 288{\cdot}49e^{-1{\cdot}7777x}$ ($R^2 = 0{\cdot}8098$). The dynamic curve represents a mean mortality rate of 0·57 and a slope described by the exponential $y = 125{\cdot}04e^{-1{\cdot}4722x}$ ($R^2 = 0{\cdot}9055$) (Text-fig. 3).

INTERPRETATION

The element-shedding hypothesis predicts the loss of S elements in the adult, thereby providing a biological explanation for the commonly observed platform over-representation in collections of conodonts. In large collections of *Idiognathodus* natural assemblages the vast majority of these are complete across a wide range of sizes and large apparatuses are complete. Evidence from these is also lacking for an absolute reduction in element size, to be expected if elements were resorbed in later growth stages (Purnell 1994; Purnell and Donoghue 1998). The element-shedding hypothesis is rejected and other taphonomic explanations have to be sought for P element overrepresentation.

Merrill and Powell (1980) used linear scaling to determine the full size of fragmentary specimens. This scaling imposes a bias on the data, consistently underestimating P_1 element size and overestimating S element size as compared with scaling relationships determined by regression methods. Comparison of recalculated average S/P_1 length ratios in *Idiognathodus* from the Drum Limestone samples with the growth model obtained from natural assemblage data indicates that in the Upper and Lower Limestone samples S/P_1 ratio is consistently higher but falls on the same growth trajectory. That the growth trajectory obtained from the natural assemblages can be mapped onto the Drum Limestone faunas, collections from different localities, facies and stratigraphical level, indicates that apparatus growth rate is likely to have been a genetically controlled character. It is the shale parting sample that shows the greatest departure from this trajectory, with S elements larger than they should be for the P_1 element size. This implies either greater removal of smaller S elements from the shale sample or that these animals grew unusually long S elements relative to P_1 element size. The shale sample has a much lower P_1 element overrepresentation than the limestone samples, implying that fewer S elements have been removed. Individuals preserved in the shale may simply have had longer S elements relative to their P_1 elements and arguably could be referred to a different taxon. There is clearly not a simple relationship between element size and preservation.

A fully rigorous test of the hypothesis that euconodonts shed their entire apparatus using survivorship analysis requires two conditions to be fulfilled: that the specimens were derived from a natural population, and that ages are known. It has been assumed that the size of the feeding elements increases with age. This is a biological imperative and is supported by limited data that larger euconodonts contained larger elements (Purnell 1994). However, it remains an open question as to whether size cohorts represent true age clusters in the *Idiognathodus* assemblages, a question that could be answered by histological analysis of specimens selected from each cluster. The survivorship analysis of the ostracod population provides a model for an organism that moults. Survivorship curves have very similar slopes and identical mortality rates between age clusters. The observed differences between the slope and mortality rates between age clusters in the survivorship curves of *Idiognathodus* indicate that this assemblage formed from the natural mortality of individuals and not from moulting.

The close spacing of the size cohorts along euconodont growth trajectories could reflect either small sample sizes or biological factors. For example, a single size cohort could include individuals with a range of ages resulting from a prolonged spawning period, a feature of many fish species (Jobling 1995), or could reflect phenotypic or ecophenotypic variation. Significantly larger collections than are currently available would be needed to distinguish between these factors. There is no evidence from the growth analysis for breaks in growth consistent with a functional or ecological shift from a pelagic mode of life in early growth stages to a nektobenthic habit in later life (Sweet 1988).

The shape of the dynamic survivorship curve for *Idiognathodus* is convex upwards [a shape also found in *O. confluens* (Jeppsson 1976)], indicating a constantly increasing probability of death with increasing age. This type of survivorship is a feature of K-strategists favouring maximum competition between adults, larger body size, late first reproduction, iteroparity and with few progeny per breeding cycle (Skelton 1993).

CONCLUSIONS

It is concluded that the axiom of platform overrepresentation in *Idiognathodus*, based largely on the evidence from the Drum Limestone samples, is the result of either a taphonomic bias (see Purnell and Donoghue 2005) or that individuals preserved in the shale parting may have been a taxon (subspecies or species) distinct from those in the limestone samples. A scatter plot of element dimensions in *Idiognathodus* shows closely spaced clusters along the growth trajectory. In the absence of an independent test for age it remains an open question as to whether these size cohorts represent true age clusters in time-averaged samples.

Differences in the slope and mortality rates between age clusters represented by the dynamic and time-specific survivorship curves indicate that the apparatus was retained in *Idiognathodus*. At least in this taxon element abundance data are a true reflection of the species' success. It is likely that *Idiognathodus* was a K-strategist.

Whilst quantitative palaeoecology and the study of population dynamics in euconodonts is possible using a range of simple biometric and statistical techniques, these can only become fully rigorous when the independent age of the taxa considered can be determined.

Acknowledgements. This work was initiated at the suggestion of Dr M. A. Purnell and has greatly benefited from his subsequent input and comments. Dr D. A. Jerram is thanked for help with the cluster analysis. The paper has also benefited significantly from the comments of Dr P. C. J. Donoghue, Dr J. Dzik and the editor, Prof. D. J. Batten.

REFERENCES

ARMSTRONG, H. A. 1997. Conodonts from the Shinnel Formation, Tweeddale Member (middle Ordovician), Southern Uplands, Scotland. *Palaeontology*, **40**, 763–799.

—— 2000. Conodont micropalaeontology of mid-Ordovician aged limestone clasts from the Lower Old Red Sandstone conglomerates, Lanark and Strathmore basins, Midland Valley, Scotland. *Journal of Micropalaeontology*, **19**, 45–59.

—— and SMITH, C. J. 2001. Growth patterns in euconodont crown enamel: implications for life history and mode-of-life reconstruction in the earliest vertebrates. *Proceedings of the Royal Society of London, Series B*, **268**, 815–820.

AVCIN, M. J. 1974. Des Moines conodont assemblages from the Illinois Basin. Unpublished PhD thesis, University of Illinois at Urbana.

BENGSTON, S. 1976. The structure of some Middle Cambrian conodonts and the early evolution of structure and function. *Lethaia*, **9**, 185–206.

—— 1983. The early history of the Conodonta. *Fossils and Strata*, **15**, 5–19.

BUSH, A. M., POWELL, M. G., ARNOLD, W. S., BERT, T. M. and DALEY, G. M. 2002. Time-averaging, evolution and morphologic variation. *Paleobiology*, **28**, 9–26.

CARLS, P. 1977. Could conodonts be lost and replaced? – Numerical relationships among disjunct conodont elements of certain Polygnathidae (late Silurian–Lower Devonian, Europe). *Neues Jahrbuch für Geologie und Paläontologie, Abhandlungen*, **155**, 18–64.

CRAIG, G. Y. and HALLAM, A. 1963. Size-frequency and growth-ring analyses of *Mytilus edulis* and *Cardium edule*, and their palaeontological significance. *Palaeontology*, **6**, 731–750.

DODD, J. R. and STANTON, R. J. Jr 1981. *Paleoecology, concepts and applications*. Wiley, New York, 599 pp.

DONOGHUE, P. C. J. 1998. Growth and patterning in the conodont skeleton. *Philosophical Transactions of the Royal Society of London, Series B*, **353**, 633–666.

—— 2001. Microstructural variaton in conodont enamel is a functional adaptation. *Proceedings of the Royal Society of London, Series B*, **268**, 1691–1698.

—— and PURNELL, M. A. P. 1999. Growth, function, and the conodont fossil record. *Geology*, **27**, 251–254.

—— and SANSOM, I. J. 2002. Origin and early evolution of vertebrate skeletonization. *Microscopy Research and Technique*, **59**, 352–372.

DUBOIS, E. P. 1943. Evidence on the nature of conodonts. *Journal of Paleontology*, **17**, 155–159.

ENEVA, M., HAMBURGER, M. W. and POPANDOPULO, G. A. 1992. Spatial distribution of earthquakes in aftershock zones of the Garm region, Soviet Central Asia. *Geophysics Journal International*, **109**, 38–53.

EVERITT, B. S. 1993. *Cluster analysis*. Edward Arnold, New York, 231 pp.

HADDING, A. 1913. Untre dicellograptusskiffen i Skåne jämte några därmed ekvivalenta bildningar. *Lunds Universitets Årsskrift, Avdelningen 2*, **915**, 1–90.

HAMMER, O., HARPER, D. A. T. and RYAN, P. D. 2001. PAST: paleontological statistics software package for education data analysis. *Palaeontologica Electronica*, **4**, Article 4.

JEPPSSON, L. 1976. Autecology of Late Silurian conodonts. 105–119. *In* BARNES, C. R. (ed.). *Conodont paleoecology*. Geological Association of Canada, Special Paper, **15**, 324 pp.

—— 1979. Conodont element function. *Lethaia*, **12**, 153–171.

JERAM, A. J., CHEADLE, M. J., HUNTER, R. H. and ELLIOTT, M. T. 1996. The spatial distribution of grains and crystals in rocks. *Contributions in Mineralogy and Petrology*, **125**, 60–74.

JERRAM, D. A. and CHEADLE, M. J. 2000. On the cluster analysis of grains and crystals in rocks. *American Mineralogist*, **85**, 47–67.

JOBLING, M. 1995. *Environmental biology of fishes*. Fish and Fisheries Series, **16**. Chapman & Hall, London, 448 pp.

KRESJA, R. J. 1990. A neontological interpretation of conodont elements based on agnathan cyclostome tooth structure, function and development. *Lethaia*, **23**, 359–378.

—— BRINGAS, P. J. and SLAVKIN, H. C. 1990. The cyclostome model: an interpretation of conodont element structure and function based on cyclostome tooth morphology, function, and life history. *Courier Forschunginstitut Senckenberg*, **118**, 473–492.

KURTÉN, B. 1953. On the variation and population dynamics of fossil and recent mammal populations. *Acta Zoologica Fennica*, **76**, 1–22.

—— 1964. Population structure in palaeoecology. 91–106. *In* IMBRIE, J. and NEWELL, N. D. (eds). *Approaches in paleoecology*. Wiley, New York, 432 pp.

MERRILL, G. K. and KING, C. W. 1971. Platform conodonts from the lower Pennsylvanian rocks of northwestern Illinois. *Journal of Paleontology*, **45**, 645–664.

—— and POWELL, R. J. 1980. Paleobiology of juvenile (nepionic?) conodonts from the Drum Limestone (Pennsylvanian, Missourian, Kansas City area) and its bearing on apparatus ontogeny. *Journal of Paleontology*, **54**, 1058–1074.

MÜLLER, K. J. 1972. Growth and function of conodonts. *International Geological Congress, 24th Session, Montreal*, **6**, 20–27.

—— and NOGAMI, Y. 1971. Über den Feinbau der Conodonten. *Faculty of Science, Kyoto University, Geology and Mineralogy Series, Memoir*, **38**, 87 pp.

PURNELL, M. A. 1993. Feeding mechanisms in conodonts and the function of the earliest vertebrate hard tissues. *Geology*, **21**, 375–377.

—— 1994. Skeletal ontogeny and feeding mechanisms in conodonts. *Lethaia*, **27**, 129–138.

—— 1995. Microwear on conodont elements and macrophagy in the first vertebrates. *Nature*, **374**, 798–800.

—— and DONOGHUE, P. C. J. 1998. Skeletal architecture, homologies and taphonomy of oarkodinid conodonts. *Palaeontology*, **41**, 57–102.

—— —— 2005. Between death and data: biases in interpretation of the fossil record of complex conodonts. 7–25. *In* PURNELL, M. A. and DONOGHUE, P. C. J. (eds). *Conodont biology and phylogeny: interpreting the fossil record*. Special Papers in Palaeontology, **73**, 218 pp.

RAUP, D. M. and STANLEY, S. M. 1978. *Principles of paleontology*. Second edition. W. H. Freeman, San Francisco, 481 pp.

RHODES, F. H. T. 1952. Classification of Pennsylvanian conodont assemblages. *Journal of Paleontology*, **26**, 886–901.

SANSOM, I. J., SMITH, M. P., ARMSTRONG, H. A. and SMITH, M. M. 1992. Presence of the earliest vertebrate hard tissues in conodonts. *Science*, **256**, 1308–1311.

SKELTON, P. 1993. *Evolution: a biological and palaeontological approach*. Addison-Wesley Co. in association with The Open University, Wokingham, 1064 pp.

SWEET, W. C. 1988. *The Conodonta: morphology, taxonomy, paleoecology, and evolutionary history of a long-extinct animal phylum*. Oxford Monographs on Geology and Geophysics, Clarendon Press, Oxford, 212 pp.

—— and BERGSTRÖM, S. M. 1962. Conodonts from the Pratt Ferry Formation (Middle Ordovician) of Alabama. *Journal of Paleontology*, **36**, 1214–1252.

VERMEIJ, G. J. and HERBERT, G. S. 2004. Measuring relative abundance in fossil and living assemblages. *Paleobiology*, **30**, 1–4.

von BITTER, P. H. and PURNELL, M. A. 2005. An experimental investigation of post-depositional taphonomic bias in conodonts. 39–56. *In* PURNELL, M. A. and DONOGHUE, P. C. J. (eds). *Conodont biology and phylogeny: interpreting the fossil record*. Special Papers in Palaeontology, **73**, 218 pp.

ZHANG, S., ALDRIDGE, R. J. and DONOGHUE, P. C. J. 1997. A Triassic conodont with periodic growth? *Journal of Micropalaeontology*, **16**, 65–72.

APPENDIX

Linear dimension and cluster analysis data

Idiognathodus data were provided by Dr M. A. Purnell.

Idiognathodus S elements

Specimen no.	No. of clusters	Length	Height	Cluster frequency (%)	d*l* − mean d*l*	Distance level
1	34	0·840	0·040	100·00	−0·023	0·00806
2	33	0·900	0·054	97·06	−0·020	0·03046
3	32	1·020	0·050	94·12	−0·037	0·03373
4	31	1·309	0·054	91·18	−0·053	0·03413
5	30	1·340	0·060	88·24	−0·059	0·039
6	29	1·548	0·072	85·29	−0·075	0·0392
7	28	1·548	0·054	82·35	−0·085	0·03996
8	27	1·560	0·060	79·41	−0·096	0·04123
9	26	1·577	0·072	76·47	−0·111	0·04295
10	25	1·606	0·054	73·53	−0·097	0·06943
11	24	1·610	0·060	70·59	−0·103	0·07841
12	23	1·625	0·060	67·65	−0·117	0·07864
13	22	1·625	0·060	64·71	−0·133	0·08612
14	21	1·660	0·070	61·76	−0·138	0·09209
15	20	1·664	0·060	58·82	−0·126	0·11673
16	19	1·750	0·070	55·88	−0·147	0·11717
17	18	1·760	0·078	52·94	−0·156	0·11763
18	17	1·794	0·060	50·00	−0·174	0·1287
19	16	1·830	0·070	47·06	−0·193	0·14068
20	15	2·000	0·060	44·12	−0·191	0·16907
21	14	2·019	0·072	41·18	−0·203	0. 1867
22	13	2·020	0·060	38·24	−0·198	0·22129
23	12	2·040	0·060	35·29	−0·233	0·2327
24	11	2·067	0·060	32·35	−0·268	0·2686
25	10	2·100	0·078	29·41	−0·282	0·29296
26	9	2·120	0·072	26·47	−0·324	0·29993
27	8	2·160	0·080	23·53	−0·355	0·33526
28	7	2·170	0·070	20·59	−0·295	0·45974
29	6	2·212	0·078	17·65	−0·303	0·55419
30	5	2·356	0·090	14·71	−0·397	0·55581
31	4	2·360	0·072	11·76	−0·336	0·83702
32	3	2·428	0·084	8·82	−0·459	0·99007
33	2	2·466	0·084	5·88	−0·060	1·63861
34	1	2·616	0·096	2·94	0·457	2·72305
35		3·026	0·096			

Random distance levels

Specimen no.	Random 1	Random 2	Random 3	Random 4	Random 5	Random 6	Random 7	Random 8	Random 9	Random 10
1	0·03478	0·04123	0·06825	0·01779	0·01886	0·02815	0·02603	0·0488	0·02558	0·00559
2	0·05723	0·05736	0·07032	0·0429	0·03877	0·05493	0·04087	0·05076	0·03311	0·06328
3	0·06587	0·10419	0·08278	0·04884	0·0674	0·0724	0·04454	0·08854	0·04959	0·08178
4	0·06653	0·10794	0·08698	0·06367	0·10005	0·08869	0·09778	0·12091	0·05135	0·08417
5	0·0813	0·12407	0·09181	0·07406	0·10301	0·09914	0·11416	0·12456	0·07876	0·09318

Random distance levels (continued)

Specimen no.	Random 1	Random 2	Random 3	Random 4	Random 5	Random 6	Random 7	Random 8	Random 9	Random 10
6	0·10737	0·14461	0·11531	0·07949	0·13889	0·1085	0·12732	0·13339	0·08286	0·10068
7	0·12537	0·15397	0·11602	0·09309	0·14316	0·12224	0·12962	0·14887	0·09692	0·12123
8	0·14129	0·18349	0·12456	0·10246	0·14616	0·13708	0·13788	0·16933	0·10669	0·12788
9	0·16932	0·19529	0·12617	0·13253	0·17909	0·15072	0·14008	0·17326	0·12544	0·14457
10	0·17147	0·22732	0·1342	0·14016	0. 1823	0·15574	0·15569	0·17415	0·15579	0·16289
11	0·17689	0·24226	0·15747	0·15424	0·19113	0·16075	0·18538	0·19649	0·17355	0·17171
12	0·22548	0·25079	0·16234	0·15776	0·21769	0·16336	0·19101	0·19802	0·20603	0·18864
13	0·24601	0·2605	0·21654	0·17609	0·26549	0·19263	0·19211	0·22067	0·20671	0·21416
14	0·2524	0·26266	0·22727	0·2239	0·27163	0·19704	0·21025	0·23207	0·2092	0·21547
15	0·27563	0·27114	0·23971	0·22798	0·27333	0·2127	0·23442	0·23322	0·24048	0·21925
16	0·30045	0·30793	0·24316	0·28071	0·28972	0·2195	0·24428	0·2767	0·24438	0·23061
17	0·30232	0·31515	0·27559	0·28207	0·2916	0·25622	0·25737	0·27978	0·24687	0·23075
18	0·32605	0·33087	0·28408	0·28773	0·34989	0·27892	0·31733	0·31444	0·26857	0·26713
19	0·369	0·33127	0·35156	0·28881	0·38057	0·30041	0·32099	0·35803	0·32315	0·31354
20	0·42711	0·33361	0·38667	0·28948	0·39731	0·30632	0·33587	0·39382	0·41136	0·31769
21	0·44801	0·33481	0·38854	0·34896	0·40008	0·34907	0·33594	0·47606	0·44914	0·36928
22	0·46129	0·36632	0·41409	0·38484	0·40955	0·38977	0·38124	0·49267	0·46439	0·42786
23	0·49774	0·37706	0·47578	0·44196	0·49118	0·49973	0·41106	0·51349	0·50684	0·43794
24	0·54665	0·60408	0·53276	0·51992	0·54096	0·54949	0·46819	0·57726	0·53362	0·49023
25	0·5495	0·60817	0·55219	0·54024	0·6175	0·55539	0·5928	0·61288	0·5551	0·56429
26	0·59579	0·73829	0·60462	0·54882	0·72067	0·60945	0·62988	0·63782	0·56784	0·58225
27	0·81959	0·73937	0·67758	0·56426	0·74865	0·67229	0·64977	0·72298	0·61775	0·69307
28	0·8488	0·75977	0·72465	0·77203	0·84972	0·68557	0·69002	0·77454	0·74625	0·69367
29	0·87243	0·8092	0·82105	0·77587	0·9539	0·937	0·69952	0·9005	0·86373	0·94222
30	0·90739	1·07864	0·83058	0·77753	1·01053	1·11747	0·75159	1·00159	1·08992	0·96565
31	1·2377	1·15164	1·03254	1·14492	1·18326	1·1455	1·14773	1·35689	1·09753	1·23651
32	1·41208	1·46523	1·29068	1·44399	1·59447	1·51311	1·29962	1·57805	1·51382	1·38028
33	1·84326	1·72417	1·55492	1·69201	1·59825	1·70089	1·57331	1·89725	1·79169	1·60927
34	2·36701	2·23037	2·11371	2·10649	2·29353	2·32573	2·05961	2·53581	2·27747	2·34684

Confidence limits

Specimen no.	Mean residual distance	Population SD	0·99	0·95	0·9
1	0·031506	0·0198259	0·011027	0·006568	0·005512
2	0·050953	0·0118164	0·006572	0·003915	0·003285
3	0·070593	0·0193028	0·010736	0·006395	0·005367
4	0·086807	0·0213947	0·0119	0·007088	0·005948
5	0·098405	0·0181191	0·010078	0·006003	0·005038
6	0·113842	0·0224193	0·01247	0·007427	0·006233
7	0·125049	0·0201921	0·011231	0·00669	0·005614
8	0·137682	0·0250657	0·013942	0·008304	0·006969
9	0·153647	0·0242289	0·013476	0·008027	0·006736
10	0·165971	0·0260909	0·014512	0·008644	0·007254
11	0·180987	0·0257187	0·014305	0·00852	0·007151
12	0·196112	0·0301584	0·016774	0·009991	0·008385
13	0·219091	0·0299185	0·016641	0·009912	0·008318
14	0·230189	0·0246538	0·013713	0·008168	0·006855
15	0·242786	0·0227797	0·01267	0·007547	0·006333
16	0·263744	0·0310252	0·017256	0·010278	0·008626
17	0·273772	0·0260447	0·014486	0·008628	0·007241
18	0·302501	0·0288358	0·016039	0·009553	0·008017
19	0·333733	0·0301152	0·01675	0·009977	0·008373

Confidence limits (continued)

Specimen no.	Mean residual distance	Population SD	0·99	0·95	0·9
20	0·359924	0·0486622	0·027066	0·016122	0·01353
21	0·389989	0·0517795	0·0288	0·017154	0·014396
22	0·419202	0·0417529	0·023223	0·013833	0·011609
23	0·465278	0·0460517	0·025614	0·015257	0·012804
24	0·536316	0·0389074	0·021641	0·01289	0·010817
25	0·574806	0·0296871	0·016512	0·009835	0·008254
26	0·623543	0·0619703	0·034468	0·02053	0·01723
27	0·690531	0·0721742	0·040144	0·023911	0·020067
28	0·754502	0·0598743	0·033302	0·019836	0·016647
29	0·857542	0·0817748	0·045484	0·027092	0·022736
30	0·953089	0·131587	0·073189	0·043594	0·036585
31	1·173422	0·0883194	0·049124	0·02926	0·024556
32	1·449133	0·1053463	0·058594	0·034901	0·02929
33	1·698502	0·1174488	0·065326	0·03891	0·032654
34	2·265657	0·1439497	0·080066	0·04769	0·040022

Random distance level – mean random distance level

Specimen no.	Random 1	Random 2	Random 3	Random 4	Random 5	Random 6	Random 7	Random 8	Random 9	Random 10
1	0·003274	0·04123	0·06825	0·01779	0·01886	0·02815	0·02603	0·0488	0·02558	0·00559
2	0·05723	0·05736	0·07032	0·0429	0·03877	0·05493	0·04087	0·05076	0·03311	0·06328
3	0·06587	0·10419	0·08278	0·04884	0·0674	0·0724	0·04454	0·08854	0·04959	0·08178
4	0·06653	0·10794	0·08698	0·06367	0·10005	0·08869	0·09778	0·12091	0·05135	0·08417
5	0·0813	0·12407	0·09181	0·07406	0·10301	0·09914	0·11416	0·12456	0·07876	0·09318
6	0·10737	0·14461	0·11531	0·07949	0·13889	0·1085	0·12732	0·13339	0·08286	0·10068
7	0·12537	0·15397	0·11602	0·09309	0·14316	0·12224	0·12962	0·14887	0·09692	0·12123
8	0·14129	0·18349	0·12456	0·10246	0·14616	0·13708	0·13788	0·16933	0·10669	0·12788
9	0·16932	0·19529	0·12617	0·13253	0·17909	0·15072	0·14008	0·17326	0·12544	0·14457
10	0·17147	0·22732	0·1342	0·14016	0. 1823	0·15574	0·15569	0·17415	0·15579	0·16289
11	0·17689	0·24226	0·15747	0·15424	0·19113	0·16075	0·18538	0·19649	0·17355	0·17171
12	0·22548	0·25079	0·16234	0·15776	0·21769	0·16336	0·19101	0·19802	0·20603	0·18864
13	0·24601	0·2605	0·21654	0·17609	0·26549	0·19263	0·19211	0·22067	0·20671	0·21416
14	0·2524	0·26266	0·22727	0·2239	0·27163	0·19704	0·21025	0·23207	0·2092	0·21547
15	0·27563	0·27114	0·23971	0·22798	0·27333	0·2127	0·23442	0·23322	0·24048	0·21925
16	0·30045	0·30793	0·24316	0·28071	0·28972	0·2195	0·24428	0·2767	0·24438	0·23061
17	0·30232	0·31515	0·27559	0·28207	0·2916	0·25622	0·25737	0·27978	0·24687	0·23075
18	0·32605	0·33087	0·28408	0·28773	0·34989	0·27892	0·31733	0·31444	0·26857	0·26713
19	0·369	0·33127	0·35156	0·28881	0·38057	0·30041	0·32099	0·35803	0·32315	0·31354
20	0·42711	0·33361	0·38667	0·28948	0·39731	0·30632	0·33587	0·39382	0·41136	0·31769
21	0·44801	0·33481	0·38854	0·34896	0·40008	0·34907	0·33594	0·47606	0·44914	0·36928
22	0·46129	0·36632	0·41409	0·38484	0·40955	0·38977	0·38124	0·49267	0·46439	0·42786
23	0·49774	0·37706	0·47578	0·44196	0·49118	0·49973	0·41106	0·51349	0·50684	0·43794
24	0·54665	0·60408	0·53276	0·51992	0·54096	0·54949	0·46819	0·57726	0·53362	0·49023
25	0·5495	0·60817	0·55219	0·54024	0·6175	0·55539	0·5928	0·61288	0·5551	0·56429
26	0·59579	0·73829	0·60462	0·54882	0·72067	0·60945	0·62988	0·63782	0·56784	0·58225
27	0·81959	0·73937	0·67758	0·56426	0·74865	0·67229	0·64977	0·72298	0·61775	0·69307
28	0·8488	0·75977	0·72465	0·77203	0·84972	0·68557	0·69002	0·77454	0·74625	0·69367
29	0·87243	0·8092	0·82105	0·77587	0·9539	0·937	0·69952	0·9005	0·86373	0·94222
30	0·90739	1·07864	0·83058	0·77753	1·01053	1·11747	0·75159	1·00159	1·08992	0·96565
31	1·2377	1·15164	1·03254	1·14492	1·18326	1·1455	1·14773	1·35689	1·09753	1·23651
32	1·41208	1·46523	1·29068	1·44399	1·59447	1·51311	1·29962	1·57805	1·51382	1·38028
33	1·84326	1·72417	1·55492	1·69201	1·59825	1·70089	1·57331	1·89725	1·79169	1·60927
34	2·36701	2·23037	2·11371	2·10649	2·29353	2·32573	2·05961	2·53581	2·27747	2·34684

[Special Papers in Palaeontology, 73, 2005, pp. 39–56]

AN EXPERIMENTAL INVESTIGATION OF POST-DEPOSITIONAL TAPHONOMIC BIAS IN CONODONTS

by PETER H. VON BITTER* *and* MARK A. PURNELL†

*Department of Palaeobiology, Royal Ontario Museum, and Department of Geology, University of Toronto, Toronto, Ontario, Canada M5S 2C6; e-mail: peterv@rom.on.ca

†Department of Geology, University of Leicester, Leicester LE1 7RH, UK; e-mail: map2@le.ac.uk

Abstract: The different types of elements that occurred together in conodont apparatuses are not recovered from the fossil record in the expected numbers. The causes of this are complex and difficult to study. Numerous complete articulated skeletons of *Ozarkodina excavata* (Branson and Mehl) have been recovered from the Eramosa Member at Hepworth, Ontario, and as a consequence, several of the potential biases affecting recovery of isolated conodont elements can be ruled out *a priori*. Based on processing ten replicate samples ('runs') of nodular carbonate and bituminous shale, we tested the role of post-mortem compaction, laboratory processing and difficulties in element identification in biasing the expected, or predicted, recovery of apparatus elements.

Although the numbers of different elements of *O. excavata* reported in the literature do not exhibit as marked a bias as do Late Palaeozoic conodont faunas, they are biased nonetheless. This is also true of elements recovered from the Eramosa Member. In both carbonate and shale samples, P_1 and $S_{1/2}$ elements are significantly under-represented, whereas P_2 and S_0 elements are significantly over-represented. In the carbonate runs, this bias is a consequence of the difficulties in differentiating between broken remains of morphologically similar elements. When this factor is taken into account in the shale runs, however, the fauna still exhibits significant bias, and we are able to rule out all potential biases except one. Surprisingly, apparent over-representation of P elements and under-representation of S elements can arise as a result of element fragmentation during sediment compaction and diagenesis alone.

Key words: conodont, elements, bias, fossil record, post-mortem, breakage, diagenesis.

WE now know that the fossil record of conodonts is fundamentally biased. Decades of research provides clear evidence that the elements of the conodont skeletal apparatus are not found in the relative abundances that they should be. But this simple statement masks an important issue: recognition that this bias exists, and assessment of the degree to which it obscures our view of conodont evolutionary history, is predicated on our understanding of the conodont skeleton as it was in life, and as other papers in this volume (Purnell and Donoghue 2005*a*) attest this is still a matter of some discussion.

The debate over the nature of the conodont skeleton was among the most important in the history of conodont research. Did each conodont, whatever organism that might be, possess a single element, multiple elements of the same type, or several different types of element? Without resolution of this fundamental question there could be no stable taxonomy, no meaningful scientific investigation of conodont evolution, and limited understanding and interpretation of stratigraphical distributions. Of central importance in this debate were fossils that have come to be known as 'bedding plane assemblages', fossils that, in contrast to the isolated elements that comprise a standard conodont sample, preserve a number of conodont elements clustered together. It was the discovery of one such fossil that led Hinde (1879) to suggest that the conodont skeleton had numerous elements of different types. Unfortunately, the assemblage described by him as *Polygnathus dubius* is a faecal or regurgitated pellet resulting from several conodonts being consumed by an unknown predator, and this is how all bedding plane assemblages were interpreted by some early conodont authorities. But as more and more were discovered through the first part of the twentieth century the recognition of patterns of element association within the clusters made this interpretation increasingly untenable. Scott (1942, p. 296) for example observed recurrent patterns in Carboniferous bedding plane assemblages from the Heath Formation of Montana and in answer to those that advocated a faecal origin noted dryly that 'it would be strange indeed to find a group of animals that had such a perfectly balanced diet the excretal material would consist time after time of one pair of prionioids [M elements], one pair of spathognaths [P_1 elements], one pair of prioniodells [P_2 elements], and approximately four pairs of hindeodells [S elements]'. It took another 30 years before the implications of this for conodont taxonomy became fully accepted (see Sweet and Donoghue 2001 for a recent review), but today nobody would suggest that conodonts bore only one element.

Ironically, it is the acceptance and implementation of multi-element taxonomy that has highlighted the bias in the conodont fossil record. Bedding plane assemblages clearly indicate that S elements should be the most commonly occurring components of the skeleton, but this is far from what we find. The Carboniferous *Idiognathodus* (*sensu* Baesemann 1973; Grayson *et al.* 1991) provides the most compelling example of this. Bedding plane assemblages have been known for decades (Du Bois 1943; Rhodes 1952) and are the most abundantly preserved of any conodont. They provided templates for the first applications of multi-element methods to Upper Palaeozoic conodonts (von Bitter 1972; Baesemann 1973), and the structure of the apparatus is also among the best known (Purnell and Donoghue 1997, 1998). Like many other ozarkodinid taxa, an unbiased collection should contain elements in the ratio of 2 M elements, 4 $S_{3/4}$ elements, 4 $S_{1/2}$ elements, 1 S_0 element, 2 P_2 elements, and 2 P_1 elements (see Purnell *et al.* 2000 for discussion of element notation), but von Bitter (1972) and subsequent authors found massive over-representation of P_1 elements (or under-representation of all other parts of the apparatus). Of the 20,244 elements of *Idiognathodus* spp. recovered by von Bitter (1972), only 379 were M elements (14% of the expected number), 330 were S_{1-4} elements (3% of expected), 56 were S_0 elements (0·04% of expected) and 1326 were P_2 elements (49% of expected). In stark contrast, 18,153 were P_1 elements (675% of expected) (Table 1).

Bias and Ozarkodina excavata

The unique nature of our data (discussed below) means that the focus of this paper is *Ozarkodina excavata* (Text-figs 1–3, Pls 1–3). In many ways this is fortuitous. As apparatus H (Walliser 1964), this species was among the first to have its complement of elements reconstructed, and it was also among the first conodonts for which the taxonomic consequences of the multi-element skeleton were properly explored (Jeppsson 1969).

Klapper and Murphy (1974, p. 36) considered *Ozarkodina excavata* to be 'among the most convincing of all reconstructions of Silurian multielement species', but in the absence of natural assemblages, opinion concerning the number of elements in the apparatus has varied. Jeppsson (1969) included *O. excavata* among members of his genus *Hindeodella*, and suggested that the apparatus probably included 16 paired elements, four P_1 elements (as two pairs), two P_2, two S_0, two $S_{1/2}$, four $S_{3/4}$ (as two pairs) and two M elements (using current element notation). Rexroad and Nicoll (1972) used the statistical methods of Kohut (1969) to look for associations of elements in their collections that might correspond to multi-element *Ozarkodina excavata.* Despite employing a statistical approach, they were among the first to articulate the view that the use of ratios of occurrence of element types to recognize apparatuses was suspect. Nevertheless, they did discuss ratios for *O. excavata*, suggesting that they cast serious doubt on Walliser's (1964) and Jeppsson's (1969) reconstructions. They tabulated ratios of elements for *O. excavata* in the Bainbridge Formation as eight P_1 elements, five P_2, six S_0, two $S_{1/2}$, zero $S_{3/4}$ and 7·5 M elements. In retrospect, it is clear that rather than undermining the reconstructions of *O. excavata* their data support their opinion that element ratios

TABLE 1. *Idiognathodus* spp., discrete elements recovered and identified from the four cyclothems of the Shawnee Group (Virgilian, Pennsylvanian) of eastern Kansas by von Bitter (1972). Percentage of each element is of the total recovered shown in parentheses under each element category. Expected percentage of each element, based on 15 element ozarkodinid model, shown in bottom row.

	P_1	P_2	S_0	$S_{1/2,\ 3/4}$	M	Total
Topeka Limestone	8088	399	23	131	166	8807
	(91·836%)	(4·531%)	(0·261%)	(1·487%)	(1·885%)	(100%)
Deer Creek Limestone	2395	250	9	67	50	2771
	(86·431%)	(9·022%)	(0·325%)	(2·418%)	(1·804%)	(100%)
Lecompton Limestone	4155	324	9	50	77	4615
	(90·033%)	(7·021%)	(0·195%)	(1·083%)	(1·668%)	(100%)
Oread Limestone	3515	353	15	82	86	4051
	(86·769%)	(8·714%)	(0·370%)	(2·024%)	(2·123%)	(100%)
mean percentage	88·77%	7·32%	0·29%	1·75%	1·87%	100%
Total Conodont Elements	18,153	1326	56	330	379	20244
% of Total	(89·671%)	(6·550%)	(0·277%)	(1·630%)	(1·872%)	(100%)
% Expected	(13·333%)	(13·333%)	(6·667%)	(53·334%)	(13·333%)	(100%)

are fundamentally biased. Table 2 summarizes literature reports of recovery for the elements of the apparatus of *O. excavata*. It is clear that although element recovery is less biased than that of *Idiognathodus*, P_1 elements are consistently over-represented, whereas S_{1-4} elements are generally under-represented.

CAUSES OF BIAS IN CONODONT ELEMENT RECOVERY

To a greater or lesser degree, the bias in element numbers discussed above affects all collections of conodonts recovered through standard sampling practice from all parts of the stratigraphical column (i.e. elements derived from laboratory dissolution or disaggregation of rock, followed by sieving, and picking). The potential causes of this bias are numerous and complex (see e.g. Carls 1977 for a review); here we consider seven potential factors.

1. Variability in the element composition of the skeletal apparatus

Do apparently biased numbers of elements reflect repeated shedding and replacement of over-represented elements or resorption and loss of under-represented elements? Hypotheses that the number of elements in conodont apparatuses was variable and under some sort of biological or

TABLE 2. Summary of element recovery of *Ozarkodina excavata* as reported in selected literature. Only elements of *O. excavata* are included in calculation of totals and percentages. Data include those of Jeppsson (1974, table 2), Klapper and Murphy (1974, tables 2–8), Simpson and Talent (1995, tables 2–5), Helfrich (1980, tables 1, 3–5), Aldridge [1972, tables 1–5, except that we have excluded samples which lack *O. excavata* P elements because, as noted by Jeppsson 1974, some '*N. excavatus*' may include M elements of *Oulodus* (his *Ligonodina*), and the same may be true of other counts] and Rexroad *et al.* (1978, table 1). Data are listed in approximate stratigraphical order, younging upwards. N, number of samples containing *O. excavata*.

Element	P_1	P_2	S_0	$S_{1/2}$	$S_{3/4}$	M	$P_1 + P_2$	$M + S_0 + S_{1/2}$	Total
Expected proportion of fauna	13·33%	13·33%	6·67%	26·67%	26·67%	13·33%	26·67%	46·67%	
Klapper and Murphy 1974									
$n = 126$									
total	1902	634	210	398	487	339	2536	947	3970
mean percentage of total	53·54%	13·71%	4·73%	9·07%	9·80%	9·14%	67·26%	22·94%	
ratio observed:expected	4·016	1·029	0·710	0·340	0·367	0·640	2·522	0·492	
Jeppsson 1974									
$n = 44$									
total	454	267	193	263	509	337	721	793	2023
mean percentage of sample	23·47%	13·18%	7·66%	9·22%	27·64%	18·84%	36·64%	35·72%	
ratio observed:expected	1·760	0·988	1·149	0·346	1·036	1·413	1·374	0·765	
Simpson and Talent 1995									
$n = 72$									
total	213	113	35	76	127	55	326	166	619
mean percentage of sample	41·88%	19·19%	2·93%	9·72%	20·13%	6·15%	61·07%	18·80%	
ratio observed:expected	3·141	1·439	0·440	0·365	0·755	0·461	2·290	0·403	
Rexroad *et al.* 1978									
$n = 5$									
total	22	19	9	8	16	24	41	41	98
mean percentage of sample	21·52%	10·25%	8·75%	6·25%	19·27%	33·95%	31·77%	48·95%	
ratio observed:expected	1·614	0·769	1·313	0·234	0·723	2·547	1·191	1·049	
Helfrich 1980									
$n = 17$									
total	78	55	105	117	27	36	133	258	418
mean percentage of sample	19·66%	24·46%	19·94%	23·18%	3·90%	8·85%	44·13%	51·97%	
ratio observed:expected	1·475	1·835	2·991	0·869	0·146	0·664	1·655	1·008	
Aldridge 1972									
$n = 15$									
total	111	98	43	41	66	39	209	123	398
mean percentage of sample	37·61%	22·61%	9·57%	6·04%	14·91%	9·27%	60·22%	24·88%	
ratio observed:expected	2·82	1·696	1·435	0·226	0·559	0·695	2·258	0·533	

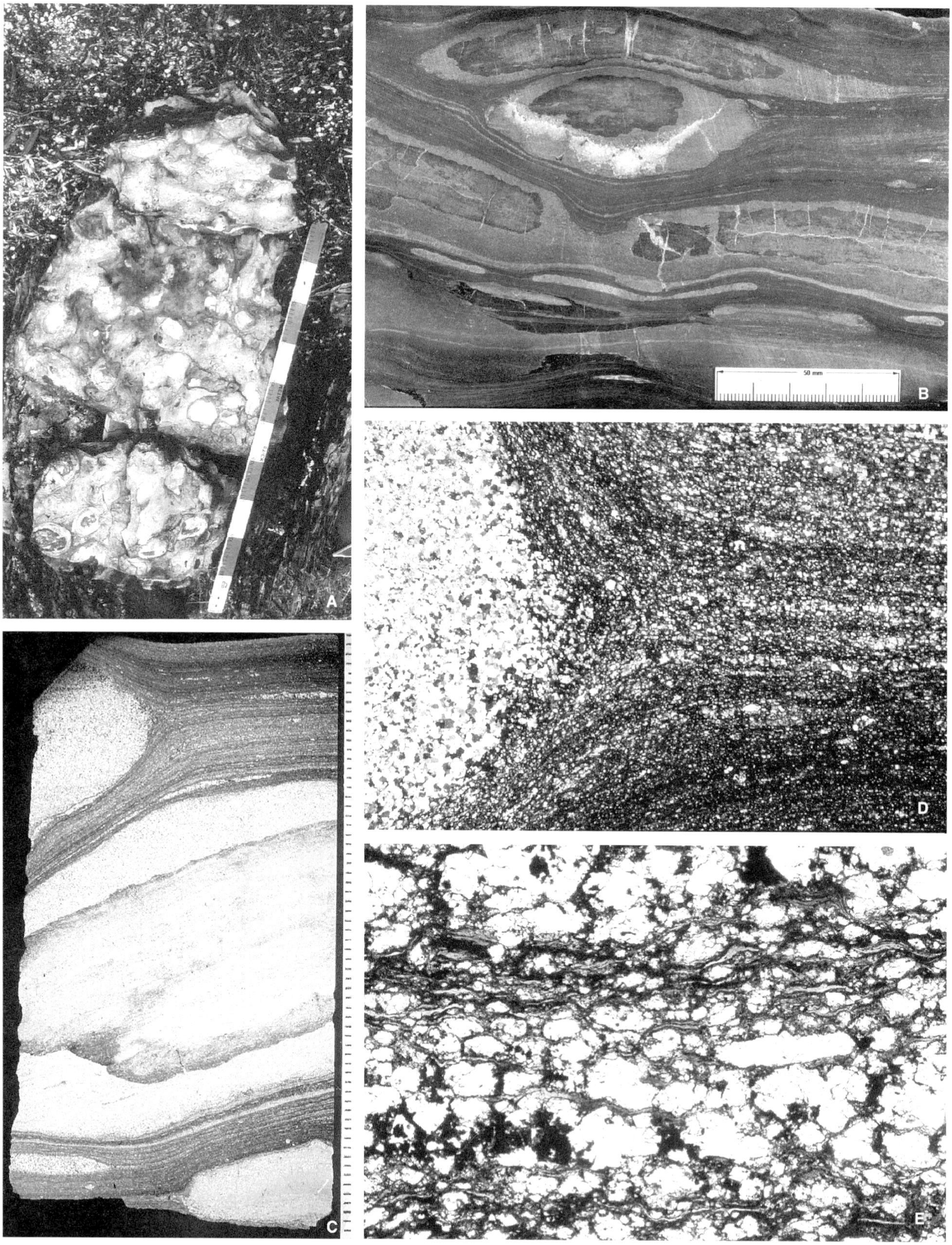
A
B
50 mm
C
D
E

developmental control have involved several interrelated ideas. The first of these, that conodonts repeatedly shed and replaced individual elements during their lifetime, began with Gross (1954), and was followed by Carls (1977) and Krejsa *et al.* (1990). Interestingly Carls' conclusions derived from a detailed analysis of element ratios and were proposed as an explanation for under-representation of S and M elements. Jeppsson (1976), however, drew on evidence of repair to damaged elements to conclude that rather than being shed, conodont elements were permanent structures. Similarly, Purnell's (1994) investigation of natural assemblages of *Idiognathodus* and *Gnathodus* found no evidence in the element composition of assemblages, or in the ontogeny of the apparatus, for element shedding or replacement. The accumulating evidence of natural assemblages of a broad range of taxa (Purnell and Donoghue 1998) provides strong support for the hypothesis that the number of elements in an apparatus was stable throughout the life of the animal that bore it, and in the face of this direct evidence it is now difficult to sustain Merrill and Powell's (1980) inference that elements were lost from the apparatus during ontogeny.

The second line of reasoning to do with variability in apparatus composition began with Müller and Nogami (1971). They interpreted discontinuities within conodont elements to be surfaces along which resorption of phosphatic crown tissue had occurred during growth. Von Bitter and Merrill (1980) built on this idea by postulating that conodonts resorbed and removed calcium phosphate from their elements as a body weight control mechanism. More recently, Zhuravlev (1999) has continued to develop these concepts. It is difficult to exclude confidently post-mortem or laboratory processes as the cause of the etched features he illustrates, but what is relevant for the present discussion is that even if the features he observed are accepted as evidence of resorption, they are too rare for resorption and shedding to account for P_1 element over-representation (Zhuravlev 1999). The hypothesis that resorption of elements has a role in explaining bias is further undermined by the recent suggestion (Purnell 1995; Donoghue and Purnell 1999) that internal discontinuities are not the result of resorption, but are wear surfaces that formed during periods of function and were subsumed during subsequent element growth. Donoghue and Purnell (1999) documented the correlation between surface wear and internal discontinuities, and noted that this provides strong evidence that elements are retained and not shed.

2. Differential breakage and dissolution during ingestion, digestion and excretion by predators

Although several authors, including von Bitter and Merrill (1998) and Tolmacheva and Purnell (2002), have studied conodont apparatuses that have gone through the gut of a predator, the effects of these ingestive, digestive and excretory processes on the completeness of individual apparatuses, or on individual elements, have apparently not been studied in a systematic way. One of the more interesting possibilities is that during these processes calcium phosphate is removed from conodont elements by bacteria (von Bitter *et al.* 1996). It is also pertinent to note, however, that complete articulated apparatuses are known from coprolites (e.g. Purnell and Donoghue 1998, text-figs 15–16) and from within the digestive tract of predators (Nicoll 1977; Purnell 1993; Purnell and Donoghue 1998).

3. Post-mortem sorting due to differential current entrainment and settling

Ellison (1968) was among the first to interpret variation in the number and kind of elements from sample to sample as evidence of sorting during deposition. Other authors (e.g. von Bitter 1972) considered this and related factors to account for under- and over-representation; however, it was not until the early 1990s that post-mortem sorting began to be studied more formally. Broadhead *et al.* (1990) studied gravitational settling of conodont elements and its implications for palaeoecological interpretations of conodont samples; similarly, McGoff (1991) determined the hydrodynamic properties of a number of different types of element. These studies confirmed previous suspicions that the susceptibility of

TEXT-FIG. 1. Conodont-bearing lithologies in the Eramosa Member, Hepworth, Bruce County, Ontario. A, bedding-plane surfaces of carbonate nodules in bituminous shale; scale represents 1 m. B, cross-section of nodular carbonate alternating with bituminous shale; carbonate has core of grey chert, and both chert and carbonate were fractured, and subsequently cemented by white, fluorescent calcite; scale represents 50 mm. C, thin-section of nodular carbonate alternating with dark, bituminous shale; darker core of chert in the largest (central) carbonate nodule, also 'pinching and swelling' of bituminous shale, ROM thin-section 99PB15A, crossed nicols; scale represents 64 mm. D, uppermost part of thin-section ROM 99PB15A, carbonate nodule on left, showing compaction of bituminous shale laminae around nodule. Carbonate nodules alternate with darker organic material that has a 'knitted', cellular texture, crossed nicols; ×30. E, enlargement of bituminous shale portion of D showing lighter carbonate in a matrix of darker layered organic material. Elongated organic material is golden yellow in colour, both under plane and polarized light. Thin-section ROM 99PB15A, crossed nicols; ×259.

TABLE 3. Discrete elements and 'clusters' of *Ozarkodina excavata* (Branson and Mehl) recovered and identified from the Eramosa Member (Silurian), Hepworth, Bruce County, Ontario, in ten carbonate (CO_3) and shale (Sh) laboratory processing runs. Some processing data and calculated conodont element yield per kilogram also shown. Asterisks refer to inclusion of counts from fused element pairs (clusters).

Run	P_1	P_2	$P_{indet.}$	S_0	$S_{1/2}$	$S_{3/4}$	M	Total	Clusters	Litho.	Amount proc.	Breakdown	Yield	Preservation
CO_3/1	5*	13	5	11	10	16	13	73	3	Mixed CO_3/Sh	2943 g	2018 g (69%)	36/kg	mod.good
CO_3/2	11*	24*	–	11	15*	20	15	96	19	CO_3	4558 g	2746 g (54%)	35/kg	mod.good
CO_3/3	3	7	–	2	3	11	3	29	3	CO_3	2082 g	1220 g (59%)	24/kg	mod.good
CO_3/4	9*	13	6	10	22	34*	17	111	~22	CO_3	4183 g	2741 g (66%)	40/kg	good to excellent
CO_3/5	15	28	4	17	41	40	34	179	9	CO_3	4275 g	3100 g (73%)	58/kg	mod.good to good
Sh/6	–	1	1	–	–	1	2	5	–	Shale	4005 g	1068 g (27%)	5/kg	poor
Sh/7	5*	42*	22	26	32	33	21	181	24	Shale	1303 g	576 g (44%)	314/kg	poor
Sh/8	2	102	48	53	62	116	56	439	36	Shale	1300 g	668 g (51%)	657/kg	poor
Sh/9	1	9	4	5	7	20	3*	49	1	Shale	2526 g	1543 g (61%)	32/kg	poor
Sh/10	–	8	3	3	2	10	6	32	–	Shale	1063 g	120 g (11%)	266/kg	poor
Summary														
5 CO_3 runs	43	85	15	51	91	121	82	488	~56	Mostly CO_3	18041 g	11,555 g (64%)	42/kg	good
% of total	8·811	17·418	3·074	10·45	18·648	24·795	16·803	100%						
5 shale runs	8	162	78	87	103	180	88	706	61	Shale	10197 g	3975 g (39%)	178/kg	poor
% of total	1·133	22·946	11·05	12·32	14·589	25·496	12·465	100%						
5 CO_3 and 5 shale runs combined	51	247	93	138	194	301	170	1194	~117	Nodular CO_3 and bituminous shale	28238 g	15,530 g (55%)	77/kg	
% of total	4·271	20·687	7·789	11·56	16·248	25·209	14·238	100%						

elements to current entrainment, transport and sorting is correlated with their size and shape. Broadhead and Driese (1994) addressed the related question of the effects of abrasion on conodont elements during current transport. They concluded that when suspended in water abrasion of elements is limited and that their destruction through abrasion is highly unlikely.

4. Lithological sampling bias

Almost all conodonts are collected blind, and although most conodont workers intuitively develop their own strategies regarding which lithologies are worth collecting, to our knowledge none of these has ever been put to the test. Aside from avoiding non-marine lithologies and

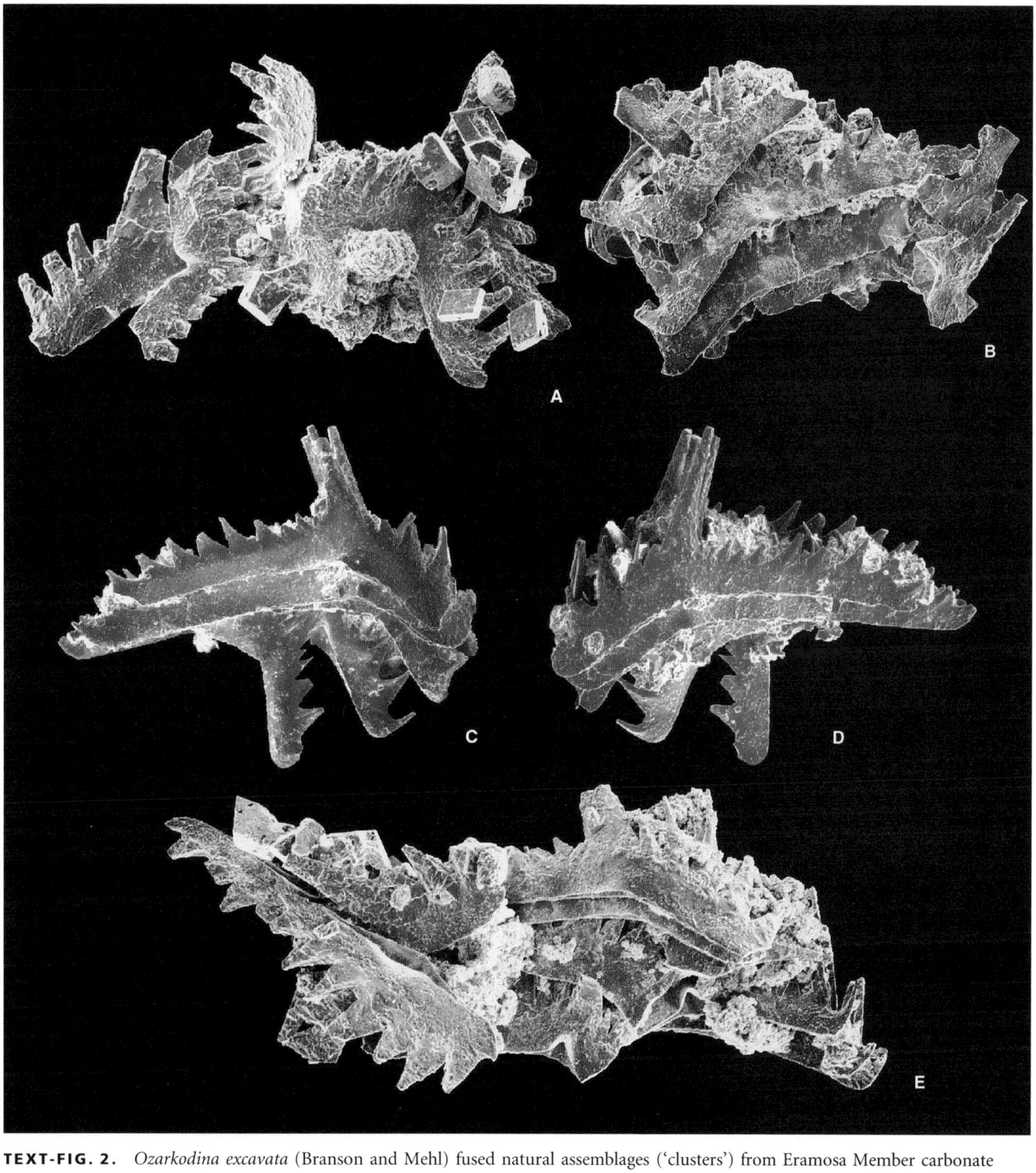

TEXT-FIG. 2. *Ozarkodina excavata* (Branson and Mehl) fused natural assemblages ('clusters') from Eramosa Member carbonate (Silurian), Hepworth, Bruce County, Ontario. A, B, E, three almost complete apparatuses, specimens ROM 56390, 56391, 56393; C–D, both sides of an almost complete S element array, specimen ROM 56392. All ×139.

those that will not yield to laboratory processing it is by no means clear that there is a general strategy that can be applied to rocks from all depositional settings through all parts of the conodont stratigraphical record. The consequence of these intuitive approaches to collecting is that lithologies that are difficult to break down, or which experience suggests are not worth the effort (too few informative conodont elements to justify the difficulty of processing), will tend not to be collected unless there is a pressing biostratigraphical (or other) need to obtain material. Among those that have recently touched on the subject of lithological sampling bias are Orchard (2002, 2005) and Broadhead and Repetski (2002); see also Purnell and Donoghue (2005*b*) for more detailed discussion.

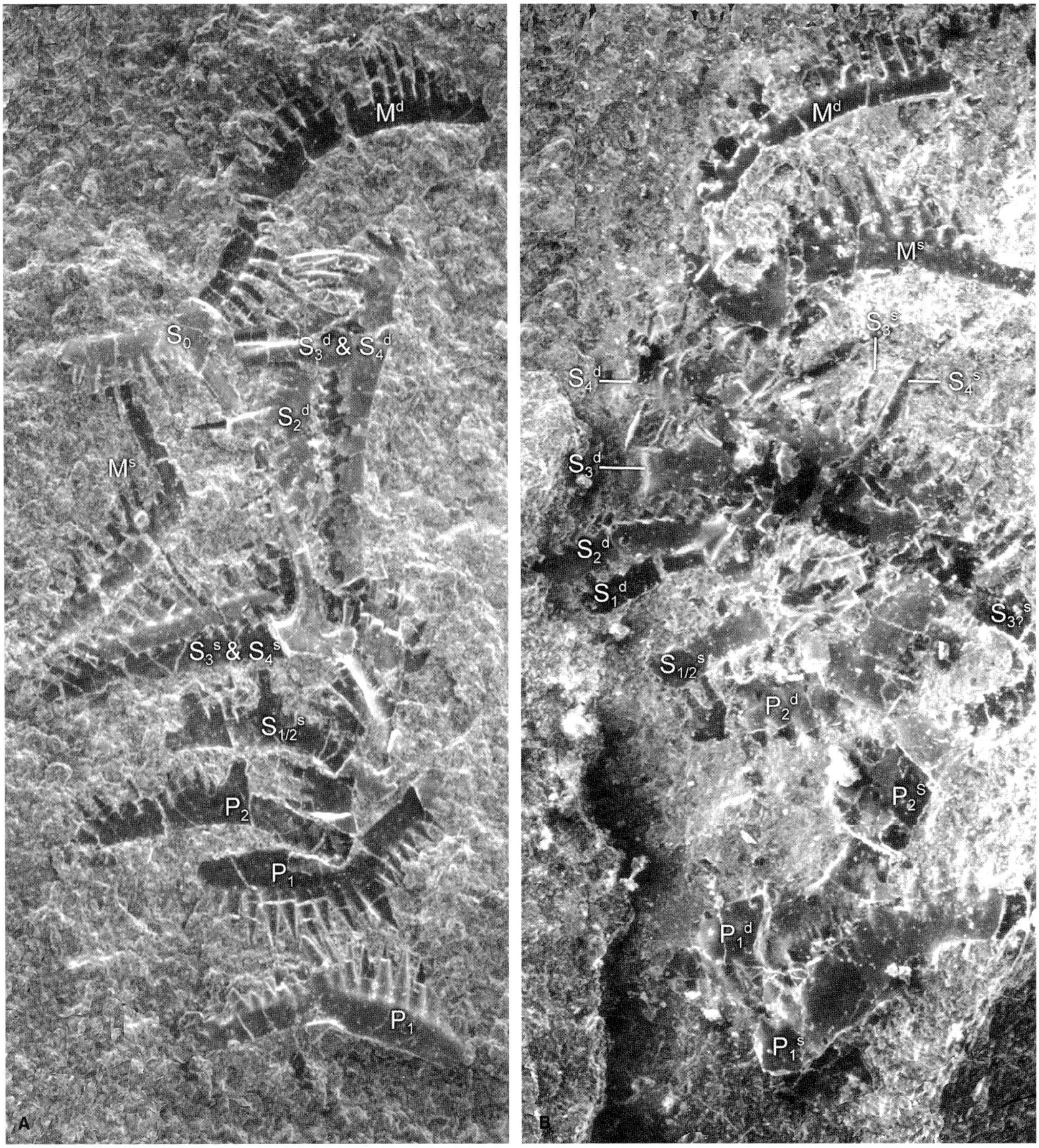

TEXT-FIG. 3. *Ozarkodina excavata* (Branson and Mehl) bedding plane assemblages from Eramosa Member shale (Silurian), Hepworth, Bruce County, Ontario. Elements show extensive fracturing. A, specimen ROM 56394; ×42; B, specimen ROM 56395; ×52.

5. Differential breakage and solution of elements during sediment compaction and diagenesis

Jeppsson (1976) and Carls (1977) were among the first to consider this factor. Questions of differential breakage and solution are an important part of our own study, and we consider these topics elsewhere in this paper.

6. Differential loss of elements during laboratory processing

Since Ziegler *et al.* (1971) and von Bitter (1972) observed that differential destruction of elements during laboratory processing may be a significant factor in the 'loss' of conodont elements, a number of papers have suggested ways of reducing, or measuring, that loss. Among the techniques for decreasing or minimizing the loss of elements, those to do with buffering the acids used during dissolution of carbonates (Jeppsson *et al.* 1985, 1999; Jeppsson and Anehus 1995) have been particularly effective and important. Introducing known 'spikes' into a sample provides a means of checking on possible processing bias as it enables phosphate dissolution during processing, element loss during separation, and effectiveness of picking to be assessed (von Bitter and Millar-Campbell 1984).

7. Bias resulting from differences in element identifiability

It is clear from natural assemblages that although the number and locations of elements in ozarkodinid and prioniodinid apparatus are stable (Purnell and Donoghue 1998), the degree to which the elements within the apparatus differ from one another varies. In many taxa S_3 and S_4 elements, for example, are extremely similar to one another. The S_1 and S_2 elements are usually more similar to one another than either is to the $S_{3/4}$ elements, but this varies considerably. Further confusion can arise in taxa which have digyrate elements in $S_{1/2}$ positions: these elements are easily distinguished from $S_{3/4}$ elements, even when broken, but may be more easily confused with digyrate elements from P positions. Clearly, if different elements vary in the ease with which they can be identified when broken, or in their similarity and potential confu-

TABLE 4. Laboratory processing data for runs 1–5 (carbonate, CO_3), and 6–10 (shale, Sh), from the Eramosa Member, Hepworth, Bruce County, Ontario.

Run	Sample	Rock type	Preparation	Acid/ chemical used	Amount processed (g)	Amount coarse fraction (g)	Weight fine fraction (g)	Amount breakdown (g)	Percentage breakdown	Comments
CO_3/1	99PB15	Interbedded carbonate and shale	Crushed	Acetic acid	2943	925	139	2018	69	
CO_3/2	01PB1	Mostly carbonate	Not crushed	Acetic acid	4558	2082	127	2476	54	
CO_3/3	01PB1	Mostly carbonate	Coarse fraction from run 2	Formic acid	2082	862	15	1220	59	coarse fraction is mostly shale
CO_3/4	01PB1	Mostly carbonate	Not crushed	Acetic acid	4183	1442	158	2741	66	coarse fraction is mostly shale
CO_3/5	02PB1	Mostly carbonate	Not crushed	Acetic acid	4275	949	214	3100	73	
Sh/6	01PB2	Shale	Not crushed	Acetic acid	4005	2603	20	1068	27	
Sh/7	01PB2	Shale	Large pieces of coarse fraction from run 6	Acetic acid	1303	727	16	576	44	poor breakdown
Sh/8	01PB2	Shale	Small pieces of coarse fraction from run 6	Sodium hypochlorite	1300	632	112	668	51	
Sh/9	02PB2	Shale	Not crushed	Acetic acid	2526	983	76	1543	61	
Sh/10	02PB2	Shale	Small pieces not crushed	Sodium hypochlorite	1063	943	28	120	11	poor breakdown

TABLE 5. *t*-test results determining whether the number of each element type recovered from the five carbonate and five shale runs processed from the Eramosa Member, Hepworth, Bruce County, Ontario, differ significantly from the number that would be expected from complete apparatuses of *Ozarkodina excavata* (Branson and Mehl). For raw data see Table 3. *t* is calculated as $(\bar{x}-\mu)/s_e$, where, in this case, $\bar{x}$ is the mean per cent recovery for an element, μ is the per cent expected recovery of that element; s_e is the standard error [calculated as $s\sqrt{(1/n)}$ where *s* is the standard deviation of the runs, and *n* is the number of runs]. The critical values of *t* for rejection of the null hypothesis at a significance level, *P*, of 0·05 are ±2·132 [for 4 degrees of freedom (= $n-1$); 5 carbonate runs, 5 shale runs) and ± 2·353 [for 3 degrees of freedom; 4 shale runs in analysis, excluding aberrant (shale) run 6]. This is a one-tailed test, because we are testing for significant over- or under-representation (positive and negative values of *t*, respectively), not simply difference from the expected value.

	P_1	P_2	S_0	$S_{1/2}$	$S_{3/4}$	M	Total P	$S_0 + S_{1/2}$	$S_0 + S_{1/2}$ + M
Carbonate runs									
% Expected	13·333	13·333	6·667	26·667	26·667	13·333	26·667	33·333	46·667
mean (%)	9·028	18·860	10·386	16·479	26·732	15·618	30·786	26·865	42·482
s_e	0·826	2·531	1·377	2·220	3·300	1·485	2·211	2·557	4·006
t	−5·211	2·184	2·700	−4·590	0·020	1·539	1·866	−2·528	−1·045
H0 rejected? ($P < 0·05\%$)	yes	yes	yes	yes	no	no	no	yes	no
Shale runs									
mean (%)	1·052	21·961	9·203	10·468	27·344	17·846	35·138	19·671	37·517
s_e	0·569	1·208	2·456	3·220	4·090	5·890	1·956	6·334	2·273
t	−21·589	7·145	1·033	−5·031	0·166	0·766	4·332	−2·157	−4·025
H0 rejected? ($P < 0·05\%$)	yes	yes	no	yes	no	no	yes	yes	yes
Shale runs (excluding run Sh/6)									
mean (%)	1·315	22·452	11·504	13·085	29·181	12·308	33·923	24·589	36·896
s_e	0·651	1·425	1·108	2·421	4·718	2·589	1·978	3·398	2·823
t	−18·454	6·400	4·366	−5·610	0·533	−0·396	3·671	−2·573	−3·461
H_0 rejected? ($P < 0·05\%$)	yes	yes	yes	yes	no	no	yes	yes	yes

sion with other elements from the apparatus, this has the potential to introduce bias in the numbers of the different elements reported.

AN EXPERIMENTAL APPROACH TO EVALUATING UNDER- AND OVER-REPRESENTATION OF CONODONT ELEMENTS

The principal difficulty in trying to assess the cause of bias in recovered collections of conodont elements is that any given sample is likely to have been affected by many or all of biases 2–7 outlined above. Because it is often impossible to disentangle the effects of one bias from another, determining the relative importance of each is problematic to say the least. We have been able to take advantage of the unique nature of the preserved conodont fauna of the Eramosa Member at Hepworth, Ontario, to provide new insights into the contribution of individual biases to the fauna recovered from this unit.

Materials, methods and experimental design

Material. The Eramosa Member of the Guelph Formation (Silurian, Wenlock) on the Bruce Peninsula of southern Ontario is commonly regarded as a dolostone (Armstrong and Meadows 1988; Armstrong 1989; Armstrong and Goodman 1990), but at Hepworth it occurs as nodular, lightly dolomitized limestone that alternates with bituminous shale. It crops out in a small roadside section, approximately 30 cm thick, with no exposed top or bottom. No excavations were possible, and all material was collected as loose material thrown up by ditch-clearing and maintenance operations. The shaly, bituminous nature of these beds, and the presence of abundant carbonate nodules, suggests that these two alternating lithologies represent the ‘Interbedded Unit’ of the upper part of the Eramosa Member (Armstrong and Meadows 1988).

The beds sampled are characterized by the presence of abundant carbonate nodules (Text-fig. 1A) around which bituminous shale thickens and thins (Text-fig. 1B). The carbonate nodules are generally between 5 and 12 cm

EXPLANATION OF PLATE 1

Figs 1–8. *Ozarkodina excavata* (Branson and Mehl) fused natural assemblages (‘clusters’) showing severe fracturing. All specimens from Eramosa Member shale (Silurian), Hepworth, Bruce County, Ontario. 1–2, fused S elements, ROM 56396; ×150. 3–4, fused S elements, ROM 56397; ×150. 5–8, a pair of P elements in apposition (i.e. fused in near functional position), ROM 56398; 5–6, ×100; 7–8, repair of fractures, possibly by mobilized phosphate; ×2035.

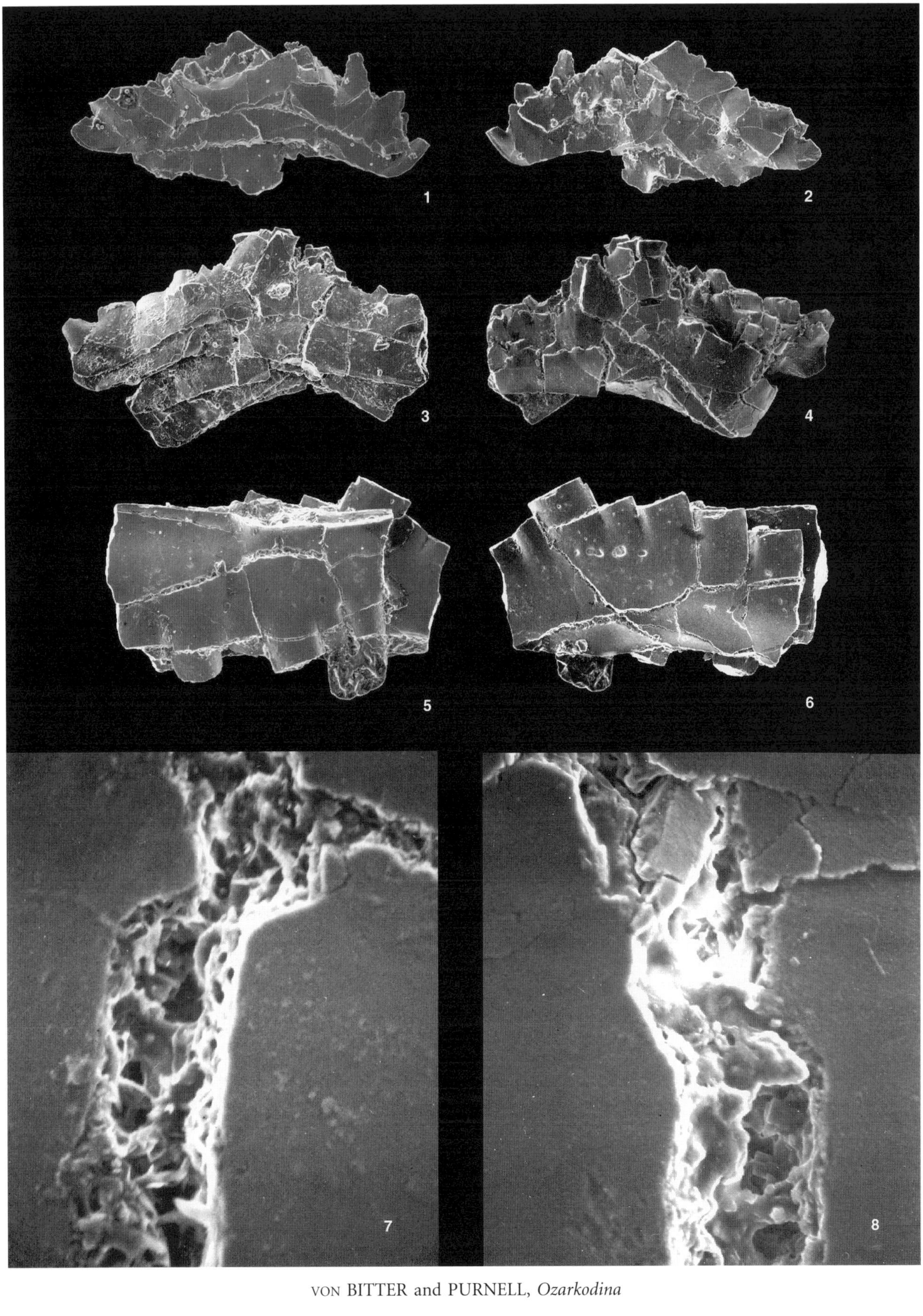

VON BITTER and PURNELL, *Ozarkodina*

in diameter, but are often considerably elongated along bedding (Text-fig. 1B); they generally possess a core of silica (Text-fig. 1B–C), sometimes replacing fossils, and they generally show evidence of early diagenetic shrinkage in the form of syneresis cracks (Text-fig. 1B). The chert is succeeded in an outward direction by lightly dolomitized limestone (Text-fig. 1D). Common accessory minerals in the nodules are sphalerite and brown fluorite, the latter often possessing an unusual acicular crystal habit. The shale enclosing the nodules is bituminous, and contains abundant elongated organic fragments parallel to bedding (Text-fig. 1E), fragments that are honey-yellow in colour under both plane polarized light and crossed nicols.

When processed for conodonts, this mixed lithology yielded a well-preserved discrete-element fauna dominated by *Ozarkodina excavata* (run 1, Table 3). From this and subsequent runs (Table 3), we also recovered clusters of elements (Text-fig. 2) representing partial to nearly complete natural assemblages in which the elements are fused together in original life position.

Subsequent to processing run 1, carbonate and shale were processed separately and this revealed that both lithologies contained conodont clusters (Table 3); however, those recovered from carbonate are noticeably better preserved than are those from shale. Unexpectedly, manual splitting of the shale and carbonate from Hepworth also produced abundant bedding plane assemblages (Text-fig. 3), natural assemblages in which the conodont elements are preserved in original, or near original, life position. These occur flattened on the convex and concave surfaces of the shale separated from around carbonate nodules. To date more than 200 bedding plane assemblages of *Ozarkodina excavata* have been recovered from Hepworth.

The fauna is dominated by *Ozarkodina excavata*, but rare associates, both as discrete elements and as natural assemblages, include *Ctenognathodus* sp., *Panderodus* sp., *Pseudooneotodus* sp. and an undescribed taxon with an apparatus of diminutive, finely denticulated elements. This small species is known from Sweden (L. Jeppsson, pers. comm. 2002) and has been illustrated from Britain (Aldridge 1985, pl. 3.4, fig. 2) as *Ozarkodina*? sp. nov. None of these elements was included in the counts that form the basis for the present study.

Experimental design and method. This conodont fauna is unique in that the dominance of *Ozarkodina excavata* and the completeness of the natural assemblages make it possible to rule out biases discussed above under headings 1–4. The natural assemblages of *Ozarkodina excavata* from nodular carbonate and from bituminous shale demonstrate unequivocally that the apparatus contained 15 elements in the usual ozarkodinid configuration. We doubt that the number of elements in the apparatus of ozarkodinid conodonts varied from 15 (see discussion of bias 1 above), but the evidence from the Hepworth fauna allows us to rule out this potential bias from the outset. The natural assemblages are generally complete and, although we recovered a few faecal element concentrations, there is remarkably little evidence of predator ingestion, digestion and excretion (allowing us to discount bias 2, the effects of predation). The natural assemblages show little or no evidence of sedimentological dismemberment and sorting (allowing us to eliminate bias 3, post-mortem sorting). Because we recovered relatively complete natural assemblages from both carbonate and shale, and because we processed both lithologies, we concluded that bias 4, lithological sampling bias, could also be safely discounted.

The elimination of these four factors leaves only differential breakage and solution during sediment compaction and diagenesis (bias 5), differential loss of elements during laboratory processing (bias 6) and identification bias (bias 7) as the possible causes of any deviation in recovery of elements from the numbers in which they occurred in the apparatus during life.

Light and scanning electron microsocopy of both natural assemblage types, and of conodont elements recovered by laboratory processing of both nodular carbonate and bituminous shale, allowed us to evaluate bias 5, differential breakage and solution during sediment compaction and diagenesis. Bias 6, loss of elements during laboratory processing, was evaluated by processing the two lithologies in a number of different ways, and processing each lithology in a number of 'runs' (see Table 4). The expected number of elements was then compared with the actual number and kinds of elements recovered. Ten runs (Table 3) were processed by Ms Kathy David of the Palaeobiology Department, Royal Ontario Museum (ROM), utilizing acetic and formic acids individually, or successively, to break down the nodular carbonates, and acetic acid and sodium hypochlorite to disaggregate the bituminous shale (Table 4). The processing of the runs was monitored using introduced 'spikes' (von Bitter and

EXPLANATION OF PLATE 2

Figs 1–17. *Ozarkodina excavata* (Branson and Mehl) from Eramosa Member carbonate (Silurian), Hepworth, Bruce County, Ontario. 1–6, P_1 elements, 'lateral' views, ROM 56355–56361. 7–8, P_1 element pair, 'lateral' views, fused in near life position; ROM 56362. 9–11, P_2 elements, 'lateral' views; ROM 56363–56365. 12–13, P_2 element pair, 'lateral' views, fused in near life-position; ROM 56366. 14–17, S_0 elements; ROM 56366–56369; 14–16, 'posterior' views, 17, oblique aboral view. All × 58.

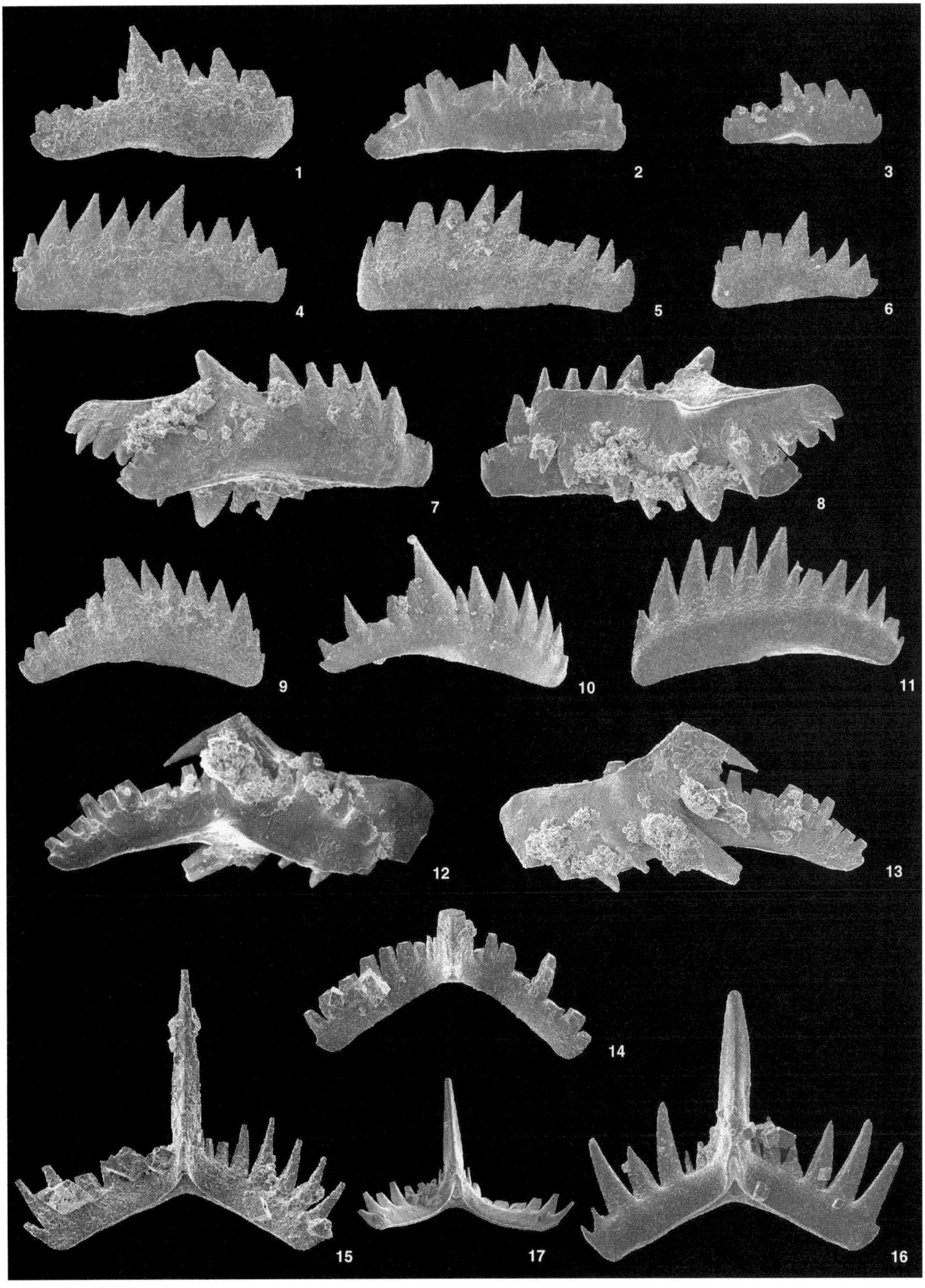

VON BITTER and PURNELL, *Ozarkodina*

Millar-Campbell 1984), controlled by the use of buffered acids, and by the daily removal of conodont elements from unspent acid. The residues from the first run were picked by Mr Christopher Stott, presently of the University of Western Ontario, whereas the remaining runs were picked by Ms David and/or PvB. Picking of samples is here regarded as part of standard laboratory processing procedure. Bias 7, identification bias, was evaluated by comparing individual counts and combined element counts to the expected frequency of occurrence of the elements (based on natural assemblage data).

Our data are summarized in Table 3. Rather than simply evaluating these data subjectively, we performed t-tests (Table 5) to determine whether the numbers of each element type recovered from the carbonate and shale fractions of the Eramosa Member differ significantly from the numbers that would be obtained from complete apparatuses [summaries of t-test methodology as applied to such problems appear in general statistical textbooks for geologists and biologists, such as Bailey (1981) or Davis (1986)]. For example, P_1 elements should make up 13·33 per cent (2/15) of an unbiased sample; does the mean value for P_1 element recovery from the Eramosa Member carbonate samples (9·028%) differ significantly from this value or not? This equates to a null hypothesis that the mean relative contribution of an element to the samples does not differ significantly from the proportion of a complete apparatus that the element makes up.

RESULTS: EVALUATION OF BIAS IN CONODONT FAUNAS FROM THE ERAMOSA MEMBER

The degree of conodont element over- and under-representation

Table 5 summarizes the results of t-tests applied to our data for recovery of elements of *O. excavata* from the Eramosa Member. The critical values of t for rejection of the null hypothesis was taken at a significance level of $P = 0{\cdot}05$ (although for several elements, t-values indicate that the null hypothesis could be rejected at a much higher level of significance).

One unusual potential bias in the data that must be dealt with arises because several runs yielded fused clusters of elements, in some cases in moderately large numbers (Table 3). The difficulty with these is that the S elements are very closely juxtaposed and it is extremely difficult to identify and count them. Consequently, the elements in clusters have not been included in the totals, except where the cluster comprises only a pair of elements. If this was a significant bias, then those runs with the fewest clusters should have S levels closest to those expected. However, this is not the case: in runs with few or no clusters (e.g. 1, 3, 9, 10) $S_{1/2}$ elements, for example, are strongly under-represented.

For the five carbonate runs, the null hypothesis is rejected for P_1, P_2, S_0 and $S_{1/2}$ elements, P_1 and $S_{1/2}$ elements being significantly under-represented, and P_2 and S_0 elements being significantly over-represented. The contribution of $S_{3/4}$ and M elements to the fauna does not differ significantly from the numbers expected in an unbiased sampling of the complete apparatuses. When S_0, $S_{1/2}$ and M element numbers are combined, however, the null hypothesis cannot be rejected, so the bias in S_0 and $S_{1/2}$ element numbers is probably due, at least in part, to broken $S_{1/2}$ elements being included in the counts for S_0 and M elements. Similarly, when P element numbers are combined, including indeterminate P elements (i.e. those that because of breakage cannot be assigned to P_1 or P_2 positions), the number of P elements does not differ significantly from the number expected. Laboratory processing of the carbonate fraction therefore yields an unbiased fauna, and other than the errors in assigning elements to positions noted above, post-depositional and laboratory processes have almost no effect on the numbers of elements recovered.

These results differ markedly from the pattern of element recovery of these elements in Late Palaeozoic faunas (Table 1), where strong positive bias in P_1 element recovery is the norm. It also differs from the published data for *O. excavata* recovery (Table 2), but it is worth noting that although P_1 element over-representation is usual, both Jeppsson (1974) and Rexroad *et al.* (1978) documented positive bias in M element recovery, and the data of Jeppsson (1974), Rexroad *et al.* (1978) and Aldridge (1972) all showed over-representation of S_0 elements (Table 2).

The results for the shale runs 6–10 from the Eramosa Member are quite different. The null hypothesis is rejected for P_1 elements (extreme under-representation), P_2 elements (over-representation) and $S_{1/2}$ elements (under-representation). The numbers of S_0, $S_{3/4}$ and M elements do not differ significantly from the numbers expected in an

EXPLANATION OF PLATE 3

Figs 1–20. *Ozarkodina excavata* (Branson and Mehl) from Eramosa Member carbonate (Silurian), Hepworth, Bruce County, Ontario. 1–9, $S_{1/2}$ elements, 'inner lateral' view; ROM 56370–56378. 10–11, 13–15, $S_{3/4}$ elements, 'inner lateral' view; ROM 56379, 56380, 56382–56384. 12, S_3 and S_4 element pair fused together in near life-position, 'inner lateral' view; ROM 56381. 16–20, M elements, 'posterior' view; ROM 56385–56389. 1–15, ×58; 16–20, ×50.

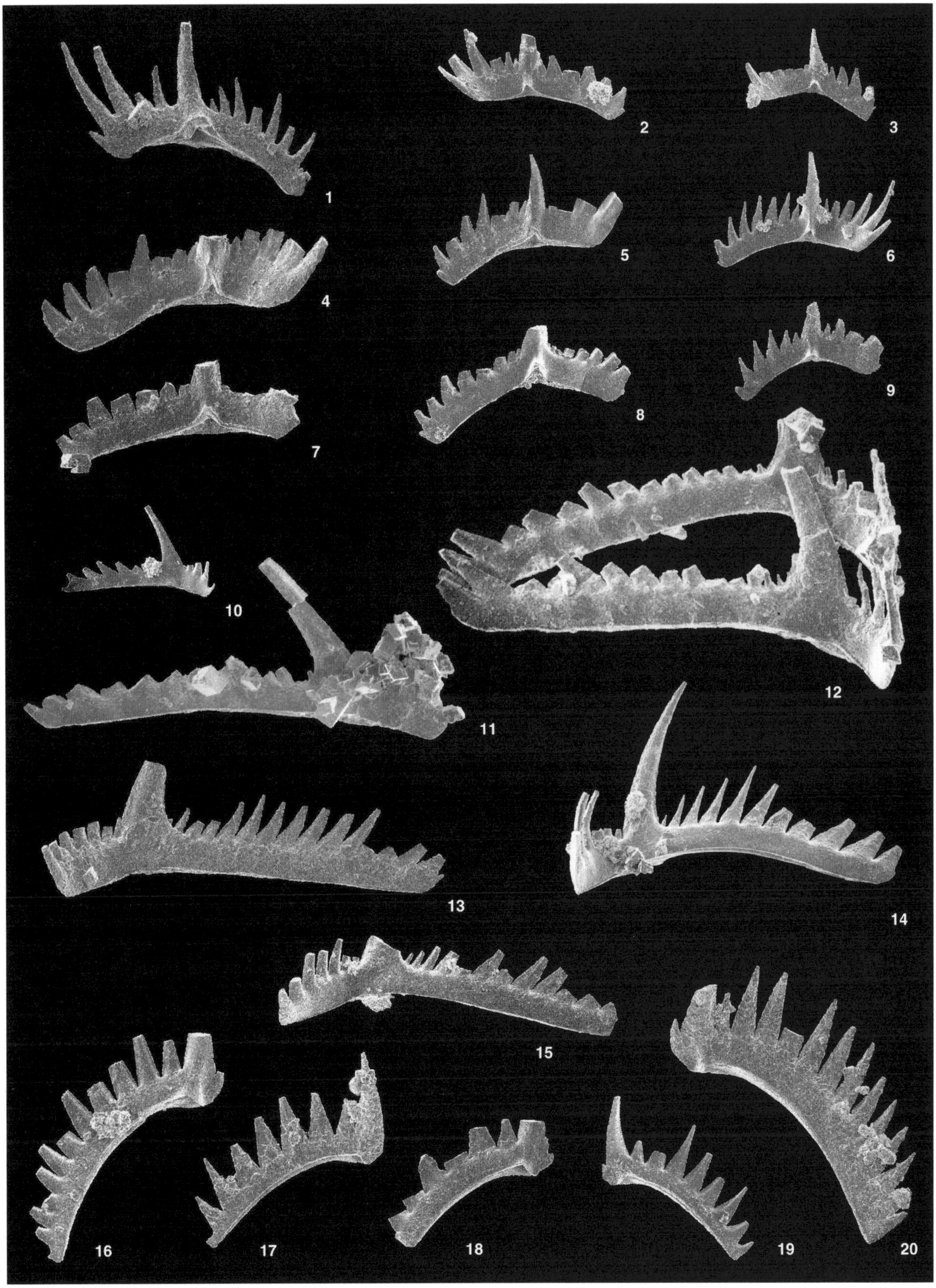

VON BITTER and PURNELL, *Ozarkodina*

unbiased sample, although when the aberrant run 6 (Table 3) is excluded from analysis, S_0 elements are significantly over-represented. The extreme under-representation of P_1 elements is due primarily to the difficulty in recognizing broken P_1 elements and differentiating them from broken P_2 elements; when P element numbers are combined they differ significantly from the expected numbers in being over-represented. The over-representation of S_0 elements and the under-representation of $S_{1/2}$ elements may in part result from broken $S_{1/2}$ elements being mistakenly included in S_0 element counts, but this is not the whole story, as when S_0, and $S_{1/2}$ counts, and S_0, $S_{1/2}$ and M counts are combined, these elements are still significantly under-represented. In the case of shale runs 6–10 (Table 3), then, post-depositional and/or laboratory processes have biased recovery of elements, leading to significant over-representation of P elements and significant under-representation of $S_{1/2}$ elements. The fact that laboratory processes have no significant effect on recovery of elements from the carbonate runs strongly supports the hypothesis that the bias in the shale runs is a result of post-depositional processes, probably connected to sedimentary compaction, leading to elements being broken to the point at which their identity cannot be determined.

Differential breakage and solution during diagenesis and sediment compaction (bias 5)

The nature of the nodular carbonate and bituminous shale at the Hepworth exposure of the Eramosa Member (see discussion above) is important in explaining one of the more startling differences between the shale and carbonate runs: the much greater abundance of conodont elements in the bituminous shale (Table 3). In the most productive shale run (Table 3), 558 conodont elements per kg were recovered, with an average yield of 159 elements per kg for the five runs. In contrast, the best yielding carbonate run produced only 49 elements per kg, with an average of 46 per kg for the five runs (Table 3). This is consistent with early diagenetic growth of the nodules preventing compaction of the carbonate. In the bituminous shales, on the other hand, loss of carbonate and later compaction will have significantly increased the number of elements per unit volume. This is also consistent with differences in preservation between the two lithologies. Compaction and possible pressure solution are both potentially important to conodont element abundance and preservation. Evidence from bedding plane assemblages on shale surfaces (Text-fig. 3), and from clusters obtained from residues of laboratory processed shale (Pl. 1), demonstrates that sediment compaction, and resultant differential breakage of conodont elements, is the norm in the bituminous shales at Hepworth, whereas elements and clusters derived from residues of laboratory-processed carbonate (Text-fig. 2, Pls 2–3) are whole, unfractured and well preserved.

Evidence for diagenetic removal of individual conodont elements or of clusters in the shale by solution is lacking. Bedding plane assemblages in the bituminous shale, although often highly fractured as a result of compaction (Text-fig. 3), are remarkably complete, and there is no evidence that individual elements, or groups of elements, were differentially removed by solution. The best evidence for solution lies in the fracture fillings in clusters (Pl. 1) and in isolated elements from the bituminous shale. The mineral(s) filling the fractures, most likely mobilized calcium phosphate, were probably introduced shortly after the differential breakage of the conodont elements, i.e. during or shortly after sediment compaction. It is probably these same fracture filling minerals that have fused together both the broken elements and the clusters.

Dolomite crystals adhering to conodont clusters (Text-fig. 2), as well as white powdery coatings on some conodont clusters from the carbonate runs, suggest that dolomitization may have been both early and a factor in the preservation of the often exquisite conodont clusters in the carbonate. Potential phosphatic fracture fillings, of the kind observed and documented in the shale-derived conodont clusters (Pl. 1), have not been detected in the unfractured conodont elements and clusters from the carbonates.

CONCLUSIONS

Because nodular carbonate and bituminous shale from Hepworth, Ontario, Canada, contain abundant natural assemblages of *Ozarkodina excavata*, we can eliminate *a priori* four of seven factors, or biases, thought to lead to conodont element under- and over-representation. In processing five runs of each lithology by standard laboratory methods, we found that there was remarkably little under- or over-representation of elements, that the overall yields of the elements of the apparatus of *Ozarkodina excavata*, the dominant taxon, were relatively balanced, and that there was little or no evidence for differential loss of elements due to laboratory processing (elimination of bias 6).

Nevertheless, statistically significant under- and over-representation was detected. In both carbonate and shale runs, P_1 and $S_{1/2}$ elements are significantly under-represented, whereas P_2 and S_0 elements are over-represented. We interpret this as being largely due to bias 5, element breakage during sediment diagenesis and compaction, compounded by bias 7, the difficulty of correctly differentiating between fragments of morphologically similar elements. In the carbonate runs, when numbers of P_1 and P_2 elements are combined and numbers of S_0, $S_{1/2}$ and M elements are combined, element numbers do not differ significantly from those in articulated assemblages. In shale samples, however, that exhibit much higher levels of element fragmentation, P elements remain significantly over-represented and S elements significantly under-represented even when individual element counts are combined.

Acknowledgements. We are indebted to Ms Kathy David of the Palaeobiology Department, Royal Ontario Museum (ROM), for her capable processing of the ten 'runs' from Hepworth, Ontario, as well as for her invaluable work in scanning electron microscopy and preparation of digital illustrations and tables. We are grateful to Mr Derek Armstrong of the Ontario Geological Survey for sending us the initial Hepworth sample that was processed as the first run, and to Mr Christopher Stott of the University of Western Ontario for 'picking' runs 1 and 6, and for some of the initial conodont identifications. Dr Desmond Collins of the ROM suggested splitting and separating the two Hepworth lithologies in order to search for bedding plane assemblages, and we gratefully acknowledge his help and encouragement. We thank Dr Mario Coniglio of the University of Waterloo for discussing the petrography of the carbonate nodules from Hepworth with PvB, and Howard Armstrong and Jerzy Dzik for their reviews of the manuscript. PvB thanks the Town of South Bruce Peninsula for permission to collect at Hepworth; he also gratefully acknowledges funding to conduct this research from a ROM Field and Laboratory Studies Operating Grant. MAP acknowledges funding from the Natural Environment Research Council (Advanced Fellowships GT59804ES and NER/J/S/2002/00673).

REFERENCES

ALDRIDGE, R. J. 1972. Llandovery conodonts from the Welsh Borderland. *Bulletin of the British Museum (Natural History), Geology*, **22**, 125–231.

—— 1985. Conodonts of the Silurian System from the British Isles. 68–92. *In* HIGGINS A. C. and AUSTIN R. L. (eds) *A stratigraphical index of conodonts*. The British Micropalaeontological Society. Ellis Horwood Limited, Chichester.

ARMSTRONG, D. K. 1988. Paleozoic geology of the southern Bruce Peninsula area, southern Ontario. *Ontario Geological Survey, Open File Report*, **5875**, 1–18.

—— 1989. Paleozoic geology of the southern Bruce Peninsula. 222–227. *In* COLVINE, A. C., CHERRY, M. E., BURKHARD, O. D., WHITE, O. L., BARLOW, R. B. and RIDDLE, C. (eds). *Summary of field work and other activities, 1989*. Ontario Geological Survey, Miscellaneous Paper, **146**.

—— and GOODMAN, W. R. 1990. *Stratigraphy and depositional environments of Niagaran carbonates, Bruce Peninsula, Ontario*. Ontario Petroleum Institute, Field Trip Guidebook, Field Trip **2**, 59 pp.

—— and MEADOWS, J. R. 1988. Stratigraphy and resource potential of the Eramosa Member (Amabel Formation), Bruce Peninsula. *Ontario Mines and Minerals Division, Ontario Geological Survey, Open File Report*, **5662**, 89 pp.

BAESEMANN, J. F. 1973. Missourian (Upper Pennsylvanian) conodonts of northeastern Kansas. *Journal of Paleontology*, **47**, 689–710.

BAILEY, N. T. J. 1981. *Statistical methods in biology*. Second edition. Edward Arnold, London, 216 pp.

BROADHEAD, T. W. and DRIESE, S. G. 1994. Experimental and natural abrasion of conodonts in marine and eolian environments. *Palaios*, **9**, 564–560.

—— —— and HARVEY, J. L. 1990. Gravitational settling of conodont elements, implications for paleoecologic interpretations of conodont assemblages. *Geology*, **18**, 850–853.

—— and REPETSKI, J. E. 2002. Taphonomic influences on the distribution of conodonts, with examples from Ordovician peritidal carbonate sequences of the eastern United States. *Strata, Serie 1. Abstracts of the Eighth International Conodont Symposium held in Europe*, **12**, 21.

CARLS, P. 1977. Could conodonts be lost and replaced? – Numerical relations among disjunct conodont elements of certain Polygnathidae (late Silurian – Lower Devonian, Europe). *Neues Jahrbuch für Geologie und Paläontologie, Abhandlungen*, **155**, 18–64.

DAVIS, J. C. 1986. *Statistics and data analysis in geology*. Second edition. Wiley, New York, 646 pp.

DONOGHUE, P. C. J. and PURNELL, M. A. 1999. Growth, function, and the conodont fossil record. *Geology*, **27**, 251–254.

DU BOIS, E. P. 1943. Evidence on the nature of conodonts. *Journal of Paleontology*, **17**, 155–159.

ELLISON, S. P. Jr 1968. Conodont census studies as evidence of sorting. *Abstracts and Program of the Annual Meeting of the Geological Society of America, North-Central Section, Iowa City, Iowa (May 8–11, 1968)*, p. 42.

GRAYSON, R. C., MERRILL, G. K. and LAMBERT, L. L. 1991. Carboniferous gnathodontid conodont apparatuses: evidence for the dual origin for Pennsylvanian taxa. *Courier Forschungsinstitut Senckenberg*, **118**, 353–396.

GROSS, W. 1954. Zur Conodonten-Frage. *Senckenbergiana Lethaea*, **35**, 73–85.

HELFRICH, C. T. 1980. Late Llandovery–Early Wenlock conodonts from the upper part of the Rose Hill and the basal part of the Mifflintown formations, Virginia, West Virginia, and Maryland. *Journal of Paleontology*, **54**, 557–569.

HINDE, G. J. 1879. On conodonts from the Chazy and Cincinnati Group of the Cambro-Silurian, and from the Hamilton and Genesee-Shale divisions of the Devonian, in Canada and the United States. *Quarterly Journal of the Geological Society of London*, **35**, 351–369.

JEPPSSON, L. 1969. Notes on some Upper Silurian multielement conodonts. *Geologisk Foreningens i Stockholm Forhandlingar*, **91**, 12–24.

—— 1974. Aspects of late Silurian conodonts. *Fossils and Strata*, **6**, 1–54.

—— 1976. Autecology of late Silurian conodonts. 105–118. *In* BARNES, C. R. (ed.). *Conodont paleoecology*. Geological Association of Canada, Special Paper, **15**, 324 pp.

—— and ANEHUS, R. 1995. A buffered formic-acid technique for conodont extraction. *Journal of Paleontology*, **69**, 790–794.

—— —— and FREDHOLM, D. 1999. The optimal acetate buffered acetic acid technique for extracting phosphatic fossils. *Journal of Paleontology*, **73**, 964–972.

——FREDHOLM, D. and MATTIASSON, B. 1985. Acetic-acid and phosphatic fossils – a warning. *Journal of Paleontology*, **59**, 952–956.

KLAPPER, G. and MURPHY, M. A. 1974. Silurian–lower Devonian conodont sequence in the Roberts Mountains Formation of central Nevada. *University of California Publications in Geological Sciences*, **111**, 1–62.

KOHUT, J. J. 1969. Determination, statistical analysis and interpretation of recurrent conodont groups in Middle and Upper Ordovician strata of the Cincinnati region (Ohio, Kentucky, and Indiana). *Journal of Paleontology*, **43**, 392–412.

KREJSA, R. J., BRINGAS, P. and SLAVKIN, H. C. 1990. A neontological interpretation of conodont elements based on agnathan cyclostome tooth structure, function, and development. *Lethaia*, **23**, 359–378.

McGOFF, H. J. 1991. The hydrodynamics of conodont elements. *Lethaia*, **24**, 235–247.

MERRILL, G. K. and POWELL, R. J. 1980. Paleobiology of juvenile (nepionic?) conodonts from the Drum Limestone (Pennsylvanian, Missourian, Kansas City area) and its bearing on apparatus ontogeny. *Journal of Paleontology*, **54**, 1058–1074.

MÜLLER, K. J. and NOGAMI, Y. 1971. Über den Feinbau der Conodonten. *Memoirs of the Faculty of Science, Kyoto University, Series of Geology and Mineralogy*, **38**, 1–87.

NICOLL, R. S. 1977. Conodont apparatuses in an Upper Devonian palaeoniscoid fish from the Canning Basin, Western Australia. *Bureau of Mineral Resources, Journal of Australian Geology and Geophysics*, **2**, 217–228.

ORCHARD, M. J. 2002. The what and where of Triassic multielement conodonts. *Strata, Serie 1. Abstracts of the Eighth International Conodont Symposium held in Europe*, **12**, 50.

—— 2005. Multielement conodont apparatuses of Triassic Gondolelloidea. 73–101. *In* PURNELL, M. A. and DONOGHUE, P. C. J. (eds). *Conodont biology and phylogeny: interpreting the fossil record. Special Papers in Palaeontology*, **73**, 218 pp.

PURNELL, M. A. 1993. The *Kladognathus* apparatus (Conodonta, Carboniferous): homologies with ozarkodinids and the prioniodinid Bauplan. *Journal of Paleontology*, **67**, 875–882.

—— 1994. Skeletal ontogeny and feeding mechanisms in conodonts. *Lethaia*, **27**, 129–138.

—— 1995. Microwear on conodont elements and macrophagy in the first vertebrates. *Nature*, **374**, 798–800.

—— and DONOGHUE, P. C. J. 1997. Architecture and functional morphology of the skeletal apparatus of ozarkodinid conodonts. *Philosophical Transactions of the Royal Society of London, Series B*, **352**, 1545–1564.

—— —— 1998. Skeletal architecture, homologies and taphonomy of ozarkodinid conodonts. *Palaeontology*, **41**, 57–102.

—— —— (eds) 2005*a*. Conodont biology and phylogeny: interpreting the fossil record. *Special Papers in Palaeontology*, **73**, 218 pp.

—— —— 2005*b*. Between death and data: biases in interpretation of the fossil record of conodonts. 7–25. *In* PURNELL, M. A. and DONOGHUE, P. C. J. (eds). *Conodont biology and phylogeny: interpreting the fossil record.* Special Papers in Palaeontology, **73**, 218 pp.

—— —— and ALDRIDGE, R. J. 2000. Orientation and anatomical notation in conodonts. *Journal of Paleontology*, **74**, 113–122.

REXROAD, C. B. and NICOLL, R. S. 1972. Conodonts from the Estill Shale (Silurian, Kentucky and Ohio) and their bearing on multielement taxonomy. *Geologica et Palaeontologica*, **1**, 57–74.

——NOLAND, A. V. and POLLOCK, C. A. 1978. Conodonts from the Louisville Limestone and the Wabash Formation (Silurian) in Clark County, Indiana and Jefferson County, Kentucky. *Indiana Geological Survey, Special Report*, **16**, 1–15.

RHODES, F. H. T. 1952. A classification of Pennsylvanian conodont assemblages. *Journal of Paleontology*, **26**, 886–901.

SCOTT, H. W. 1942. Conodont assemblages from the Heath Shale Formation, Montana. *Journal of Paleontology*, **16**, 293–300.

SIMPSON, A. J. and TALENT, J. A. 1995. Silurian conodonts from the headwaters of the Indi (upper Murray) and Buchan rivers, southeastern Australia, and their implications. *Courier Forschungsinstitut Senckenberg*, **182**, 79–215.

SWEET, W. C. and DONOGHUE, P. C. J. 2001. Conodonts: past, present, future. *Journal of Paleontology*, **75**, 1174–1184.

TOLMACHEVA, T. Y. and PURNELL, M. A. 2002. Apparatus composition, growth, and survivorship of the Lower Ordovician conodont *Paracordylodus gracilis* Lindström, 1955. *Palaeontology*, **45**, 209–228.

von BITTER, P. H. 1972. Environmental control of conodont distribution in the Shawnee Group (Upper Pennsylvanian) of eastern Kansas. *University of Kansas, Paleontological Contributions*, **59**, 1–105, 16 pls.

—— DAVID, K. and FERRIS, G. 1996. Fossil bacteria on faecal conodont assemblages from the Upper Carboniferous of England, Scotland, and the United States. 62. *In* DZIK, J. (ed.). *Sixth European Conodont Symposium, Abstracts*. Instytut Paleobiologii PAN, Warszawa, 70 pp.

—— and MERRILL, G. K. 1980. Naked species of *Gondolella* (Conodontophorida): their distribution, taxonomy, and evolutionary significance. *Royal Ontario Museum, Life Sciences Contributions*, **136**, 1–56.

—— —— 1998. Apparatus composition and structure of the Pennsylvanian conodont genus *Gondolella* based on assemblages from the Desmoinesian of northeastern Illinois, USA. *Journal of Paleontology*, **72**, 112–132.

—— and MILLAR-CAMPBELL, C. 1984. The use of 'spikes' in monitoring the effectiveness of phosphatic microfossil recovery. *Journal of Paleontology*, **58**, 1193–1195.

WALLISER, O. H. 1964. Conodonten des Silurs. *Abhandlungen des Hessischen Landesamtes Bodenforschung*, **41**, 1–106.

ZHURAVLEV, A. V. 1999. Resorption of conodont elements. *Lethaia*, **32**, 157–158.

ZIEGLER, W., LINDSTRÖM, M. and McTAVISH, R. 1971. Monochloroacetic acid and conodonts – a warning. *Nature*, **230**, 584–585.

[Special Papers in Palaeontology, 73, 2005, pp. 57–71]

BIASES IN THE RECOVERY AND INTERPRETATION OF MICROPALAEONTOLOGICAL DATA

by LENNART JEPPSSON

Department of Geology, Sölvegatan 12, SE-22362 Lund, Sweden; e-mail: Lennart.Jeppsson@geol.lu.se, Lennart.Jeppsson@telia.com

Abstract: Bias caused by collecting and processing bulk samples is largely independent of what fossil clade or mineral is searched for. Instead, different methods bias the data to a different, frequently very large, degree. Furthermore, biases accumulate with each recovery step, and the sum may be extreme unless appropriate methods to minimize it are employed. The effects depend on what the data are used for, e.g. establishing range ends (zonal boundaries), taxonomy (frequencies as an aid to, for example, conodont apparatus reconstruction) and ecology (relative frequency in a fauna, frequency/kg, faunal diversity). However, the best published methods remove calcium carbonate and dolomite without bias. All rocks with such cement can be broken down without bias, and so can some claystones with little lime. The bias caused by concentration can be measured, kept low, and documented. Removal of clay is an exception: screening or decanting removes all small elements. Extraction methods should be stated in all publications so that the data can be assessed more fully and quoted properly. Information about the acid methods, screen hole diameter and collection size are especially important because these usually cause the greatest bias. Reliability of observed range ends increases with increasing number of specimens and with decreasing sample distance (recollecting near the boundary). Samples that are too small, yielding subadequate collections, can strongly bias placement of zonal boundaries and implied diversity. Not taking the uncertainty intervals of zonal boundaries into account may result in artificially extended observed ranges of other species. Methodological progress over the last 25 years has increased the potential average yield per hour of manpower over 100 times for samples yielding fewer than 100 elements/kg. This has made it possible to overcome most of the biases outlined herein. Similarly, taking the biases that are known or are likely to have affected data into account allows levels of precision in data to be evaluated and published.

Key words: bias, microfossils, conodont, collecting, laboratory work, analysis, methods.

FROM sampling to picking to final analysis, the methods by which conodont elements are recovered for study influence what is recovered. The observed record thus deviates more or less from the preserved one; that is, it is biased. Some kinds of bias are indiscriminate, affecting all elements to the same degree, but most discriminate strongly. In order to minimize bias and to document bias that is unavoidable, it is necessary to identify the cause and magnitude of bias. Most studies of conodont recovery have been directed towards maximizing yield, and our knowledge of bias is concerned largely with minimizing loss during recovery rather than as a topic that has been studied in its own right. Those parts of this account that deal with the destructive effects of unbuffered acids are based on controlled experiments and are thus empirically scientific. Anyone doubting the conclusions drawn from these experiments can and should repeat them (the critical test takes less than half an hour of work to repeat). In accordance with this principle, other workers (e.g. R. J. Aldridge, pers. comm. *c.* 1981/82) quickly confirmed my discovery of the destructiveness of unbuffered acetic acid (Jeppsson in Sweet 1982; Jeppsson *et al.* 1985). Similarly, Mawson (1987) tested and confirmed the increase in yield with buffered acetic acid. Dismissing the results of these experiments without repeating them is unscientific, as is the use of untested methods of acid processing.

Here I describe how the loss of conodont elements can be monitored and corrected at most stages of laboratory processing. A useful spike test (von Bitter and Millar Campbell 1984) measures element loss except that due to removal of fines. Other biases include fieldwork bias and small-collection bias. Apart from the acetic acid method of Jeppsson *et al.* (1999) and the buffered formic acid method of Jeppsson and Anehus (1995) the methods needed to cope with the various difficulties we face in extracting conodonts from rocks have not been subjected to rigorous testing. In the absence of such empirical studies, descriptions of bias in some methods are based on long personal experience, often including resampling and reprocessing of the same stratum. During 38 years of research I have tried many different methods: most of them can be very efficient in destroying or losing conodont elements.

I also discuss how to determine the amount of bias in different methods. Some tests may be found useful as a daily routine. Others are important for quantifying the amount of bias in the routine application of a method and may, perhaps, not be used for every sample. Many published methods are not mentioned, either because they have been superseded by better techniques or because I lack experience of them. The methods recommended here have functioned equally well for samples from all geographical areas and geological intervals processed in the Lund conodont laboratory, not just Silurian residues from Gotland and Skåne. Hence, I expect that my conclusions will be applicable widely. Similarly, although this study is published in a conodont context, some of the types of bias identified are found in all kinds of collecting work, be it fossils or other particles, micro- or macrofossils of phosphate or other minerals or material.

Some of my conclusions are drawn from discussions with colleagues or resulted when we compared our collections and I tried to determine why they differed. Other conclusions are based on analysis of published data. I have treated any bias discovered in this way as due to honest mistakes, and I thus consider it inappropriate to cite the source of the data. Some examples of what can occur using the wrong method may therefore appear artificial and exaggerated. Unfortunately they are not.

Bias can be minimized but not avoided entirely: hence one should always discuss the methods used and evaluate the degree of bias caused by different techniques. In practice, the pressure for space in publications will usually limit any such discussion to references to standard published methods, with notes of any deviations from these methods and their effects on the data. However, using standard methods has a major advantage in that data in different papers can be compared. Different kinds of data will be biased to a different degree by different fieldwork and laboratory methods. Some of the more frequent causes of bias are listed in Table 1, together with what can and should be undertaken to minimize them and document their effects.

The methods advocated in this paper are those that I have found to be optimal in reducing the biases inherent in collecting and processing samples for conodonts (and other microfossils). I recognize that constraints of field logistics, laboratory time or other resources may conspire

TABLE 1. Some sources of bias, effects, possible improvements, and what documentation should be included in any publication; see text for details.

Source of error (and unit)	Effect	Improvements	Documentation
Sampling distance (years; metres are only a local proxy)		recollecting critical gaps	sampling specified, e.g. in a stratigraphic column
Collection size	abundance cut-off-limit (are rarer taxa recovered?)	adequate sample size; recollecting intervals where collections are too small	sample and collection size specified
Element destruction	distorted yield/kg; smaller collections	employ scientifically tested methods	references given for all methods; any deviation described
Selective destruction of elements	distorted relative frequencies	gentle handling: screening in water, etc.	references given for all methods; any deviation described
Selective recovery	distorted relative frequencies	routine testing of methods and result	references given for all methods; any deviation described
Screening (mm)	no specimens smaller than the cut-off-limit recovered	finer screen	sieve mesh diameter specified
Decanting	element size cut-off fuzzy and operator-dependent	use of fine screen	the smallest abundant specimen size discussed
Removal of non-phosphatic minerals	skewed proportions; smaller collections	use only the best methods available; testing all fractions; redo separations that are too biased	references given for all methods; any deviation described
Picking	skewed proportions; smaller collections	testing every tray picked	details of method discussed
Analytical error, e.g. no uncertainty intervals for frequencies, zonal boundaries	incorrect conclusions; unnecessary conflicts	analysis of confidence limits	confidence limits discussed

to prevent optimal work in some circumstances. However, I hope that this paper can be used as a guide to how element recovery will be biased in such cases, and what steps can be taken to minimize bias or take it into account when interpreting the collections that result. Some descriptions of methodology may seem unnecessarily detailed, or too obvious to be repeated here. However, in order to compare the relative biasing effects of alternative methods it is essential that processing details are made clear.

BIAS IN THE RECOVERY OF CONODONT ELEMENTS

The accumulation of bias during sampling, processing and interpretation

If something goes wrong during processing a sample, it is usually impossible afterwards to identify at what stage it occurred (or even realize that something went wrong). In order to illustrate the progress in processing methods therefore two hypothetical samples are used in Table 2 to provide a summary of the biases introduced in processing and interpreting the resulting collections. The first of these hypothetical examples is based on methods of collection, processing and analysis that were the state of the art 30 years ago; the second is based on the best methods available today. The biostratigraphic implications of bias for each of the resulting collections ('Interpretation and zonal identification') compare well with two real sets of samples collected in 1973 and from 1989 onwards, respectively, from the Slitebrottet 1 and 2 sequence on Gotland, now *c.* 50 m of strata. The 1989 samples varied in size as a result of collecting constraints, and variation in what was found to be an adequate size for each interval (some samples exceeding 100 kg). Nevertheless, the two sets of samples differed in size by about two orders of magnitude. Similarly, muddiness, yield, etc., varied through the sequence.

To avoid exaggerating the perils of the 5-kg hypothetical sample, the losses described for each step in Table 2 are at the lower end of the probable range. The differences in the real results were even greater than those in the hypothetical set: in the Slitebrottet samples from 1973 none of the zones named by Walliser (1964) at Cellon was identified (Jeppsson 1983*a*). The second set of samples, however, yielded both *K. patula* (at an average frequency of one platform element per 30–40 kg) and *Ozarkodina sagitta sagitta* (at a frequency in the per mille range). The latter taxon has P_1 elements (sp) that are thin-walled, and all its other elements are small and fragile, so recovery requires very gentle handling and very low discriminating bias in spite of the large volumes of mud etc. that must be removed during processing. With their ranges known reasonably well, two other zones could be characterized, the Sheinwoodian conodont zonation revised, and the new standard Sheinwoodian zonation applied to published sequences, including identifying one of the new zones and a major gap at Cellon (Jeppsson 1997*b*).

Another case of resampling illustrates the improvement in yield up to the 1990s: samples Kl67-32 and Kl67-33 from Skåne gave eight and 12 specimens from 0·468 and 0·485 kg, i.e. 17 and 25 elements kg^{-1}, respectively (Jeppsson 1975; the methods used are described). Samples Sk93-4LJ, 18·0 kg, and Sk93-3LJ, 37·2 kg, from the same beds (treated with most of the methods recommended herein) yielded *c.* 1200 and *c.* 4000 conodont elements, respectively. This equates to 67 and 107 elements kg^{-1}, respectively; compared with the earlier samples this is four times as many per kilogram for both beds and over 100 times as many specimens per sample.

Fieldwork

Sample size and yield. There is a relationship between sample size and yield that through the effect of 'rounding off' can have an important bearing on sampling strategies. For example, a unique specimen in 100 g does not equate to ten such specimens kg^{-1}. Because of the statistics of small numbers, the true frequency is probably much less, and in my experience a new sample ten times as large would not be expected to yield more than approximately two or three specimens (Anders Martinsson, working with ostracodes, had a similar 'rule'). This has implications where resampling is undertaken in order to find more specimens of biostratigraphically important taxa, for example to increase the reliability of ranges (see below and Table 2). Recollecting a sample less than ten times as large will in most cases be a waste of time and other limited resources.

Sample collection. Collecting standard size samples from a lithologically diverse sequence (ranging, for example, from low-energy fine-grained marly lithologies to higher-energy, rapidly deposited coarse reef-detritus limestones) results in collections of different size. The yield can vary by a factor of much more than ten, and all or most of the rarer taxa will consequently be absent from the smaller collections. Strong apparent variations in diversity will arise as an artefact of the sampling, and stratigraphic ranges for most taxa will be artificially shortened. Only with correspondingly larger bulk samples can the expected lower yields be compensated for, although bias due to winnowing, crushing and other factors associated with samples from high-energy environments will remain and will need to be taken in account.

Collecting in quarries is usually easiest near the drill holes resulting from past blasting. However, blasting not only splits the rock but also creates microfractures within it and the fossils it contains (building-stone quarries use other methods). Hammering also causes microfracturing of rock and fossils, introducing a bias that is, for a number of reasons, unnecessary. Firstly, during collection and transport it is much more efficient to clean two small surfaces with a wire brush, and, when dry, label them with a marker pen, than to break a sample and pack a lot of

TABLE 2. Comparison between two hypothetical samples to illustrate the differences in bias introduced during collection, processing and analysis using methods that were the state of the art 30 years ago and those advocated herein. The characteristics of the two hypothetical samples are as follows: both from the same stratum and of average lithology [a limestone containing 10–20% mud (but not so much mud that less than 90% of the sample breaks down in acid), 2‰ quartz sand, 5‰ ore minerals (mostly pyrite and weathered former pyrite)]; containing 100 conodont elements per kg, representing 16 species, the most abundant at 30 per cent, and each of the subordinate taxa at *c.* 30 per cent of the remainder (i.e. 30, 21, 15, 11, 7, 5, 4, 3, 2, 0·8, 0·5, 0·3, 0·2, 0·1 and 0·05%). This is more favourable than most of my collections in the frequency of the tenth species being as high as 0·8 per cent, the drop-off being regular and the diversity higher. The two hypothetical samples belonged to two hypothetical sets of samples, the first set spaced *c.* 1 m apart, the second at *c.* 3 m apart except where recollected (to compensate for the doubled processing time of the larger samples and to leave resources for recollecting critical intervals).

Methods of 1973	Methods advocated here
Sample size and theoretical max. element recovery	
5 kg	50 kg
500 elements	5000 elements
Rock processing	
Crushed (breaking larger elements and many others beyond recognition). Sample suspended in colander in a bucket.	Not crushed. Sample suspended in a 'colander' in a 165-litre vessel (Jeppsson 1987, p. 48).
Ten changes of acid solution (100 min working time).	Ten changes of acid solution (150 min working time).
Spent acid decanted to point when fine sediment almost spilling out, leaving all insoluble residue and elements in bucket (possibly a few of the smallest elements lost; if clay removed by decanting, loss would have been higher).	Spent acid solution siphoned off through 53-μm screen (except for amount needed as buffer) leaving all insoluble residue and elements in vessel.
Vigorous addition of water from tap to dilute the solution to *c.* 10 per cent, resulting in a complete mixing during nine of the changes. Spent acid remaining in the bucket resulted in a ±buffered acid solution.	Vigorous addition of new acid and water through a hose reaching down into the bottom sediment.
pH not measured (but happened to be on safe side of 3·6 after nine of the ten changes).	pH measured, and any deviation from a *c.* 7 per cent solution and a pH slightly above 3·6 adjusted with more buffer or acid (Jeppsson *et al.* 1999).
	When repeated pH measurements indicate that remaining lumps are decalcified, most clay and silt automatically siphoned away under water through a 63-μm screen.
Residue processing	
Acid-insoluble residue hand-washed on a 63-μm screen.	Acid-insoluble residue hand-washed gently on a 63-μm screen, dried, treated with boiling water with some sodium carbonate, and washed. Treatment repeated until all clay removed.
Reduced using Franz Isodynamic Separator Model L1 (not available in Lund in 1973).	Reduced using Franz Magnetic Barrier Separator Model LB1. Traces of tap water carbonate and any dolomite removed with buffered formic acid (Jeppsson and Anehus 1995).
Density-separated using bromoform (a quartz crystal was used to judge density, density was high enough when it was floating on the surface).	Two-stage density-separation using sodium polytungstate at densities of 2·84 and 3·04, respectively (Jeppsson and Anehus 1999).
Result: large residue consisting mainly of pyrite that was not oxidized enough to be removed magnetically.	Result: residue consisting essentially of phosphatic fossils only; smaller amount than that from the 5-kg sample.
Picking	
Large residue spread thickly; each tray wet picked once (here in Lund, electrostatically).	Rich residue spread thinly; each tray picked electrostatically (Barnes *et al.* 1987) twice.
Loss of elements	

TABLE 2. Continued

Methods of 1973	Methods advocated here
Estimated at *c.* 63 per cent. After fifth acid change (when 50% of the sample was dissolved) pH in insoluble residue dropped to 3·0 and all elements in that residue were dissolved, i.e. 50 per cent of the original elements. Without feedback from bias control and with a less efficient magnetic-separator an additional *c.* 5 per cent of elements were lost in each of magnetic separation, density separation and picking. Small and thin-walled elements preferentially lost (due to e.g. rafting with quartz sand during density separation).	Estimated at *c.* 4 per cent total. No elements were lost in the bias-free acid treatments. With feedback from bias control during all other steps, correction of any too biased step, and an efficient magnetic-separator, less than 1 per cent the elements was lost in each of magnetic separation, density separation, and picking). Small and thin-walled elements preferentially lost.
Processing summary	
187 specimens recovered (37 per cent; but varying strongly according to which acid change buffering failed).	At least 4802 specimens recovered (96 per cent).
Working time: 5 units of time (1 unit of time per kg), 37 specimens per unit time (= 1 unit of yield per unit time).	Total working time twice that of 5-kg sample, i.e. 0·2 units of time per kg; 480 specimens per unit time (13 units of yield per unit time).
Faunal recovery (based on hypothetical frequency distribution)	
11 species (three rarest absent due to crushing, biased recovery of small elements, and the randomness affecting small numbers, respectively).	16 species (rarest represented by two elements).
Interpretation and zonal identification	
None of the defined zones could be identified in the collections from the 5-kg samples; all collecting and processing effort wasted if that was the only purpose.	All index species recovered; recollecting gave the zonal boundaries with a precision of better than 0·3 m. Further, the collections together yielded 14 specimens of the rarest species, indicating that all species had been found.

sharp-edged pieces in several bags (rock travels better than bags). More importantly, however, breaking a sample into small pieces or crushing before dissolution results in only slight improvements in processing efficiency. These are due not, as one might expect, to increasing reactive surface area but to decreasing the maximum distance from the centre of the thickest piece to the nearest surface (Jeppsson *et al.* 1999, p. 969). Crushing results in pieces of different sizes, and the increase in bias due to breakage is disproportionately much larger than the decrease in thickness. The position and orientation of the sample in the vessel (slabs should be approximately vertical) influence processing time more than crushing (Jeppsson *et al.* 1999). Today my samples are cleaned, weighed and dissolved as they are, except when a piece does not fit into the 'colander' (0·6 × 0·4 m). Occasionally, evident bedding plane weakness indicates that the thickest pieces in a sample will split easily and without much hammering into slabs, and these provide the only exception.

Although many conodont workers may collect samples much smaller than those I suggest here, this is by no means standard among microvertebrate specialists. Ward (1984), for example, drew the boundary between small and large samples at 100 kg and described methods for efficient handling of samples above that limit. Similarly, the vessels for dissolution of limestone described by Cooper and Grant (1972) and those used by Barry Fordham (pers. comm. 1996) are larger than mine, although mine are approximately 75 times larger than the standard vessels I used before 1972. My point here is that many of the methods for handling large samples were in part developed and tested some time ago. The cost of scaling up processing can be small (most of the required equipment can be obtained second hand, from supermarkets, or home made), and doubling the size of a dissolution vessel effectively halves the amount of time spent changing acid, so after only a few samples the vessel has paid for itself. With such adjustments in methods and equipment the increase in total manpower per sample only doubles for a ten-fold increase in collection size (= yield). I have found this scaling relationship to hold true over two orders of magnitude (samples 100 times larger requiring only four times the work).

Sampling strategies. Most palaeontologists know that different lithologies can yield different faunas, and collect accordingly. Therefore, such facies-controlled faunal differences are not of major concern here (although where facies are preferentially sampled or avoided this should be documented). Lithologies that are difficult to process are often under-sampled, because processing can be prohibitively expensive and frustrating (how this is exacerbated by using methods that yield inconsistent results is discussed below). The resulting bias will be most severe when working with condensed sequences or in quickly changing intervals where facies-related faunas and taxa may easily be missed. Similarly, external forcing may trigger a change to both a unique fauna and a lithology that is generally low-yielding or otherwise difficult. Using a model for what happened during a time interval, how that affected different taxa, and how later diagenesis and weathering have affected the fossils as a guide in collecting can yield new data by highlighting layers that would otherwise probably have been left unsampled. In my own work, I have found my model for oceanic cycles (e.g. Jeppsson 1990, 1998) extremely useful. For example, together with colleagues I have identified an interval representing probably only a few thousands of years at several localities on two palaeocontinents.

Owing to economic constraint, not every bed or even every 10,000-year interval can be collected adequately. A more or less unconscious eye for subtle lithological differences can yield better collections and better results, and sampling at standard distances may not be the best alternative in most work. We need to be aware of this and to find ways to document it. Apart from this, sampling for microfossils does not include any of the biases inherent in picking fossils by hand in the field.

Laboratory methods

The effectiveness of a laboratory method can be rated according to how well it removes a given mineral and what percentage of the original conodont element content remains. An unbiased method removes 100 per cent of a specific mineral and leaves 100 per cent of the fossils that are searched for; in short, a 100/100 method. Some widely used concentration methods yield <100/≈100 and repeatable results; that is, the degree of bias is about the same for every sample if the worker and laboratory set up are unchanged. Other methods yield strongly varying results (<<100/0–90), even if worker and laboratory variables are constant. Some methods still in use may even destroy all fossils (Jeppsson *et al.* 1999).

Rock dissolution and disaggregation. Chemical methods tested and found to be safe for phosphate reach 100/100 effectiveness in routine work. The buffered pH-measured acetic acid method (Jeppsson *et al.* 1999) and the buffered formic acid method (Jeppsson and Anehus 1995) are of this kind. Hence, limestones, dolostones and all lithologies with such cement can be broken down with 100/100 efficiency, if: (1) the method is tested for all kinds of phosphate that one wants to recover (some taxa require raising the minimum pH; Jerre 1994; Siverson 1995; D. Fredholm pers. comm. *c.* 1986–1990); (2) the laboratory work is done as prescribed; for example, the pH meter is correctly calibrated and measurements are undertaken properly. Hydrochloric acid may also be used (Jeppsson *et al.* 1985), but the method is not yet well enough developed to be efficient. Thioglycollic acid dissolves iron compounds (Ward 1984, p. 256), and a method based on controlled experiments needs to be developed. Many claystones break down with sodium carbonate or petroleum ether/kerosene methods (Jeppsson and Anehus 1999, p. 58); both approaches are very efficient, but the former is preferable for reasons of health, environment, economy and versatility.

Only these chemical methods reach 100/100 efficiency and the sum of processing bias will consequently be lowest if these methods are used whenever possible. It is possible to remove any remaining grains of calcite and dolomite using physical methods, but this will increase bias, and it is better to use buffered formic acid as the final acid step. Similarly, it is often best to finish all chemical preparations before applying physical methods other than screening.

Some methods yield inconsistent results. This includes unbuffered or incompletely buffered acids (herein referred to as ±unbuffered, for short) and methods involving 'gentle kneading' or freeze–thaw action to break down difficult sediment. The bias due to ±unbuffered acids varies strongly with minor methodological details and minor differences such as the muddiness of the sample, and the amount of carbonate per litre of pure acid used. There is, however, little size-related variation in the bias. If unbuffered acid is used but not exchanged and if the ratio of 'pure lime to amount of acid' happens to be optimal, then the efficiency can approach >>90/≈90 for a nearly pure limestone. Such an optimal ratio requires weighing the sample, determining its lime content and calculating how much acid to use. A complete exchange of the spent acid results in a total loss of all previously freed conodont elements and of those nearest the surface of the remaining rock pieces (some limestone must be dissolved before the acid is sufficiently buffered). Even an incomplete exchange of the acid may be disastrous if, for example, the acid was not completely spent (pH should be 4·8–5 or higher). In practice, such methods result in a strong increase in the average element loss with decreasing lime content, and element loss is probably essentially complete even in 'medium difficultly' lithologies (typically high-yielding with pH-measured methods). Results can be as poor as <<100/≈90, i.e. lots of limestone and other grains remaining, but very few or no conodont elements found after days of picking through the residue (pers. obs.).

Unbuffered formic acid reacts much faster than acetic acid, and results are generally worse and more unpredictable.

Destruction by acid is insidious in that damaged specimens are often rare in comparison with those dissolved or destroyed beyond recognition. Furthermore, damaged specimens are easily missed in the variation of preservation we find. I have repeatedly encountered colleagues who were unaware that their own material included such specimens and were hence confident that their acid method was safe. This is probably the main reason that some colleagues continue to use untested acid methods. (For further aspects of the variation in bias caused by such methods, see Jeppsson *et al.* 1985, 1999; Jeppsson and Anehus 1995.) Another frequent objection is that I have not taken into account the lateral variation in conodont faunas within a bed, the implication being that improved yield on resampling is the result of encountering part of the bed with more abundant conodonts, and not due to improved techniques and buffered acid. Although it is possible that conodont faunas may vary significantly in adjacent parts of a heterogenous bed, this cannot be the cause of my results. In all cases the better yield is obtained using the improved technique and enough localities have now been resampled to rule out statistically the alternative possibility. Furthermore, any conclusions regarding large lateral variation in conodont abundance within a bed that derives its data using processing methods that are demonstrably inconsistent in their results is, to say the least, open to question.

Physical methods of residue reduction. Physical methods, including sieving, decanting, density separation and magnetic separation, never achieve 100/100 performance in routine work. Some grains of the mineral that should have been removed will remain, and some of the phosphatic fossils will end up in the fraction removed. The degree of bias is strongly related to how these methods are used and is highest for small specimens (Jeppsson and Anehus 1999, p. 61). However, with proper care, the efficiency of some methods will be ≈100/≈100 and the bias will be more or less consistent. The degree of bias in the work can be evaluated by checking both fractions once processing has advanced so far that phosphatic particles can be seen in the residue. For example, if 100 phosphatic particles are seen per minute in one fraction but none in the other, over a similar period, then the loss is below or about 1 per cent if both fractions are of a similar size. If that is considered acceptable, the latter fraction can be stored without further processing. If loss is at an unacceptable level, processing is modified and evaluation repeated until an acceptable result is achieved. In this context, other phosphatic fossils are useful in two ways. Firstly, they increase the number of phosphatic particles seen in the conodont-containing fraction when testing a separation, and secondly, being on average more porous than conodont elements, such fossils tend to have more impurities and are more likely to be separated incorrectly. Their absence from the separated fraction thus provides an extra safety margin (see below).

A prerequisite when using physical methods of residue reduction is that the fossils are essentially free from adhering grains. Using sodium carbonate routinely during the washing of residues can help to achieve this.

Sieving and decanting. Removing the unwanted fine fraction from a residue by sieving or decanting causes the largest bias among all standard methods of residue reduction. It is the major reason why, to my knowledge, no collections contain the earliest growth stages of conodont elements. Furthermore, some taxa, e.g. *Decoriconus*, have elements that are so small that elements are found only rarely unless fractions below *c.* 150 μm are picked and many records are based on elements less than 125 μm in diameter.

Sedimentary, diagenetic and weathering processes may destroy the smallest specimens in a fauna, but generally the lower size limit of elements recovered is set by the choice of sieve. Decanting is often claimed to be better than sieving, but I have seen many resulting collections and the recovery limit is markedly operator dependent, rarely as low as 125 μm and usually considerably higher (the recovery limit is here defined so that essentially all specimens larger than that size are recovered; below that size the recovery percentage drops gradually to zero). The limit varies with the shape of the element and with their 'bulk density' (Jeppsson and Anehus 1999, p. 61). In contrast, it is easy to remove tens of litres of clay in routine work even if a sieve as fine as 63 μm is used (see below; a finer sieve has also been tested with good results). The recovery limit is much sharper than that for decanting, but occasional, smaller elements may also be recovered. Sieving has another major advantage: the bias can be expressed exactly, as in 'only elements over xx μm in diameter were recovered'.

Another kind of bias is also due mainly to the removal of the fines. The central parts of mature elements are relatively resilient, but other parts, especially denticles, distal processes and juvenile elements, are very brittle. Rough or extended washing intended to break down lumps of decalcified rock physically is much more destructive than generally thought. The effect is worsened if the residue contains heavy minerals such as pyrite, the effects of which resemble those of a ball mill or a tumbler. Methods for minimizing loss at this stage are detailed in the Appendix (section 1).

Washing under water is frequently used to recover brittle fossils. An 'explosive expansion' in the knowledge of Mesozoic metatherians (Mammalia) was 'in large part' due 'to widespread use of underwater screen-washing and associated techniques' (Cifelli 2001, p. 1219). I have had similar experience with methodological improvements in conodont processing (Jeppsson *et al.* 1985, 1999; Barnes *et al.* 1987; Jeppsson 1987; Jeppsson and Anehus 1995, 1999; herein) resulting in a large increase in the number of conodonts extracted [average recovery of elements during 1967–1983, 1000 elements per year (1967–1973, see Jeppsson 1975); improvements during 1984–1995 gradually raised recovery to a few hundred thousand elements per year, including highly significant new data and previously unknown taxa]. This increase is similarly due in part to an underwater screen-washing machine (Jeppsson 1987, p. 49). It functions very well, routinely washing away most of the clay through a 63-μm screen. It was primarily constructed to save expensive working time and to decrease the amount of tedious work. This saving, invested in larger samples, together with the reduced element breakage, increased the yield. Very gentle removal of the remaining clay has now made it possible to benefit more fully from the latter effect (see Appendix, section 1, points 6 and 7) and has helped to maintain the yearly yield despite decreasing availability of manpower.

Density separation. Density separation in liquids with densities below and above that of conodont elements (Jeppsson and Anehus 1999) is the only concentration method needed to produce routinely a phosphate fraction that is *c.* 99 per cent clean. Sometimes such a residue includes fewer 'other phosphate' than conodont elements, making it faster to pick away that and pour the remaining conodonts into the slide. Densities very close to that of conodont elements give the cleanest residue but increase the risk of losing some elements, such as those with impurities in their basal cavity. If picking a residue directly after the first density separation would take more time than a second density separation, then picking it directly is a waste of limited time or money. Bias caused by density separation and how to minimize it were described by Jeppsson and Anehus (1999). Low-viscosity liquid causes an unacceptable loss at a density of 3040 kg m^{-3} (G. K. Ahlberg, pers. comm. 2004), most probably because that density is so close to the density of the conodont elements that the rapid downward movement of heavier minerals, such as pyrite, is enough to drag them down, even when very small quantities are added at the time.

Magnetic separation. Magnetic separation removes paramagnetic material. Yellowish and grey pieces of hard clay are easiest to remove. (Different kinds of extra equipment permit removing ferromagnetic and diamagnetic minerals.) Both precision, in terms of degree of bias (i.e. the deviation from 100/100%), and the efficiency in terms of throughput vary with the magnetic-separator model. In the past I have obtained reasonable results using a Franz Isodynamic Separator L-1, but rebuilding it into a Franz Magnetic Barrier Separator Model LB1 increased both precision and efficiency enormously. If properly adjusted for each size fraction of a sample, bias can be kept low. However, with eight continuous variables to adjust on an LB1, the number of combinations is infinite and finding the best combination (with lowest bias) by random adjustments is impossible. It is necessary to know the effect of each possible adjustment in order to obtain good results (see Appendix, section 2).

Density separation is done in batches and the manpower requirement increases stepwise with size. Magnetic separation is therefore very useful for reducing large residues rich in paramagnetic particles prior to density separation. However, if residues are small or their unwanted parts consist mainly of quartz or other diamagnetic material, then magnetic separation may not save manpower. The cost per man-hour determines how many hours need to be saved to compensate for the investment in a separator because double density separation alone can yield a phosphate concentrate that is almost clean.

Sometimes even the clean phosphate fraction is too large to pick through and we have to choose between maximizing removal of unwanted material or minimizing the loss of conodonts. Impurities may make at least some of the other phosphatic fossils slightly denser and slightly more magnetic than the conodont elements, and fine-tuning the magnetic separator (see Appendix, section 2) or density separating at a density closer to that of the conodont elements can yield a fraction enriched in conodonts.

Picking. Picking is not free of bias. If particles are of different size, smaller specimens may be concealed beneath larger particles, but this bias can be reduced through size fractionation of a residue in a sieve-stack. Picking effectiveness and bias can be evaluated as follows: some material is spread thinly and evenly on a picking tray, which is then held over a sheet of paper and gently tapped horizontally a few times so that any perched particles are moved down by the horizontal vibrations; the tray is picked under a microscope at low magnification; the tray is again tapped over the sheet and picked at higher magnification; comparison of the numbers of elements obtained from first and second pickings provides a measure of picking bias. For example, if the two pickings gave 1000 and 100 elements, respectively, then a third would be expected to give *c.* ten specimens. If, on the other hand, the two pickings gave 1000 and 400 elements, 160 would be expected in the third, and so on. Even after the fifth picking, the remaining bias would be nearly twice that of the second picking in the first example, suggesting that bias would have been reduced (and picking time decreased) if the material had been spread more thinly in the tray.

The electrostatic picking method (Barnes *et al.* 1987) allows 50 or more conodont elements min^{-1} to be picked from a more or less clean phosphatic residue (R. Anehus and G. K. Ahlberg, pers. comm. 2002). The technique takes only a few minutes to learn and in addition to increased picking speed allows even the smallest specimens to be handled without breaking them. Sorting of the resulting stack of loose specimens is also much faster. Picking with a wet brush is inferior in all these respects and increases the bias, especially for small specimens (which are more brittle and easily destroyed during wet handling).

Other causes of bias

Previous work has discussed upwards reworking and identification of such range end bias (Jeppsson 1997*a*, pp. 472, 488), and contamination (Jeppsson 1997*a*, p. 472). Mislabelling or mixing sample numbers during handling is part of an old problem the cause of which is well known: *errare humanum est.*

Observed breakage of elements is the sum of breakage caused by usage during life (Jeppsson 1976), sedimentary processes, compaction of sediments, tectonics, sampling, processing, concentration and handling of the residue (von Bitter and Purnell 2005 discuss biases caused by some of these factors in more detail). Breakage of different origin has different characteristics, allowing appropriate measures to be taken to minimize that caused by sampling and processing (see Appendix, section 3). [Wear (Purnell 1995) and abrasion due to sedimentary processes (Broadhead and Driese 1994) also affect conodont elements but, in my experience, breakage is the main destructive effect of our handling of conodont remains.]

BIAS IN INTERPRETATION OF MICROPALAEONTOLOGICAL DATA

Interpreting how the observed micropalaeontological record deviates from that preserved in the studied strata is a two-stage process. Firstly, the various biases discussed above must be taken into account, their relative importance depending on the nature of the conclusions being drawn. Secondly, common sense or intuition is used to decide what conclusions can or cannot be drawn from the available data, or data are subjected to more rigorous analysis, including statistical techniques, to determine the degree of error or uncertainty in the results. This more rigorous approach is clearly superior in that results are not dependent on the subjective evaluation of the operator. Marshall (1990, 1994, 1997) has described methods for statistical evaluation of discontinuous records, and these aspects of analysis are not discussed here. His 1995 paper showed the power of such methods and clearly illustrated the danger in taking range ends at face value. Here I concentrate on problems that are more or less specific to bulk sample collecting and also comment on how the number of specimens in a collection influences the robustness of an identified range end.

Relative frequencies

Conclusions regarding the relative frequency of different species in a community, the ratio between different elements in an apparatus, the ratio of juvenile to mature elements, etc., are strongly affected by sieve mesh diameter. Similarly, using any method in a way that discriminates against small specimens has strong effects. Two examples illustrate this effect. All species of *Panderodus* have a short, small, symmetrical element (with a true furrow on both sides). In collections of the more gracile species the frequency of this element may be as low as 1 per cent (Jeppsson 1983*b*) even when a 63-μm screen is used. In the comparatively robust *P. greenlandensis*, on the other hand, recovery of this element approaches the expected frequency of approximately 7 per cent (assuming an apparatus of 15 elements). This is because the smallest individuals will always be represented only by their largest elements, even with a fine sieve, and the bigger elements will be over-represented as a consequence.

A similar example is provided by some Ordovician 'platform conodont genera', which until recently were thought to lack ramiform elements, despite being studied by many different authors through the years. As a result, they have even been separated from their closest relatives as a distinct order. Adopting new methods has gradually resulted in better collections of these taxa (Löfgren and Zhang 2003), and these authors could show how a varying combination of decanting, different sieve sizes and different acetic acid solutions yielded collections of very different composition [A. Löfgren, pers. comm. 2003; Löfgren's acid dissolutions performed in Lund used the buffered acetic acid method (Jeppsson *et al.* 1985) until the pH-measured technique was developed (Jeppsson *et al.* 1999)]. By processing with an acetic acid solution with a pH known to be more than 3·60, and sieving with a 63-μm screen Löfgren and Zhang (2003, p. 722 and pers. comm. 2003) were able to obtain the collections 'essential for firmly establishing the apparatus reconstruction' and for separation of all ramiform elements of closely related species.

Range ends, faunal diversity and variations therein

The perceived range end position is based on the absence of the fossil on one side of the observed end, and its presence on the other. Its presence is relatively unproblematic; once the possibility of reworking, contamination, incorrect labelling, etc., has been ruled out, no bias need be taken into account. However, the adjacent sample that does not contain the fossil is more difficult; absence of evidence is not evidence of absence, and it follows that the probability limits of the absence must be analysed. Many taxa are more or less rare, and most publications, for various reasons, include some collections that are too small to give a complete species list (my own papers are no exception). Graphic correlation has its own special biases (for discussion see Jeppsson 1997*b*, p. 104).

The first step in documenting the degree of bias in the position of a range end is to record every find. If only first and last occurrences are marked, the reader can only use these in an analysis of range end bias (= the difference between true and observed range ends), and the result is a range end bias many times larger than a calculation based on all occurrences. The result of such an analysis can be given as statistical confidence limits (Marshall 1994, pp. 467–468). The reason for such an analysis can be illustrated with an example: taxon A occurs in every sample from its known range but taxon B in every tenth sample. Both statistical methods and common sense tell us that the true range ends of B are probably many samples outside the known range and may well be much more than ten samples away (see Marshall 1995 for a very illustrative example). In contrast, the range ends of A are probably at the most only one or two samples away.

Narrower confidence limits require the number of specimens to be known. For example, with over ten specimens of taxon A in every sample, the range end uncertainty is not wider than the interval between samples, with the same provisions.

The next step in analysis is to determine the robustness of the observed range end based on frequency changes near the true range end. For example, a species with an average of 100 specimens per sample has an observed range end that is an order of magnitude less sensitive to frequency fluctuations near the range end than a species averaging only ten specimens per sample. Large collections from the Ireviken Event interval have shown that some species maintain a roughly similar frequency through the youngest collections (spaced by <0·1 m, calculated to correspond to < *c.* 2000 years), until they go extinct. Others, such as *Distomodus staurognathoides* and *Pterospathodus procerus*, 'faded away' and the position of the observed range end differs markedly when based on smaller and larger collections (Jeppsson 1997*a*: compare data in the main text and the supplementary note written 2·4 years later when larger collections were available).

Zonal boundaries

First and last records of index fossils are widely used for defining and identifying zonal boundaries [Lazarus gaps (Flessa and Jablonski 1983) can be important but have been used only rarely]. Hence, analysing the degree of bias in the position of a zonal boundary is a special case of range end bias. No taxon has the same frequency across all facies and all latitudes, and although all index fossils should be very widespread and may be frequent in some areas, they will be rarer in many areas [most of the Wenlock and Ludlow index fossils proposed by Walliser (1964), for example, have frequencies below 1% in most areas, in some places below 0·1%]. Other rare taxa may be ignored without significant consequences, but because the burden of providing the best zonation frequently falls on the conodont specialist, we are under pressure to do our best, even if an index fossil is rare. The significance of this is clear: any kind of bias that influences zonal boundaries has far-reaching consequences. It is also worth noting that analysis of range end bias in only a single section is not enough; any correlated section should be subjected to similar analysis (unless previous authors have carried out such an analysis).

In addition to erroneous correlations, placing zonal boundaries without consideration of the statistical uncertainty interval of first and/or last occurences can have other serious effects. For example, the ranges of taxa are related to what zone they are found in, so if a given taxon appeared everywhere at the same time, but in one area a zonal boundary is drawn too high (or too low), then that taxon will be erroneously recorded as appearing in another zone in that area. A correct analysis of the uncertainty in the zonal boundary position there would place the end record of the taxon in the uncertainty interval, resulting in the correct conclusion: that its range is not significantly different from the record elsewhere.

For a variety of reasons (biological, historical, bias against small specimens in many published collections) index taxa tend to have relatively large elements, and bias resulting from discrimination against small specimens is not a major problem for most index taxa. As shown above, however, the only insurance against a large range end bias is that the critical collection just outside the perceived zonal boundary is adequately large. Unless recovery bias is large (e.g. due to a destructive acid method), a smaller range end bias usually requires larger collections.

In principle, any zonation of a sequence includes an uncertain interval intercalated between any two intervals confidently referred to zones, and this ought to be marked as such (for an example, see Kiipli *et al.* 2001). Once collection size is adequate therefore the next step to decrease zonal boundary bias is to resample the interval between the documented range end of the index fossil and the nearest adequate collection. The first fieldwork may often include only ten samples per 10^6 years or even fewer. Thus, the range end uncertainty will be *c.* 10^5 years or more. Strategic resampling can minimize the amount of effort required to reduce this significantly. For example, 15 samples are collected between the range end and the next adequate sample from the first sample set; these are processed and analysed one at the time, selecting the sample that lies in the middle of the remaining uncertainty interval for processing, so that each new collection reduces it by half. In this way, processing only four samples reduces the uncertainty by more than one order of magnitude (to $0{\cdot}5^4$, equivalent to 0·0625 of the initial uncertainty). In my experience, planning so that resampling of critical intervals is possible before publication decreases the bias in the data enormously, because it allows most kinds of bias encountered to be addressed. Thus, range end bias is decreased not by closely spaced small collections but by fewer, adequately large collections and resampling: more of the same *is* much better.

SUMMARY

The hundreds of specialists on phosphatic fossils use many tens of more or less different methods of extraction. For every method in use there is at least one specialist convinced of the superiority (e.g. regarding efficiency) of that method for her or his own work. I am no exception. I am convinced about the superiority of bias-free and bias-controlled methods for all kinds of work, by all specialists. Hitherto nobody has proven me wrong, but, in contrast, many have reported considerable gains using one or more of these methods. I have also met objections and explanations from colleagues as to why they have not or did not plan to test my claims, most frequent among which are that my methods are idealistic and impractical. In short, they have assumed that such methods take too much time per sample to be useful in practice when hundreds of samples are processed. However, thousands of samples have been processed in the Lund conodont laboratory, using different methods. The bias-controlled methods now in use are by far the most efficient of those tried and my conclusion is that these are the most efficient ones existing (in addition, they yield reproducible results).

Each processing step adds to the total bias; hence, the sum can be considerable at the end. Bias is either systematic or random. Systematic bias can be statistical or complete. Not saving and picking a fraction finer than xx μm is close to a complete loss of every element below xx μm in diameter. Most physical methods discriminate more against small thin elements than against large ones. The degree of such statistical bias is strongly tied to the laboratory regime, including protocols for routine checking of the degree of bias. The person's degree of knowledge, awareness and carefulness, both regarding the proper use of each method and its peculiarities regarding bias, influence the degree of bias enormously. The statistical probability of finding a species, present in the sediment that was formed at a specific point in space and time, increases with the collection size to the level at which the collection can be described as 'adequate'. Some ways of using ±unbuffered acid methods result in random bias, and in such collections anything between 10 and 100 per cent of the original fossil content may have been lost in processing. There is also a statistical pattern in the degree of bias caused by ±unbuffered methods: samples that are muddier or otherwise more difficult to dissolve have higher probabilities of high bias.

The difference in working time between a fully bias-controlled approach and the most efficient uncontrolled approach is only a few per cent. In contrast, the many different less efficient alternative methods that have been suggested and probably are in use at many places require much more time per sample than a fully bias-controlled approach using the most efficient methods.

By choosing the best methods now available, including a protocol for routinely monitoring the amount of bias introduced, the degree of bias will be only a small fraction of that caused by methods not tested with controlled experiments. 'Adequate' samples are necessary to overcome the otherwise large bias of zonal boundaries and other range ends. Paying attention to every collecting step, from fieldwork to analysis of the resulting data, is, however, very profitable in another respect, too; the average yield per hour of work can easily be raised at least ten- to 100-fold when processing low or moderately yielding lithologies.

RECOMMENDATIONS

1. Where alternative methods exist, bias should be kept as low as possible by using the method causing the lowest bias.
2. All methods used should be referred to, or described.
3. Data should be presented in a form permitting the reader to judge the amount of bias (the practice among conodont specialists of including a table with number of specimens of each element of each species in each sample is a very important step).
4. The total bias in the data must be taken into account when the results are analysed.

Bias control is, to a large extent, like any other new technique. The instructions may at first seem very impractical, but the control quickly becomes a routine. Furthermore, it soon becomes apparent which steps are the most risky for the kind of samples being processed (on average, the largest biases are caused by unbuffered acid, samples that are too small, incompletely buffered acid, and discrimination against small/fragile elements, in that order, but be prepared for some surprises). Gradually the results of the control at each stage of processing become more intuitively predictable, and at that point risk can be assessed and informed decisions made concerning where time and effort can be conserved without significant risk to element recovery (and when no corners can be cut). On the other hand, at that point the operator will probably, like me, have found that the bias control has not only given so much better results that the time taken to learn the new methods was well spent, but also that, once it is a routine, most of it takes so little time that it is like a low-cost insurance against a high risk, and that it continues to pay off. For me, developing bias control has also resulted in development of methods and a laboratory set-up that have saved far more working time. I suppose that paying attention to potential bias also results in paying attention to other possible improvements.

Acknowledgements. It is well known that one must learn from other people's mistakes because one lacks enough time to repeat all of them oneself. I have repeated too many, but the fact that some methods are very strongly person-related can only be found with help from colleagues. The content of this paper was influenced by what I have learnt during contacts with many colleagues, including Richard J. Aldridge, Rikard Anehus, Ondrej Babek, Chris Barnes, Stig Bergström, Mikael Calner, Phil Donoghue, Michael J. Engelbretsen, Barry Fordham, Git Klintvik Ahlberg, Peep Männik, Anders Martinsson, John Talent, Viive Viira and Otto Walliser. Ann-Sofi Jeppsson typed the first version of the manuscript and corrected many linguistic errors. Mark Purnell, Phil Donoghue, Peter Molloy and John Repetski (referee) provided useful comments on the manuscript and Mark and David J. Batten carefully re-edited it. Grants from NFR, and its successor, The Swedish Research Council, have financed my experiments with methodological improvements and other work and thereby most of the many years of experience upon which this paper is based. Last but not least, this paper would not have been written had I not been invited to do so by Phil and Mark. Only recently did I realize that I was not the first in Lund to use data from large samples. Hadding (1958, p. 23) reported a general analysis of major elements, in a 'boatload of limestone' from Wenlock strata in the Smojen quarry on Gotland.

REFERENCES

BARNES, C. R., FREDHOLM, D. and JEPPSSON, L. 1987. Improved techniques for picking of microfossils. 55, 74–76. *In* AUSTIN, R. L. (ed.). *Conodonts: investigative techniques and applications.* Ellis Horwood, Chichester, 422 pp.

BROADHEAD, T. W. and DRIESE, S. G. 1994. Experimental and natural abrasion of conodonts in marine and eolian environments. *Palaios*, **9**, 546–560.

CIFELLI, R. L. 2001. Early mammalian radiation. *Journal of Paleontology*, **75**, 1214–1226.

COOPER, G. A. and GRANT, E. E. 1972. Permian brachiopods of west Texas, I. *Smithsonian Contributions to Paleobiology*, **14**, 1–231.

FLESSA, K. W. and JABLONSKI, D. 1983. Extinction is here to stay. *Palaeobiology*, **9**, 315–321.

HADDING, A. 1958. The Pre-Quaternary sedimentary rocks of Sweden VII. Cambrian and Ordovician limestones. *Lunds Universitets Årsskrift. N. F., Avd. 2*, **54** (5), 262 pp.

JEPPSSON, L. 1975. [1974]. Aspects of Late Silurian conodonts. *Fossils and Strata*, **6**, 79 pp.

—— 1976. Autecology of Late Silurian conodonts. 105–118. *In* BARNES, C. R. (ed.). *Conodont paleoecology.* Geological Association of Canada, Special Paper, **15**, 324 pp.

—— 1983*a*. Silurian conodont faunas from Gotland. *Fossils and Strata*, **15**, 121–144.

—— 1983*b*. Simple-cone studies, some provocative thoughts. *Fossils and Strata*, **15**, 86.

—— 1987. Some thoughts about future improvements in conodont extraction methods. 35, 45–51, 52, 53. *In* AUSTIN, R. L. (ed.). *Conodonts: investigative techniques and applications.* Ellis Horwood, Chichester, 422 pp.

—— 1990. An oceanic model for lithological and faunal changes tested on the Silurian record. *Journal of the Geological Society, London*, **147**, 663–674.

—— 1997*a*. The anatomy of the mid-Early Silurian Ireviken Event. 451–492. *In* BRETT, C. and BAIRD, G. C. (eds). *Paleontological events: stratigraphic, ecological, and evolutionary implications.* Columbia University Press, New York, 604 pp.

—— 1997*b*. A new latest Telychian, Sheinwoodian and Early Homerian (Early Silurian) Standard Conodont Zonation. *Transactions of the Royal Society of Edinburgh, Earth Sciences*, **88**, 91–114.

—— 1998. Silurian oceanic events: summary of general characteristics. 239–257. *In* LANDING, E. and JOHNSON, M. E. (eds). *Silurian cycles: linkages of dynamic stratigraphy with atmospheric, oceanic, and tectonic changes.* New York State Museum, Bulletin, **491**, 327 pp.

—— and ANEHUS, R. 1995. A buffered formic acid technique for conodont extraction. *Journal of Palaeontology*, **69**, 790–794.

—— —— 1999. A new technique to separate conodont elements from heavier minerals. *Alcheringia*, **23**, 57–62.

—— —— and FREDHOLM, D. 1999. The optimal acetate buffered acetic acid technique for extracting phosphatic fossils. *Journal of Paleontology*, **73**, 957–965.

—— FREDHOLM, D. and MATTIASSON, B. 1985. Acetic acid and phosphatic fossils – a warning. *Journal of Paleontology*, **59**, 952–956.

JERRE, F. 1994. Anatomy and phylogenetic significance of *Eoconularia loculata*, a conularid from the Silurian of Gotland. *Lethaia*, **27**, 97–109.

KIIPLI, T., MÄNNIK, P., BATCHELOR, R. A., KIIPLI, E., KALLASTE, T. and PERENS, H. 2001. Correlation of Telychian (Silurian) altered volcanic ash beds in Estonia, Sweden, and Norway. *Norsk Geologisk Tidskrift*, **81**, 179–194.

LÖFGREN, A. and ZHANG JIANHUA 2003. Element association and morphology in some Middle Ordovician platform-equipped conodonts. *Journal of Paleontology*, **77**, 723–739.

MARSHALL, C. R. 1990. Confidence intervals on stratigraphic ranges. *Paleobiology*, **16**, 1–10.

—— 1994. Confidence intervals on stratigraphic ranges: partial relaxation of the assumption of randomly distributed fossil horizons. *Paleobiology*, **20**, 459–469.

—— 1995. Distinguishing between between sudden and gradual extinctions in the fossil record: predicting the position of the Cretaceous-Tertiary iridium anomaly using the ammonite fossil record on Seymour Island, Antartica. *Geology*, **23**, 731–734.

—— 1997. Confidence intervals on stratigraphic ranges with non-random distributions of fossil horizons. *Paleobiology*, **23**, 165–173.

MAWSON, R. 1987. Documentation of conodont assemblages across the Early Devonian-Middle Devonian boundary, Broken River Formation, North Queensland, Australia. *Courier Forschungsinstitut Senckenberg*, **92**, 251–273.

PURNELL, M. A. 1995. Microwear on conodont elements and macrophagy in the first vertebrates. *Nature*, **374**, 798–800.

SIVERSON, M. 1995. Revision of the Danian cow sharks, sand tiger sharks, and goblin sharks (Hexanchidae, Odontaspididae, and Mitsukurinidae) from southern Sweden. *Journal of Vertebrate Paleontology*, **15**, 1–12.

SWEET, W. C. (ed.) 1982. *Pander Society Newsletter 14*. Department of Geology and Mineralogy, Ohio State University, Columbus, Ohio, 40 pp.

VON BITTER, P. H. and MILLAR CAMPBELL, C. 1984. The use of 'spikes' in monitoring the effectiveness of phosphate microfossil recovery. *Journal of Paleontology*, **58**, 1193–1195.

—— and PURNELL, M. A. 2005. An experimental investigation of post-depositional taphonomic bias in conodonts. 39–56. *In* PURNELL, M. A. and DONOGHUE, P. C. J. (eds). *Conodont biology and phylogeny: interpreting the fossil record*. Special Papers in Palaeontology, **73**, 218 pp.

WALLISER, O. H. 1964. Conodonten des Silurs. *Abhandlungen des Hessischen Landesamtes für Bodenforschung zu Wiesbaden*, **41**, 1–106.

WARD, D. J. 1984. Collecting isolated microvertebrate fossils. *Zoological Journal of the Linnean Society*, **82**, 245–259.

APPENDIX

1. *Manual for routine samples*

In sequences with stable yield, adequate samples may be of equal size and a standardized treatment possible. Efficient handling requires tight control on the man-hours spent on each sample (independent of who processes the sample). For details of methods, see Jeppsson and Anehus (1995, 1999), Jeppsson *et al.* (1999) and herein.

1. Collect a sample of adequate size from the lithology expected to produce the best yield.
2. Clean, store a reference piece, and weigh the amount to be dissolved (needed for calculation of yield), do not crush (saves time and reduces bias).
3. Place sample in a colander in a vessel large enough to hold enough acid solution for a complete dissolution (20 litres per 1 kg sample). (Saves time otherwise spent changing acid; regarding smaller vessels see 5 below.)
4. Add acetic acid, acetate and water (vigorously to achieve a complete mixing); measure pH (at least until you have succeeded to mix the right pH, slightly above 3·60 for at least ten consecutive samples; and then now and then for succeeding samples to check that, for example, the concentration of the acetate solution has not changed). Provided that the amount of solution does not exceed what is needed for dissolving the whole sample the risk from the pH being too low at the start of processing a sample is moderate even if the solution is only partly buffered because every conodont is still encased by limestone. [If the solution is completely unbuffered, the pH will have increased to 3·60, the critical point, when *c.* 15 per cent of the sample has been dissolved (Jeppsson *et al.* 1999, fig. 2).] Further, the initial reaction is fast with a clean limestone sample, hence I estimate that less than 10 per cent of the conodonts freed before the reaction ends would be destroyed beyond recognition. However, measuring pH routinely takes so little time that it remains a routine here, even at the start of a sample.
5. If smaller vessels are used, the solution needs to be changed. This adds working time (5–15? minutes for each change). Even if clean limestones are processed and the pH has been slightly above 3·60 every time after ten such changes, I do not recommend abandoning measuring pH after a change of acid because the time saved per change (*c.* 1 min, when many samples are handled at a time) is small compared with the risk that, say one in 20 or 100 changes would go wrong, destroying all previously freed conodonts. ('Recover the fines after each change' is not an alternative since that requires 10 – >>100 min for cleaning the sieve, washing out the fines, cleaning them, transferring them to a filter paper and handling two or more residues with lots of calcite grains.)
6. Once dissolution is essentially complete, remove the larger particles first with a 1-mm sieve, perhaps also a 0·5-mm sieve. Then sieve the residue on a fine screen (I use 63 μm) and dry. To minimize loss use 'gently flowing water' and minimize washing time by only washing away the acid/acetate and the clay that easily pass the sieve. Store the remaining residue overnight with some sodium carbonate (or detergent) and wash again; repeat if useful.
7. If lumps of clay or dirty material remain: pour boiling hot water with some sodium carbonate over the residue,

wash and dry. Next, dry the residue thoroughly (e.g. in an oven at 50°C), pour boiling water (with some sodium carbonate) over the dry warm residue (if inefficient, gently boil it directly) wash, repeat if useful. Marl samples are frequently moderately difficult, requiring several cycles, interspersed with acid treatment to decalcify remaining lumps of rock. More difficult lithologies require many cycles. End sieving with rinsing in distilled water if you expect that density separation may be needed. Dry. Use a microscope to decide what is needed to get an easily picked residue.

8. If calcite or dolomite grains remain: buffered formic acid dissolves both minerals. If the microscope control indicated that density separation would also be needed, rinse with distilled water directly after washing, otherwise tap-water washing is enough.
9. Use a microscope to decide if one or more further concentration methods are required, density separation at 2·84 (2·90 if white mica) or 3·04, or magnetic separation (see below), saves more picking time than the concentration method requires.
10. Pick, sort, identify and evaluate the result: are there enough conodonts for a statistically reliable conclusion or should the conclusion include 'perhaps' or 'probably' and a larger sample be collected?

2. *Separation with a Franz Magnetic Barrier Separator Model LB1, adjustments and evaluation of results*

Sliding a paper below and above the chute before starting the machine best checks its position in the pole gap (two nuts below the chute). If the chute is in contact with the magnet, the vibrations of its lower end are dampened, and transport out from it is obstructed (and there is more noise). The lateral position of the chute in that gap (screws on the underside of the chute holder) is best checked when separation results in a distinct stream of magnetic particles: the splitter should guide that stream into the upper (inner on an LB1) fraction. Forward slope and vibrations regulate the transport forward. Side slope regulates how large a part of the gravitation is used to get the phosphate out of the stream of magnetic material. The lowest degree of bias results from an adequate part of the gravitation being used with a magnetic field of adequate strength.

To minimize bias of different kinds:

1. Screen the residue in fractions: if small, 63–250 and 250–1000 μm, further if large, e.g. 63–180–250–500–1000 μm.
2. Use a high side slope to get enough gravitational effect on the phosphate; I usually use 30 degrees, i.e. 50 per cent of the gravitation.
3. Conodont elements are brittle, so the less vibration needed for moving the particles forward, the less breakage is expected. Thus, adjust forward slope so that a minimum of vibration gives a smooth flow through the chute: at least 25–30 degrees for the finest fraction and slightly less for the coarsest is often best; less only if the particles are round enough to roll. Adjust the lip of the feeder accordingly.
4. Start with a low magnetic field. If 0·1 A removes some highly magnetic particles, use that. If such a residue is started at a higher magnetic field, such particles will block through-flow, causing conodont elements to end up in the wrong fraction.
5. Once the machine is adjusted, put all trial fractions back into the feeder.
6. Run some material and analyse the result under a microscope (see below). If satisfactory, run the rest of the residue.
7. Particles magnetic enough to be removed with only a small increase will repeat the problem described in 4, if the field is increased more than *c.* 75–100 per cent. Increase the strength of the magnetic field so much (less if needed to keep the magnetic fraction below 80%) and run the conodont fraction again. There is a considerable risk that some phosphate is trapped in a steady stream of magnetic particles if, e.g., 99 per cent is removed in one run. Repeat up to maximum magnetic field.
8. If needed, continue by lowering side slope stepwise using a very low rate of feeding (to compensate for the lower gravitational effect) and maximum magnetic field.

In order to keep bias low, both fractions must be checked under the microscope after every run (see laboratory methods in the text). Three causes for a higher bias may be identified and remedied. If the magnetic fraction contains:

1. Preferentially small, light phosphatic particles. The cause is either too high a feeding rate or too little of the gravitation being used to pull them out from the stream of magnetic particles, or both. Remedy: increase side slope (= increase the gravitational pull) and/or decrease feeding rate. (Removal of too much material in one run may cause a similar bias.)
2. Phosphatic particles with magnetic grains adhering to, or lodged in them. Remedy: increase side slope. (Inefficient washing or washing methods may cause some of this bias.)
3. All sizes of phosphatic particles. Stop only the feeder and wait until the flow out from the chute has ceased. Decrease magnetic field (the ampere) to zero, and, if needed, increase vibration. Check if a group of further particles leave the chute. If not, the remedy is probably to decrease feeding rate. If the test revealed such magnetic particles, see 4 and 7 above. If some particles block transport even at zero amperes, the sample must be spread out on a paper and such ferromagnetic particles removed by a hand magnet. Keep the magnet within another sheet of paper (to permit easy cleaning) and pass it some millimetres above the particles. If such pretreatment does not function, increase side slope. (If the chute is not properly positioned in the pole gap, similar transport problems will occur; see above.)

3. *Separation of different causes of breakage*

Breakage during life was repaired, or the lost piece partly regenerated, except when breakage occurred immediately before the animal died.

Sedimentary processes caused the same kind of rounding as on other sediment grains. Breakage due to compaction was sometimes followed by deposition of phosphate between the fragments fusing them in oblique positions. Similarly, tectonic breakage and pull apart was often 'healed' by inorganic deposition of phosphate. The original pieces show standard CAI-darkening, but the inorganic phosphate lacks organic matter and thus remains hyaline (white if pitted/'frosted'). Breakage surfaces, resulting from any of these processes, exhibit the same set of microdamage (e.g. phosphate loss, crystal growth or imprints) as the outside of the last lamella; hence such breakage can be identified.

Breakage due to collecting and processing differs from that described above. Broken surfaces are clean and sharp-edged. (However, using improperly buffered acid etches the surface on all affected elements, including those due to such breakage, blurring this distinction. Only later breakage is 'unfrosted'.) Microfracturing (due to blasting or hammering) of fossils still encased in the rock can easily be identified; easiest on large blades or platforms. When the rock splits through a fossil, the latter can equally well happen to be split along any plane and most breakage is more or less oblique. Each specimen tends to be split in a unique way, and the pieces of large or relatively rare forms can be identified (and glued together). Sometimes the fracture plane has cropped the free part of two or more denticles along an oblique plane. In contrast, breakage of specimens no longer encased in the rock will be orientated along a preferred crystal plane. At least in those ramiform elements I am familiar with, breakage of freed specimens is usually perpendicular to the length axis of the cusp, denticle or process. Denticles are usually broken at different heights or, if very badly damaged, at the base.

When the pieces of the sample are dissolved, conodont elements will begin to protrude until, for example, only a denticle tip remains in the limestone. Thus, any movement of the lumps towards each other or the support will easily break the specimen. Breakage will be partly that of encased elements, partly that of freed ones. In order to avoid such damage on silicified brachiopods, Cooper and Grant (1972, p. 17) coated the lower surface of their blocks with cellulose acetate. Conodont elements are smaller; hence, the extent of such breakage will be less. Placing the slabs nearly upright, leaning against the side of the colander, decreases the surface in contact with the support and thereby the amount of damage (in addition, dissolution is much faster, because such an orientation results in a much more vigorous circulation).

Rough or extended washing is extremely destructive. The (sub)final result is characteristic; instead of say the 2000–5000 well-preserved elements expected from an argillaceous sample, the yield may be a few tens or hundreds of fragments (the rest were broken into fragments small enough to be washed away). Such destruction can be minimized; see 6 and 7 in section 1.

Magnetic separation is frequently seen as a major source of breakage. However, my experience has been that it has very small effects compared with other causes, even on high-quality collections with long, thin denticles, if properly adjusted (see section 2).

Breakage due to crushing is probably not affected much by the purity of the rock. However, breakage due to washing and other concentration work increases easily with decreasing purity. The laboratory-induced breakage is partly method-related but also to a very large extent person-related (laboratory assistant, professor, research student). Two persons might get very different results, especially when processing difficult samples, even if both used the same pH measured, methods and screen. The relative destructiveness of different methods can be evaluated by determining what percentage of the denticle tips remains in a collection from a fine-grained argillaceous limestone.

[Special Papers in Palaeontology, 73, 2005, pp. 73–101]

MULTIELEMENT CONODONT APPARATUSES OF TRIASSIC GONDOLELLOIDEA

by MICHAEL J. ORCHARD

Geological Survey of Canada, 101–605 Robson St., Vancouver, BC, V6B 5J3 Canada; e-mail: morchard@nrcan.gc.ca

Abstract: New multielement reconstructions of 26 Triassic conodont species representing 26 genera assigned to the superfamily Gondolelloidea are presented. Most of these have formerly been interpreted as single element taxa based on biased collections of disjunct elements, but here all complete apparatuses are regarded as consisting of 15 elements. Eight genera are new: *Conservatella*, *Columbitella*, *Discretella*, *Meekella*, *Novispathodus*, *Spathicuspus*, *Trammerella* and *Wapitiodus*. The genera are classified into seven subfamilies (five new: Cornudininae, Epigondolellinae, Mullerinae, Novispathodinae, Paragondolellinae). One new species is described: *Wapitiodus robustus*. All taxa are characterized by a pair of enantiognathiform S_1 elements, the diagnostic character for the superfamily. Assembly of genera into subfamilies is based on similarity in the morphology of homologous ramiform elements, with emphasis on the position of the antero-lateral processes relative to the cusp in the alate S_0 elements; the development of a second antero-lateral process in the digyrate S_2 elements; and the occurrence and position of an accessory antero-lateral process in the bipennate S_3 elements.

Key words: conodont, Triassic, apparatus, multielement taxonomy.

Two views of Triassic conodont apparatuses exist in the literature. The first is that originally presented by Sweet (1970, p. 210) and Kozur and Mostler (1971, pp. 11–12) who regarded many Triassic pectiniform conodonts as having apparatuses composed of a single element type. This view arose from both empirical and statistical analysis of disjunct element collections, which commonly have a preponderance of platforms. The implications of these authors' reconstructions are that no bias exists in such collections. With the discovery of fused clusters and a natural assemblage, the multielement view saw pectiniform elements in natural association with the ramiform elements that had hitherto been regarded as separate and rather conservative apparatuses. Kozur (1989, pp. 409–410) now viewed the apparatuses of Pennsylvanian *Gondolella* (see von Bitter and Merrill 1998), the younger *Neogondolella* (see Orchard and Rieber 1999), and many other gondolelloids as both multilelement and identical. The underrepresentation of ramiform elements was now seen as a significant bias in the fossil record, but the value of these non-pectiniform elements in deciphering relationships was largely dismissed. Consequently, a further bias was introduced into the actual study of the conodonts with ramiforms being largely neglected. Hence, both views of the Triassic apparatuses encouraged the emphasis on pectiniform element morphology as the primary basis for conodont systematics and interpretations of phyletic relationships. The present paper models many Triassic multielement apparatuses, and concludes that Triassic Gondolelloidea were indeed multielement and were essentially composed of seven paired element types and one unpaired element, making a total of 15. Furthermore, many species bore distinctive ramiform elements that lead to very different phylogenetic hypotheses from those based on the pectiniform alone.

RECONSTRUCTING TRIASSIC APPARATUSES

The Gondolelloidea are the predominant conodont group from the mid-Permian to end Triassic. Attempts at reconstructing their apparatuses date from the early empirical methods of Huckriede (1958, p. 164). Later, Sweet (1970) undertook statistical analysis of Lower Triassic material from the Salt Range of Pakistan and demonstrated natural groupings of ramiform elements that he referred to species of *Ellisonia*; however, he regarded species of *Hindeodus*, *Neospathodus* and *Neogondolella* as consisting only of pectiniform elements. These results did not support the natural association of pectiniform elements with the ramiform complex referred to multielement *Ellisonia* but natural clusters of gondolelloids (e.g. Ramovs 1977, 1978) and a natural assemblage of *Neogondolella* (Rieber 1980) demonstrated otherwise.

Empirical reconstruction of Triassic multielement species has continued in Europe (Kozur and Mostler 1971; Ramovs 1977; Mietto 1982; Bagnoli *et al.* 1985), China (Zhang and Yang 1991) and Japan (e.g. Koike 1996, 1999), but most of these works propose apparatuses that appear incomplete and/or lack modern notation. On the basis of *Neogondolella* natural assemblages, Orchard and Rieber (1999) provided a 15-element template for reconstructions of the Gondolelloidea, including a notation for the diverse ramiform elements. The notation adopted here is updated from Orchard and Rieber (1999), with Pa, Pb, M, Sa, Sb_1, Sb_2, Sc_1 and Sc_2 being replaced, respectively, by P_1, P_2, M, S_0, S_1, S_2, S_3 and S_4 (Purnell *et al.* 2000).

The following interpretations of the apparatuses of many common Triassic multielement genera draw on abundant well-dated conodont material from North America, Europe and Asia. In undertaking this exercise, it became apparent that many taxa presently combined in a single form-genus, for example in '*Neospathodus*', have sufficiently different apparatuses that they should certainly be assigned to separate genera, and even separate subfamilies. Conversely, some species formerly assigned to separate genera are judged to be congeneric based on virtually identical apparatuses (e.g. some species assigned to *Clarkina*, *Neogondolella*, *Pridaella* and *Paragondolella*). The evolution of apparatuses seems to have been very rapid during the Early Triassic following the Permian/Triassic boundary extinctions and the key to unravelling relationships within the Gondolelloidea should be sought in Induan collections: by the early Olenekian, apparatuses were substantially differentiated.

The thesis presented here is that the ramiform elements carry a high taxonomic signal, and that three elements, S_0, S_2 and S_3, have prime importance in suprageneric classification and the tracking of relationships within the Gondolelloidea. Key criteria are (1) the presence and position of a secondary anterior process in the S_3 element, (2) the character of the shorter antero-lateral process in the S_2 element, and (3) the position of the antero-lateral processes relative to the cusp in the S_0 element. It should be noted that there is some evidence that the pair of elements identified as occupying the S_1 and S_2 (= Sb) positions in fact occupy, respectively, the S_2 and S_1 positions but the evidence is inconclusive. Similar uncertainty surrounds the relative position of some element pairs identified as S_3 and S_4 (= Sc). Although this impacts on architectural considerations, it makes no practical difference to the classification introduced here.

The methodology adopted here starts with the *Neogondolella* natural assemblage template (Orchard and Rieber 1999). Low-diversity collections in which a single species predominates permit reconstruction of *Jinogondolella* (Guadalupian, Texas); *Neogondolella* (Changxingian, South China; Induan, Canadian Arctic; Olenekian, Idaho and British Columbia; Anisian, Nevada; Ladinian, British Columbia), *Neospathodus* and *Novispathodus* (Scythian, British Columbia, California), *Triassospathodus* (Spathian, Nevada, British Columbia), *Metapolygnathus*, *Norigondolella*, *Epigondolella* and *Cypridodella* (Carnian–Norian, British Columbia). Diverse collections with one or several of these genera have in turn facilitated recognition of other apparatuses. For the most part, apparatus reconstruction has been based on elements from single collections, but it has been necessary sometimes to select elements from different collections for optimal illustration.

Figured specimens are housed in the National Type Collection of Invertebrate and Plant Fossils at the Geological Survey of Canada, 601 Booth Street, Ottawa, Ontario K1A 0E8. Supplementary material is housed in the collections of GSC Vancouver.

BIAS AND COMPLETENESS

Most of the apparatuses presented here have been reconstructed from collections originating in relatively offshore, basinal and/or low-energy environments where bias arising from post-mortem sorting and selective destruction is less of a problem. These collections commonly contain all the elements of an apparatus and are more balanced in terms of pectiniforms vs. ramiforms, although the delicate nature of the ramiforms invariably results in some imbalance. Collections consisting exclusively or predominantly of pectiniform elements (typically robust *Neogondolella*, *Metapolygnathus*, *Epigondolella* and *Cypridodella*) are more commonly found in grainstones and high-energy deposits in which post-mortem sorting may account for the concentration of these elements and the destruction of the ramiforms. It may also be true that in such stressed environments ramiform elements were not mineralized (discussion in Sweet 1988, pp. 87–88, 143–144). However, the conclusion is that the apparatus blueprint for all the Gondolelloidea consists of 15 elements: paired P_1, P_2, M, S_1, S_2, S_3 and S_4, and an unpaired S_0.

SYSTEMATIC PALAEONTOLOGY

Order OZARKODINIDA Dzik, 1976
Superfamily GONDOLELLOIDEA (Lindström, 1970)

Remarks. The Gondolelloidea bore a 15-element apparatus: seven paired element types (P_1, P_2, S_1, S_2, S_3, S_4) and an unpaired bilaterally symmetrical element (S_0). In particular, the superfamily is characterized by a breviform digyrate, 'enantiognathiform' S_1 element. Subfamilial classification emphasizes the character of S_0, S_2 and S_3 elements within different apparatuses, as summarized below.

1. Subfamily Cornudininae subfam. nov. S_3–S_4 elements with arcuate, variably downturned anterior processes; M elements with two short straight processes; S_0 has antero-lateral processes far to the anterior of the cusp. Genera: *Cornudina* Hirschmann, 1959, *Spathicuspus* gen. nov.
2. Subfamily Epigondolellinae subfam. nov. S_3 has no accessory anterior process; S_2 has single denticulated antero-lateral process; S_0 element with anterior processes branching from a denticle immediately anterior of the cusp. Genera: *Epigondolella* Mosher, 1968, *Cypridodella* Mosher, 1968
3. Subfamily Gladigondolellinae Hirsch, 1994. S_3 has accessory anterior process; S_2 has two well-developed denticulated antero-lateral processes; S_0 element with anterior processes generally branching from the cusp. Genera: *Cratognathodus* Mosher, 1968, *Gladigondolella* Müller, 1962
4. Subfamily Mullerinae subfam. nov. S_3 elements are tertiopedate with an accessory antero-lateral process branching from the cusp. Genera: *Conservatella* gen. nov., *Discretella* gen. nov., *Guangxidella* Zhang and Yang, 1991, *Meekella* gen. nov.
5 Subfamily Neogondolellinae Hirsch, 1994. S_3 has accessory anterior process; S_2 has single denticulated antero-lateral process; S_0 element with anterior processes branching from the cusp or from a denticle immediately anterior of it; S_0, S_3 and S_4 elements have posterior processes that are commonly detached. Genera: *Neogondolella* Bender and Stoppel, 1965, *Chiosella* Kozur, 1989, *Columbitella* gen. nov., *Metapolygnathus* Hayashi, 1968, *Neospathodus* Mosher, 1968, *Norigondolella* Kozur, 1989.
6. Subfamily Novispathodinae subfam. nov. S_3 has no accessory anterior process; S_2 has a single well-developed denticulated antero-lateral processes; S_0 element with anterior processes branching from first or second denticle anterior of the cusp. Genera: *Novispathodus* gen. nov., *Budurovignathus* Kozur, 1988, *Mosherella* Kozur, 1972, *Pseudofurnishius* van den Boogaard, 1966, *Triassospathodus* Kozur, 1998.
7. Subfamily Paragondolellinae subfam. nov. S_3 has an accessory anterior process branching from its anterior end; S_2 has a single well-developed denticulated antero-lateral process and a second shorter process; S_0 element with anterior processes branching far in front of the cusp. Genera: *Paragondolella* Mosher, 1968, *Trammerella* gen. nov.
8. Subfamily Uncertain Genera: *Icriospathodus* Krahl, Kauffmann, Kozur, Richter, Foerster and Heinritzi, 1983, *Scythogondolella* Kozur, 1989, *Wapitiodus* gen. nov.

Family GONDOLELLIDEA Lindström, 1970

Characteristics. As for superfamily. Pectiniform P_1 elements are typically segminate or segminiplanate, less commonly carminate or carminiplanate; P_2 elements are angulate or exceptionally anguliplanate; M elements are breviform digyrate; S_3–S_4 elements are generally bipennate with variably inturned and downturned anterior processes, one of which (S_3) may be bifid; in some constituent genera, the occupants of the S_3–S_4 positions exhibit three similar bipennate element morphologies. In the Mullerinae, the S_3 element is tertiopedate. Digyrate S_1 elements are as for the superfamily. Digyrate S_2 elements include distinctive elements that have one well-developed anterolateral process and a second that may be either rudimentary or as well developed as the first; S_0 elements are alate with two antero-lateral processes diverging from the cusp or from far anterior of it.

Remarks. Apart from the diagnostic P elements, those displaying the most distinctive morphology are the S_0, S_2 and S_3 elements. This is manifest as variation in the position of the anterior processes relative to the cusp in the S_0 element, the character of a second antero-lateral process in the S_2 element; and the presence and position of the bifurcation in the anterior process of the S_3 element. In some subfamilies two morphotypes of bipennate S_4 element may be differentiated based on anterior curvature. On the basis of these differences, the Gondolellidae are here subdivided into seven subfamilies.

Subfamily CORNUDININAE subfam. nov.

Characteristics. The 15-element apparatus includes P_1 elements that are typically short, segminate or rarely segminiplanate, and have a prominent cusp; P_2 elements are angulate with short subequal processes and prominent cusps; M elements are breviform digyrate with two straight, denticulated, relatively short, and downwardly directed processes; S_3–S_4 elements are bipennate with variably inturned and downturned anterior processes and denticulation on the anterior process composed of long arcuate denticles and on the posterior processes of denticles that increase in size distally; S_1 elements are breviform digyrate; S_2 elements are (in one genus at least) small digyrate elements with some denticulation on two antero-lateral processes; S_0 element is modified alate with two antero-lateral processes diverging from a point far anterior of the cusp, from which it is separated by 4–5 denticles. The posterior process of the S_3 element and the inner lateral process of the S_1 element may be sinuous in upper view.

Remarks. Presence of, and variation in the inner curvature of, the S_3–S_4 anterior process is more pronounced than in most other gondolellids and a third element morphology may be separable in this position: its significance is uncertain. These elements have the characteristic strong anterior basal curvature which sets them apart from most other gondolelloids: the distal end of the anterior process projects posteriorly and its upper side may form an acute angle with the posterior process. This is a character seen in the ramiform elements (*'Oncodella' sensu formo*) of *Neostrachanognathus*, which is almost certainly related. A relationship with the Paragondolellinae is suggested because of their very similar S_0 elements, which were referred to the form genus *Hibbardelloides* by Kozur and Mostler (1970, pp. 438–439); it is not clear whether that name has priority over others used herein.

Genus CORNUDINA Hirschmann, 1959
Text-figure 1

Type species and holotype. Ozarkodina breviramulis Tatge, 1956, p. 139, pl. 5, fig. 12a–b.

Type locality and stratum. Lower Muschelkalk, Kalkwerk Quarry, Trubenhausen, Germany.

Original diagnosis. Based on a small angulate P_2 element with a large and prominent medial cusp and very short upturned processes (Hirschmann 1959, p. 44).

Multielement diagnosis. As for the subfamily, with a P_1 element with a long cusp, twice the length of the adjacent denticle, a very short anterior process, and a broadly excavated basal cavity. The M element has two very short, straight processes. Nature of S_2 element uncertain.

Remarks. Cornudina has been reconstructed by Koike (1996) as an apparatus consisting only of elements like that described and illustrated here as P_1. Both Kozur and Mostler (1971, p. 11) and Sweet (in Clark *et al.* 1981, p. W155) placed *Cornudina* with *Chirodella* as a multielement *Chirodella*, but Koike (1996) regarded the two genera as unrelated. The holotype of *Cornudina* (*O. breviramulus*) appears to be a P_2 element, and that of *Chirodella* (*Metalonchodina triquetra*) an S_1 element; both were from Muschelkalk collections made by Tatge (1956), who described no elements like those described here as P_1. Conversely, *'Chirodella' sensu formo* does not occur in the present collections, nor apparently in those from Japan. It is therefore uncertain that the present genus should be called *Cornudina* even though the type species of that genus resembles the P_2 element illustrated here.

In the present reconstruction of *Cornudina*, the identity of the S_2 element is uncertain: it may be a bipennate element (as illustrated), but this would be unlike its digyrate homologues in other gondolelloids. The genus probably evolved from *Spathicuspus* near the Lower/Middle Triassic boundary through overall shortening of the P_1 element

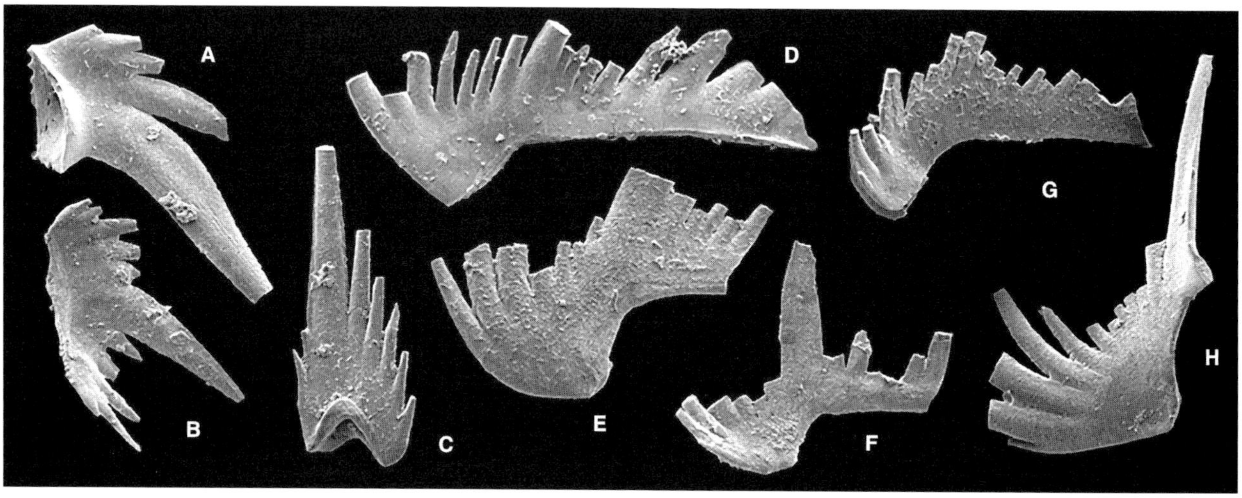

TEXT-FIG. 1. *Cornudina? igoi* (Koike). A, P_1, GSC 119939. B, P_2, GSC 119940. C, M, GSC 119941. D, S_2?, GSC 119942. E, S_4, GSC 119943. F, S_3, GSC 119944. G, S_0, lateral view, GSC 119945. H, S_1, GSC 119946. All Upper Guangdao section (UGD), Nanpanjiang Basin, South China. A, sample 15 (GSC loc. C-306504); B–D, sample 19 (GSC loc. C-306508); E–G, sample 44 (GSC loc. C-306517); H, sample 42 (GSC loc. C-306516). Anisian. All ×80.

and the enlargement of the cusp, and by process reduction in the M element.

Genus SPATHICUSPUS gen. nov.
Text-figure 2

Derivation of name. From the type species and its prominent cusp.

Type species and holotype. Neospathodus spathi Sweet, 1970, pp. 257–258, 260, pl. 1, fig. 5.

Type locality and stratum. Mittiwali Member, Mianwali Formation, Narmia, Pakistan.

Multielement diagnosis. As for the subfamily, with a P_1 element in which the posteriorly located cusp is large but not markedly higher than the denticles of the anterior process. The M element has straight denticulate inner and outer lateral processes. S elements may have sinuous posterior processes (seen in upper view), and flat to inturned or costate anterior processes.

Remarks. Sweet (1970) described *Neospathodus spathi* as a single element apparatus of segminate elements in which a midlateral ridge or platform-like brim was developed in later growth stages. Ramiform elements like those included in *Spathicuspus* were not illustrated by Sweet (1970) from Pakistan, although a few such elements do occur in his collections. Similarly, although Sweet (1970) regarded the incipient platform as typical of later growth stages, some relatively large elements may entirely lack it, as is the case in my material.

Subfamily EPIGONDOLELLINAE gen. nov.

Characteristics. Apparatuses with segminiplanate P_1 elements having a free blade and a platform that bears strong marginal denticles, at least anteriorly; P_2 elements are relatively elongate with closely spaced, laterally compressed denticles and a long posteriorly inclined subcentral cusp; S_3 and S_4 elements have a variably inturned and downturned anterior process and may have joined posterior processes; S_3 has no secondary anterior process; S_2 elements have one long denticulate process and a short adenticulate one; breviform digyrate S_1 elements have the distal part of the larger process variably developed; modified alate S_0 element has antero-lateral processes that branch from the first or second denticle anterior of the cusp, and a long posteriorly downturned and disjunct posterior process. M elements are breviform dygyrate elements corresponding to the form-genus '*Cypridodella*'.

Remarks. The Epigondolellinae arose from the Neogondolellinae (i.e. *Metapolygnathus*) by loss of the bifid S_3 element, and the reduction of the anterior processes and joining of the posterior processes in the S_3 and S_4 elements. The S_0 element is similar to that of many species of *Neogondolella*, including the possession of a disjunct posterior process.

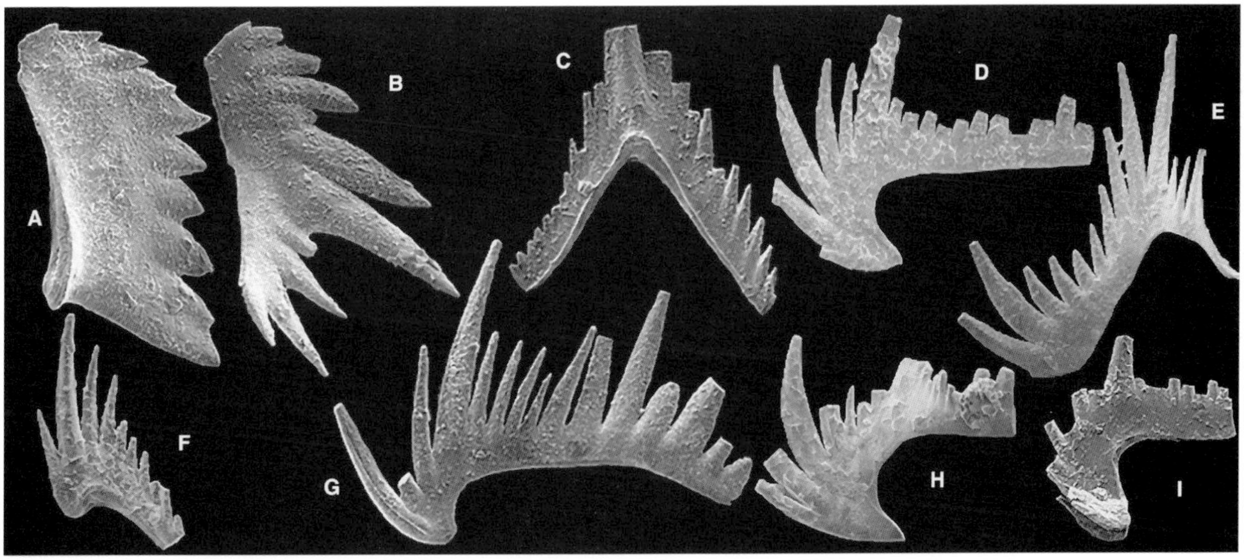

TEXT-FIG. 2. *Spathicuspus* sp. A, P_1, GSC 119954. B, P_2, GSC 119955. C, M, GSC 119956. D, S_4 variant, GSC 119957. E, S_1, GSC 119958. F, S_2, GSC 119959. G, S_4, GSC 119960. H, S_3, GSC 119961. I, S_0, lateral view, GSC 119962. A, sample 40 (GSC loc. C-306515), Upper Guangdao section (UGD), Nanpanjiang Basin, South China; B–E, G–H, sample 103b (GSC loc. C-117673), Jabal Safra, Oman; F, I, sample W13 (GSC loc. C-301226), Columbites beds, Hot Springs, Idaho, USA. Spathian. All ×80.

Genus EPIGONDOLELLA Mosher, 1968
Text-figure 3

1970 *Tardogondolella* Bender, 1970, p. 530.
1972 *Ancyrogondolella* Budurov, 1972, pp. 855–857.

Type species and holotype. Polygnathus abneptis Huckriede, 1958, pp. 156–157, pl. 14, fig. 16.

Type locality and stratum. Beds assigned to *Cyrtopleurites bicrenatus* Zone (Alaun), Someraukogel near Hallstatt, Austria.

Original diagnosis. Species have been based on ornate segminiplanate P_1 elements that commonly have a quadrate platform ornamented with strong marginal denticles, and a free blade extending onto the platform as a low carina that terminates in front of the posterior end (Mosher 1968, pp. 935–936).

Multielement diagnosis. As for subfamily, with the P_1 as described above and with the S_4 element having a longer, less inturned anterior process than the S_3 and an apparently detached posterior process. The distal end of the larger process of the S_1 element, commonly found disjunct, has high denticles set perpendicular to the bar.

Remarks. The genus was originally established to include most Norian ornate platform species, which included quadrate forms with no posterior carina (typically Lower Norian) and narrower, pointed forms with carinas extending to the posterior tip (typical of the Middle and Upper Norian) (see Orchard 1983, 1991*b*). Kozur (1989) later introduced *Mockina* to accommodate the carinate group, although *Cypridodella* is a senior synonym. Apart from their P_1 elements, which may be very similar, *Epigondolella* differs from *Cypridodella* in the longer anterior processes of the S_3 and S_4 elements, and the likely disjunct nature of the latter. In this respect, the S_4 element of *Epigondolella* resembles that seen in neogondolellins. S_1 and S_2 elements may also be distinguished from *Cypridodella* (*q.v.*).

Genus CYPRIDODELLA Mosher, 1968
Text-figure 4

1989 *Mockina* Kozur, p. 423.

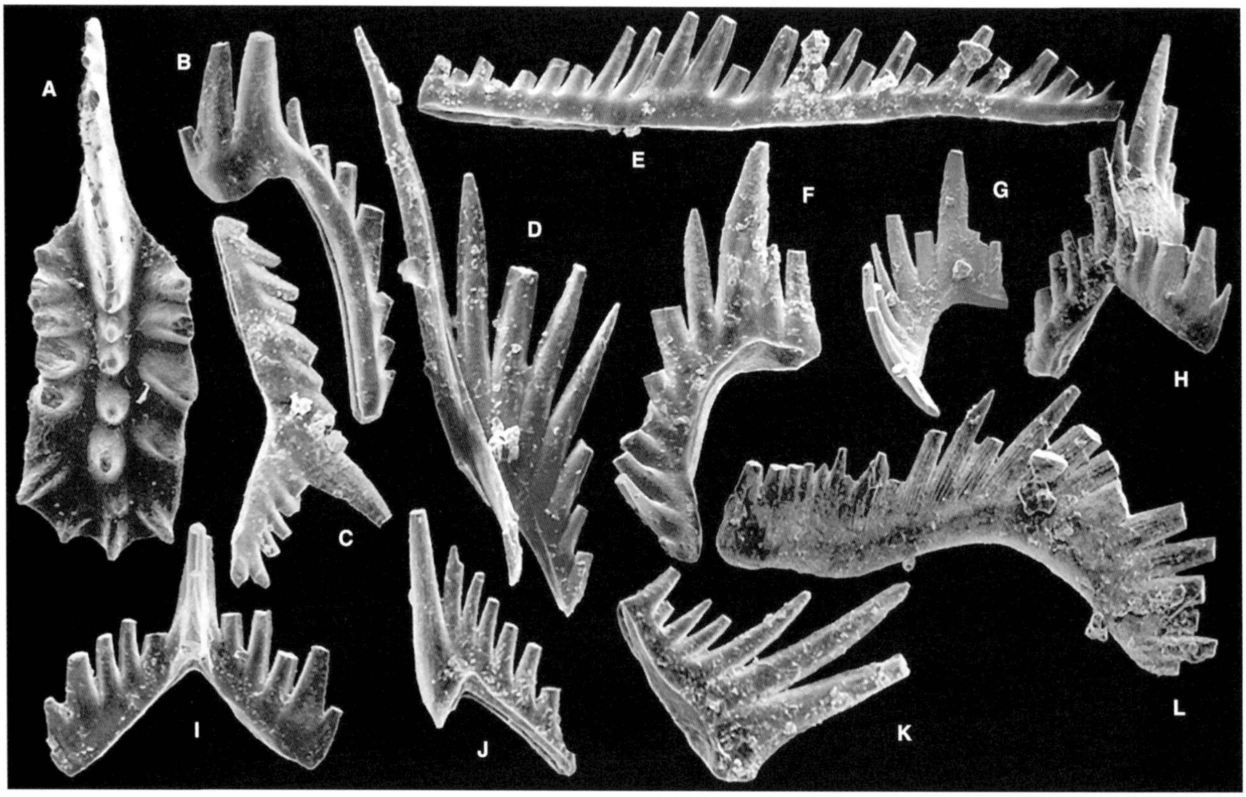

TEXT-FIG. 3. *Epigondolella triangularis* (Budurov). A, P_1, GSC 120002. B, M, GSC 120003. C, P_2, GSC 120004. D, S_1, GSC 120005. E, S_3 or S_4, posterior process, GSC 120006. F, S_4, GSC 120007. G, S_3, GSC 120008. H, S_0, antero-lateral view, GSC 120009. I, S_0, posterior view, GSC 120010. J, S_2, GSC 120011. K, S_1, distal part of long process, GSC 120012. L, S_0, posterior process, GSC 120013. From the Pardonet Formation in north-east British Columbia. All sample MS_1 (GSC loc. C-101034), McLay Spur, except E, sample PMB21 (GSC loc. C-303898), Pink Mountain. Lower Norian. All ×80.

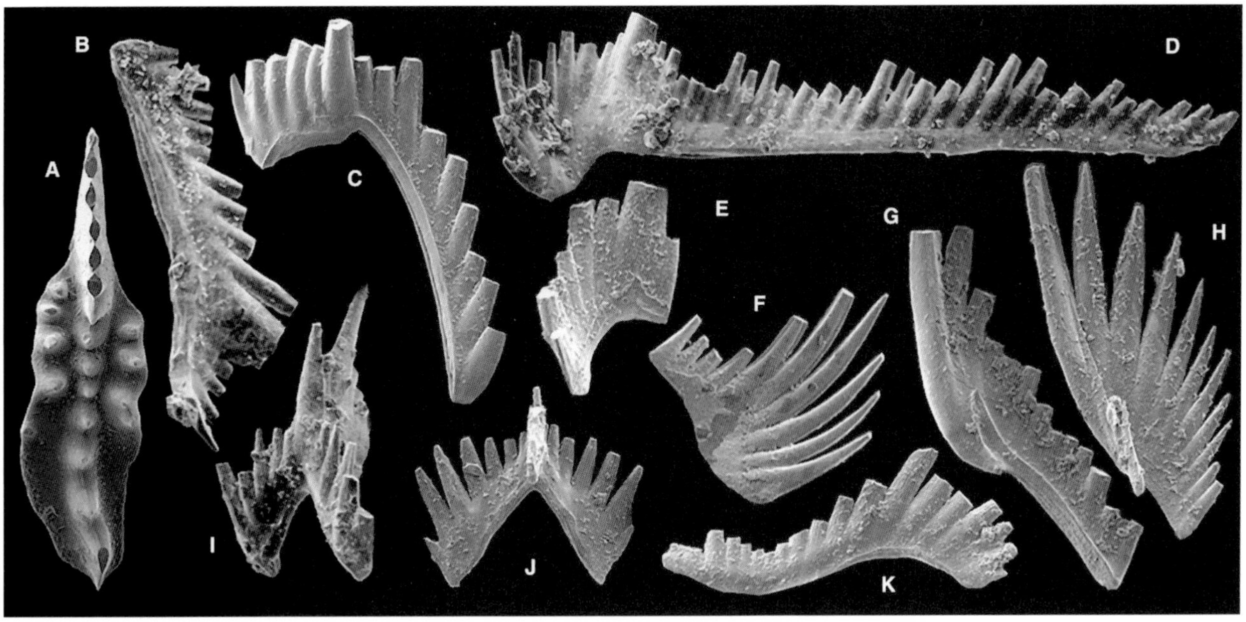

TEXT-FIG. 4. *Cypridodella multidentata* (Mosher). A, P_1, GSC 119991. B, P_2, GSC 119992. C, M, GSC 119993. D, S_4, GSC 119994. E, S_3, GSC 119995. F, S_1, distal part of long process, GSC 119996. G, S_2, GSC 119997. H, S_1, GSC 119998. I, S_0, antero-lateral view, GSC 119999. J, S_0, posterior view, GSC 120000. K, S_0, posterior process, GSC 120001. All from the Pardonet Formation in north-east British Columbia. A, C, E, sample PMK29 (GSC loc. C-304865), and K, sample PM21 (GSC loc. C-304220), Pink Mountain; B, D, I, sample 306B (GSC loc. C-101116), Black Bear Ridge; F, MS_4 (GSC loc. C-101037), McLay Spur; G, SCII-1 (GSC loc. C-305290), Sikanni Chief River; H, Sample 217e-4 (GSC loc. C-202750), and J, sample CG10 (GSC loc. C-305843), Crying Girl Prairie Creek. Middle Norian. All ×80.

Type species and holotype. Cypridodella conflexa Mosher, 1968, p. 920, pl. 113, fig. 8.

Type locality and stratum. Hallstatter Kalk, Steinbergkogel, Hallstatt Salzberg, Austria.

Original diagnosis. The genus was based on a breviform digyrate M element, and later re-described as a seximembrate (11-element) apparatus lacking a platform element (Mosher 1968, p. 920; Sweet in Clark *et al.* 1981).

Multielement diagnosis. As for subfamily, with a P_1 element characterized by a platform with high anterior marginal denticles and a variably elevated posterior carina that extends to, or close to, the posterior tip of the element. Both S_3 and S_4 elements have a short, variably inturned anterior process and a joined posterior process. The distal end of the larger process of the S_1 element, commonly recovered detached, is characterized by a crest of high, apically curved denticles. In some species, the S_2 element has a basal attachment scar extending apically towards the cusp.

Remarks. Cypridodella species are here recognized as an apparatus with a P_1 element corresponding to form-species formerly included in *Epigondolella* or, more recently, *Mockina*. *Cypridodella* has priority over *Mockina* which was introduced for species with strongly ornamented segminiplanate P_1 elements in which the carina typically reaches the posterior end of the platform. The holotype of *Cypridodella* originated in a sample with *Epigondolella bidentata* and it is here interpreted as the M element of that apparatus. In the interpretation of Sweet (1988, and in Clark *et al.* 1981), the P_2 element of *Cypridodella* was regarded as P_1, and the S_1 element as P_2.

Subfamily GLADIGONDOLELLINAE Hirsch, 1994

Characteristics. A group of conodonts with skeletal apparatuses characterized by carminiplanate, anguloplanate, or segminiplanate P_1 elements, and ramiforms that have relatively robust, commonly discrete denticulation; P_2 elements are moderately to strongly arched with an inturned posterior process bearing smaller denticles than the anterior process; S_3–S_4 elements have joined posterior processes; S_3 elements have a secondary antero-lateral process that branches from the third to fifth denticle anterior of the cusp; S_2 elements are extensiform digyrate, and the S_1 elements are breviform digyrate. The S_0 element is alate with lateral processes diverging from the cusp.

Remarks. The subfamily was introduced without clear definition, but simply to accommodate *Gladigondolella* and affirm the different character of its apparatus compared with the Neogondolellinae, at that time conceived as including most other Triassic genera (Hirsch 1994, p. 956). As shown here, there is considerable convergence between the apparatus of *Gladigondolella* and that of other gondolelloids. The most conspicuous and stable feature of the gladigondolellin apparatus is the extensiform digyrate S_2 element with its two strongly developed antero-lateral processes.

Genus GLADIGONDOLELLA Müller, 1962

Text-figure 5

1968 *Dichodella* Mosher, p. 923.

Type species and holotype. Polygnathus tethydis Huckriede, 1958, pp. 157–158, pl. 12, fig. 38a–b.

Type locality and stratum. Trachyceras austriacum bed (Julian), Feuerkogel near Röthelstein, Austria.

Original diagnosis. The name was first used for a carminiplanate P_1 element with a narrow platform, a low carina, and a relatively short posterior process and corresponding keel posterior of the pit (Müller 1962, p. 116).

Multielement diagnosis. Early growth stages of the P_1 element have a narrow platform and prominent cusp and carina; with growth, the platform broadens and the carina/cusp become increasingly suppressed. The P_2 element is angulate with an inturned, short and slender posterior process bearing, like all the ramiforms, discrete peg-like denticles. The M element has two distinct antero-lateral processes, the longer one of which bears a few long denticles that are twice the length of the triangular apical cusp; the outer lateral process bears much shorter denticles. All the S elements have a characteristic distal denticulation on the anterior (or antero-lateral) process: the largest denticle, usually the third one from the anterior end, and those adjacent to it form a high crest. The S_3 element is bifid from the fifth denticle

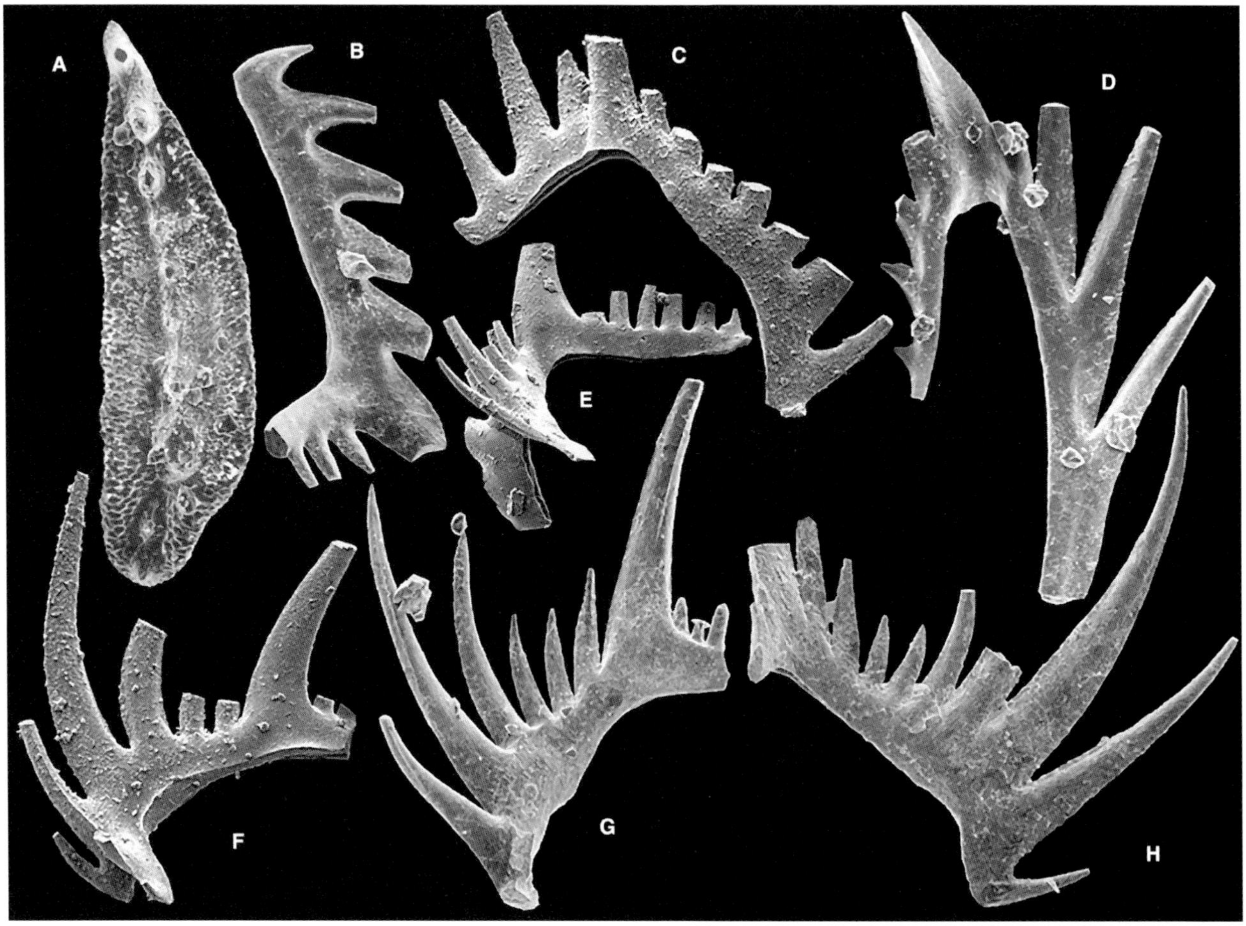

TEXT-FIG. 5. *Gladigondolella tethydis* (Huckreide). A, P_1, GSC 120022. B, P_2, GSC 120023. C, S_2, GSC 120024. D, M, GSC 120025. E, S_0, lateral view, GSC 120026. F, S_3, GSC 120027. G, S_4, GSC 120028. H, S_1, GSC 120029. A–B, D, G–H, sample 5-2-16 (GSC loc. C-301151), Vászoly, Hungary; C, F, sample 614 (GSC loc. C-306591); E, sample D9043 (GSC loc. C-304295), Desli Caira, Romania. Middle Triassic. All ×80.

anterior of the cusp; the fourth denticle immediately behind the bifurcation is the longest and rivals the cusp in size. The S_2 element is extensiform digyrate with a posteriorly inflated basal cavity.

Remarks. The 15-part apparatus is entirely compatible with that of other Gondolellidae. Although the genus has a very distinctive apparatus, it has never been fully described. Huckriede (1958, p. 164) listed 11 constituent form-species but he included two types of P_2 element ('*Ozarkodina saginata*' and '*Prioniodina kochi*') and two 'species' ('*Hindeodella petraeviridis*' and '*Prioniodella pectiniformis*') that are fragments of other elements. A form subsequently named *Dichodella* by Mosher (1968) is also a fragment of the apparatus. Kozur and Mostler (1971, pp. 10–11) listed ten elements that make up the '*Gladigondolella tethydis* multielement', but included all three 'fragment species'; their reconstruction had all the elements included here except for the bifid S_3. The anterior portion of the bifid S_3 element has commonly been incorrectly regarded as an S_0 element, for example by Hirsch (1994, p. 956, text-fig. 3), whose notation differs substantially from the homology suggested here. Koike (1999, pp. 237–238) drew attention to the fact that both Kozur and Mostler, and Hirsch may have confused the *Gladigondolella* apparatus with that of *Cratognathodus* although he did not offer a reconstruction of the former. Early Triassic *Gladigondolella* evolved from *Cratognathodus* and initially contained ramiform elements that were very similar to that genus (e.g. '*Gladigondolella*' *carinata*); however, the Middle Triassic *G. tethydis* is significantly different.

Genus CRATOGNATHODUS Mosher, 1968
Text-figure 6

Type species and holotype. Prioniodina kochi Huckriede, 1958, p. 159, pl. 12, fig. 11.

Type locality and stratum. Trachyceras austriacum Zone, Fuerkogel am Röthelstein, Austria.

Original diagnosis. Genus based on posteriorly downturned segminate P_1 elements with a large (sub)terminal cusp and smaller subequal denticles on a long anterior process (Mosher 1968, pp. 918–919).

Multielement diagnosis. Apparatus with paired segminate P_1, angulate P_2, breviform digyrate M and S_1, extensiform digyrate S_2, bipennate S_3 and S_4, and an alate S_0 element. See Koike (1999) for details.

Remarks. Koike (1999, p. 240) reconstructed the apparatus of a Middle Triassic species he referred to *C. multihamatus* and described it in detail. He noted that the type species of this genus was *Hindeodella multihamata* Huckriede, 1958, with which he synonymized *Prioniodina kochi.* However, the type species selected by Mosher (1968) stands.

Compared with the apparatus of *Gladigondolella*, that of *Cratognathodus* species has a P_1 element that lacks a platform as well as a much simpler M element. Compared with the present reconstruction, the apparatus reconstructed by Koike (1999) included S_3 and S_4 elements with longer anterior processes, an S_2 element with a more posteriorly expanded basal pit, and an S_0 element with the antero-lateral processes branching from a denticle anterior of the cusp rather than the cusp. These differences may be attributable to the fact that the species illustrated here is from the Lower Triassic rather than the Middle Triassic.

Subfamily MULLERINAE subfam. nov.

Characteristics. Constituent genera generally have a 15-element apparatus with a unique asymmetrical tertiopedate S_3 element in which a third downwardly and inwardly directed process arises directly from the cusp. The alate S_0 element also has three processes meeting at the cusp but in this case they are symmetrically arranged. Pectiniform P_1 and P_2 elements are segminate, carminate or carminiplanate. The digyrate S_2 element has two well-developed processes and either a well-defined carina on the posterior face of the cusp or a posteriorly expanded basal cavity beneath it; bipennate S_4 elements have a variably inturned and downturned anterior process; M and S_1 elements are digyrate. Many elements have partial inversion of the basal cavity.

Remarks. This subfamily is introduced for several species originally included in the form-genus *Neospathodus* but shown here to have significantly different apparatuses both from that genus and from others formerly submerged within it (*Novispathodus, Triassospathodus*). The characteristic S_3 element with its third process may lead to its misinterpretation as an S_0 element but it is clearly homologous to the accessory process developed from the anterior process in other gondolelloids. Constituent genera have elements with both discrete and rounded, or confluent and compressed denticulation.

Genus CONSERVATELLA gen. nov.
Text-figure 7

Derivation of name. From the type species.

Type species and holotype. Ctenognathus conservativa Müller, 1956, p. 821, pl. 95, fig. 25.

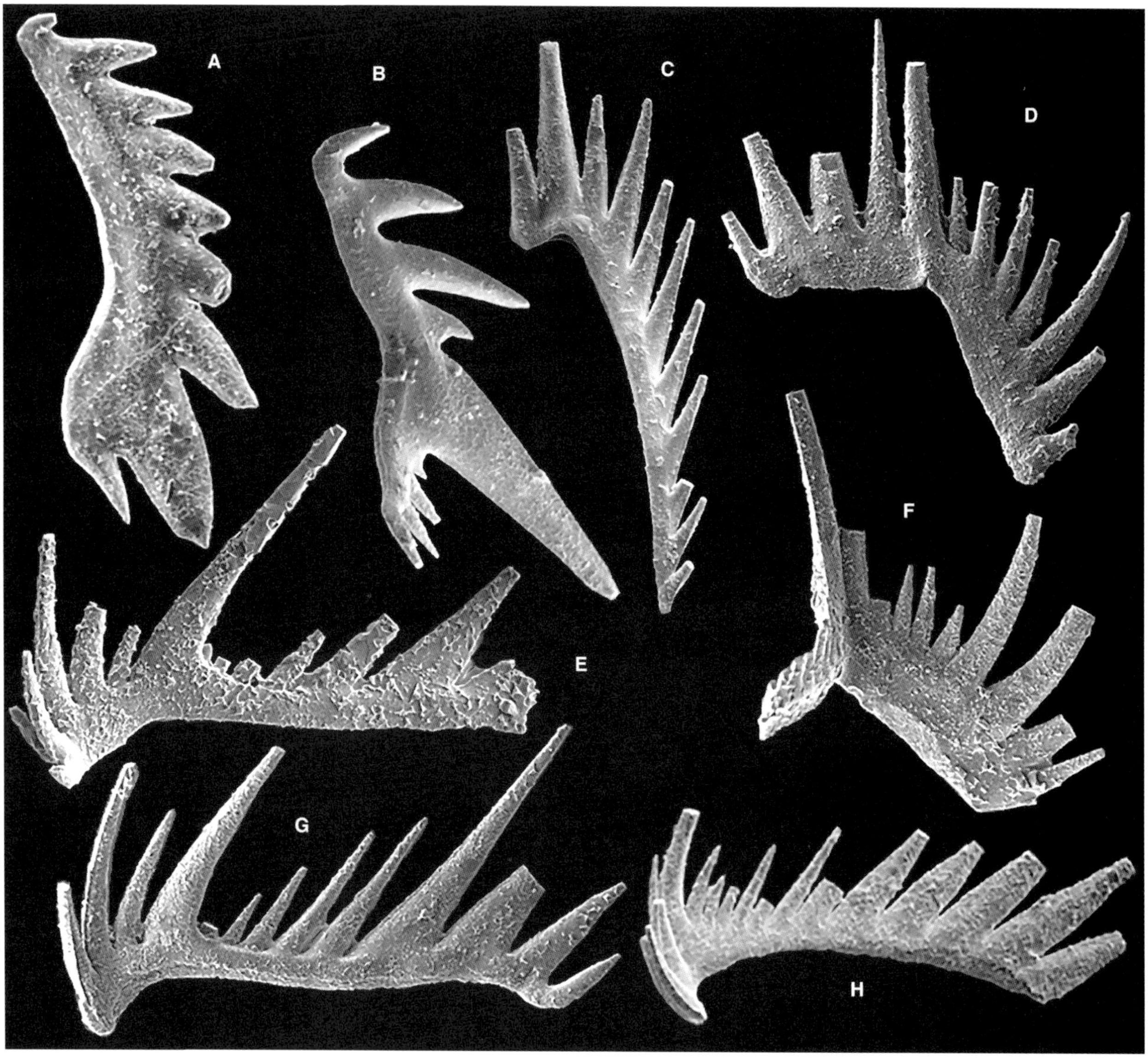

TEXT-FIG. 6. *Cratognathodus* sp. A, P_1, GSC 120014. B, P_2, GSC 120015. C, M, GSC 120016. D, S_2, GSC 120017. E, S_3, GSC 120018. F, S_1, GSC 120019. G, S_4, GSC 120020. H, S_0, lateral view, GSC 120021. All Jabal Safra, Oman. A–C, sample 103A-1 (GSC loc. C-177651); D, F–G, sample 103b (GSC loc. C-117673); E, H, sample 85314 (GSC loc. C-177663). Spathian. All ×80.

Type locality and stratum. Smithian ammonoid bed, Crittenden Springs, Elko Co., Nevada.

Multielement diagnosis. P_1 elements carminate with upturned posterior basal margin and inversion of the basal cavity; the cusp is generally larger than other confluent and compressed denticles but it is not conspicuous. P_2 elements slightly arched with cusp of variable size and an inturned posterior process. S_3 elements have a relatively short third process. S_4 elements show variation in the degree of inturning of the anterior process. S_2 elements have two processes, one half the length of the other, and a pronounced carina on the posterior face of its cusp.

Remarks. The apparatus of the type species of this genus has also been reconstructed from less well-preserved topotype material than the material illustrated here from Oman. The former differs in having less discrete denticulation, particularly in the P_2 element, which furthermore has no well-differentiated cusp; otherwise the two are very similar.

Budurov *et al.* (1988) referred '*Neospathodus*' *conservativus* to *Smithodus*, a new genus based on '*Neospathodus longiusculus* Buryi, 1979'. This latter (P_1) element differs from those of the present genus, and most other Mullerinae, in possessing an expanded basal cavity and a downarched posterior process: its apparatus is unknown, so for now I restrict *Smithodus* to its type species.

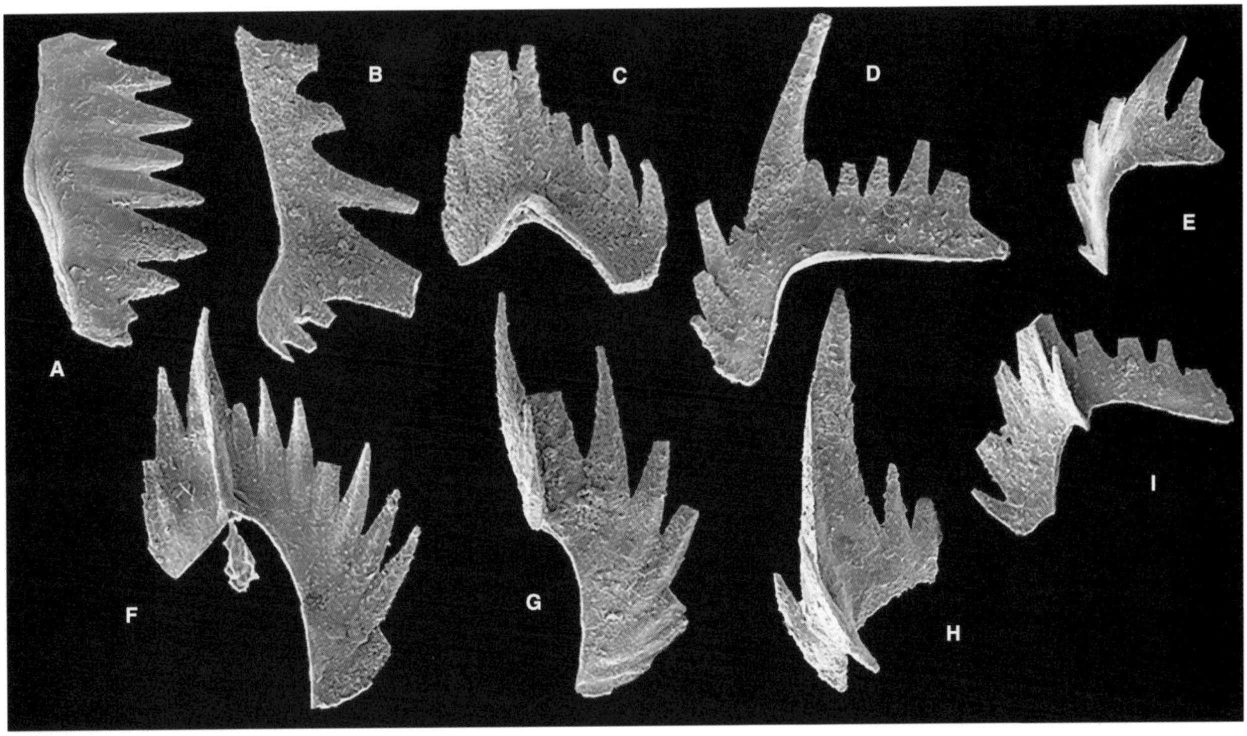

TEXT-FIG. 7. *Conservatella* aff. *conservativa* (Müller). A, P_1, GSC 120137. B, P_2, GSC 120138. C, M, GSC 120139. D, S_4 variant, GSC 120140. E, S_4, GSC 120141. F, S_2, GSC 120142. G, S_1, GSC 120143. H, S_0, lateral view, GSC 120144. I, S_3, GSC 120145. A–B, E–F, I, sample TE106B (GSC loc. C-177656); C–D, G–H, sample TE109B (GSC loc. C-177657). Jabal Safra, Oman. Smithian. All ×80.

Genus DISCRETELLA gen. nov.
Text-figure 8

Derivation of name. From the type species.

Type species and holotype. Ctenognathodus discreta Müller, 1956, pp. 821–822, pl. 95, fig. 28.

Type locality and stratum. Smithian ammonoid bed, Crittenden Springs, Elko Co., Nevada.

Multielement diagnosis. As for subfamily and characterized by elements with discrete, widely spaced, peg-like denticles. P_1 element is carminate with the basal margin upturned and inverted in the posterior one-third to one-half of the element; the cusp is generally larger than other discrete and upright denticles but it is not conspicuous. P_2 element angulate with a relatively expanded and recessive basal attachment surface. S_3 element has a short third process. S_2 element has two well-developed, denticulated processes, an excavated base ('lonchodinid'), and a posteriorly expanded basal cavity. Alate S_0 element with an apparently short posterior process.

Remarks. '*Neospathodus discreta*' was also referred to *Smithodus* by Budurov *et al.* (1988) (see above). The apparatus is very similar to that of *Guangxidella* (see below). Pending reconstruction of the type species from its type locality, the illustrated species is kept in open nomenclature.

Genus GUANGXIDELLA Zhang and Yang, 1991

Type species and holotype. Neoprioniodus bransoni Müller, 1956, p. 829.

Type locality and stratum. Smithian ammonoid bed, Crittenden Springs, Elko Co., Nevada.

Original diagnosis. The type species was based on a segminate P_1 element with a very large and inclined posterior cusp. Zhang and Yang (1991) introduced the genus as a seximembrate (11-element) apparatus (Zhang and Yang 1991, pp. 33–36).

Multielement diagnosis. As for subfamily, with a distinctive P_1 element as described above. Most other elements were described in detail by Zhang and Yang (1991).

Remarks. The type species was named *Guangxidella typica* by Zhang and Yang (1991) but that is a junior synonym of '*Neospathodus bransoni* (Müller)'. The genus was established as a multielement one consisting of six element

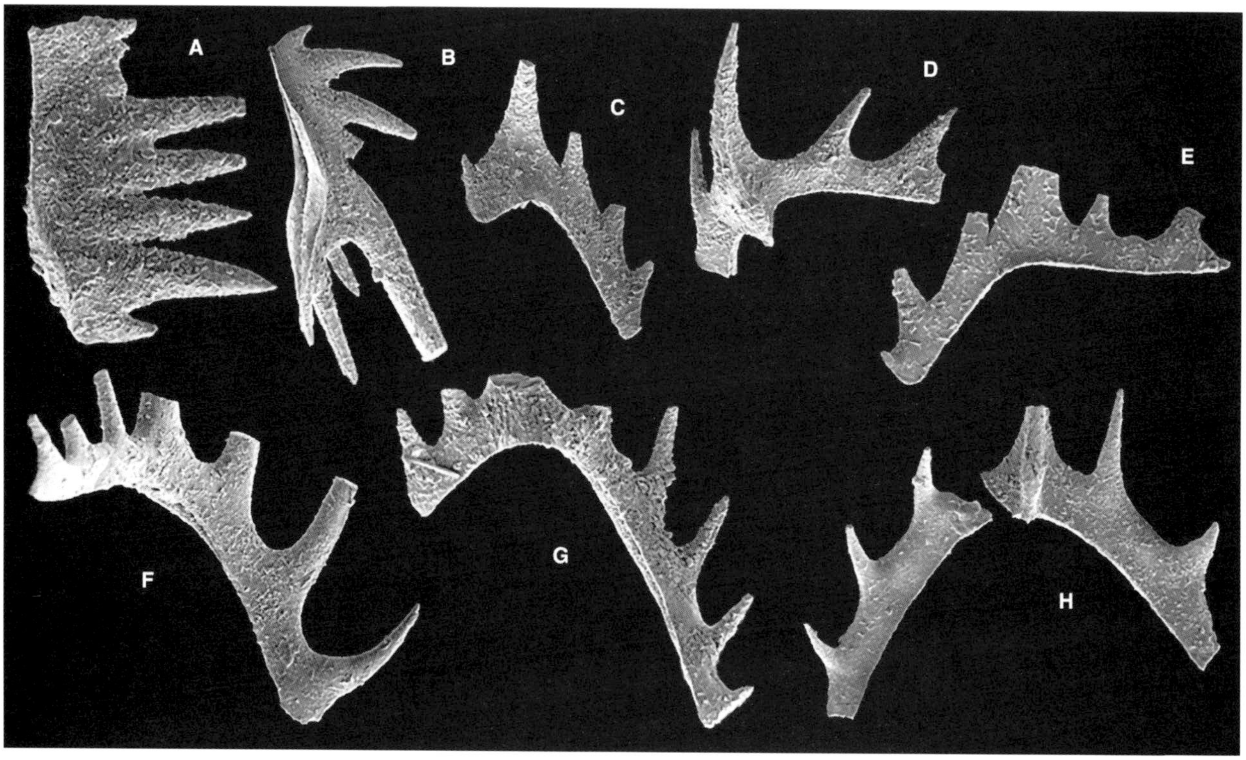

TEXT-FIG. 8. *Discretella* sp. A, P_1, GSC 120146. B, P_2, GSC 120147. C, M, GSC 120148. D, S_3, GSC 120149. E, S_4, GSC 120150. F, S_1, GSC 120151. G, S_2, GSC 120152. H, S_0, posterior view, GSC 120153. A, D, G, sample TE109B (GSC loc. C-177657); B–C, E–F, H, sample TE106b (GSC loc. C-177656). Jabal Safra, Oman. Smithian. All ×80.

types. Reinterpretation of the elements described by Zhang and Yang (1991) identifies their P_2 element as an S_1, their M as an S_2, and their Sb as an S_3–S_4; their Sc could be an S_1 element. These elements are very similar to those of *Discretella* but the P_1 elements are quite different.

Genus MEEKELLA gen. nov.
Text-figure 9

Derivation of name. From the type species.

Type species and holotype. Gladigondolella meeki Paull, 1983, pp. 189–191.

Type locality and stratum. Meekoceras beds, Thaynes Formation, Bear Lake Hot Springs, Idaho.

Multielement diagnosis. As for subfamily, with a distinctive carminiplanate P_1 element having a narrow, laterally and downwardly deflected posterior process. Ramiform elements are robust and characterized by moderately discrete denticles. The arched P_2 element has long unequal processes bearing confluent denticles and sometimes develops a flange on its flanks. Digyrate M elements have a costa on the cusp and slightly flexed processes. Tertiopedate S_3 elements have a more strongly developed third process compared with other members of the subfamily. S_2 elements have two well-developed, denticulated processes, an excavated base ('lonchodinid') and a posteriorly expanded basal cavity. Digyrate M and alate S_0 elements have a distinctive attachment scar.

Remarks. The P_1 element of the type species was originally referred to *Gladigondolella* on the basis of its posterior process but it otherwise has little in common with that genus.

Subfamily NEOGONDOLELLINAE Hirsch, 1994

Characteristics. The apparatus of species included in this subfamily (Hirsch 1994, p. 952) is characterized by a P_1 element that is generally segminiplanate, less commonly segminate, and a relatively conservative M element that in most genera corresponds to the form species *Lonchodina mülleri* Tatge. Bipennate S_3–S_4 elements have incurved and down-curved anterior processes that form an angle of 90 degrees or more with the posterior process; the S_3 element has an antero-lateral process that generally branches from the fifth (rarely the third) denticle anterior of the cusp. The S_2 element generally has a single straight, denticulated

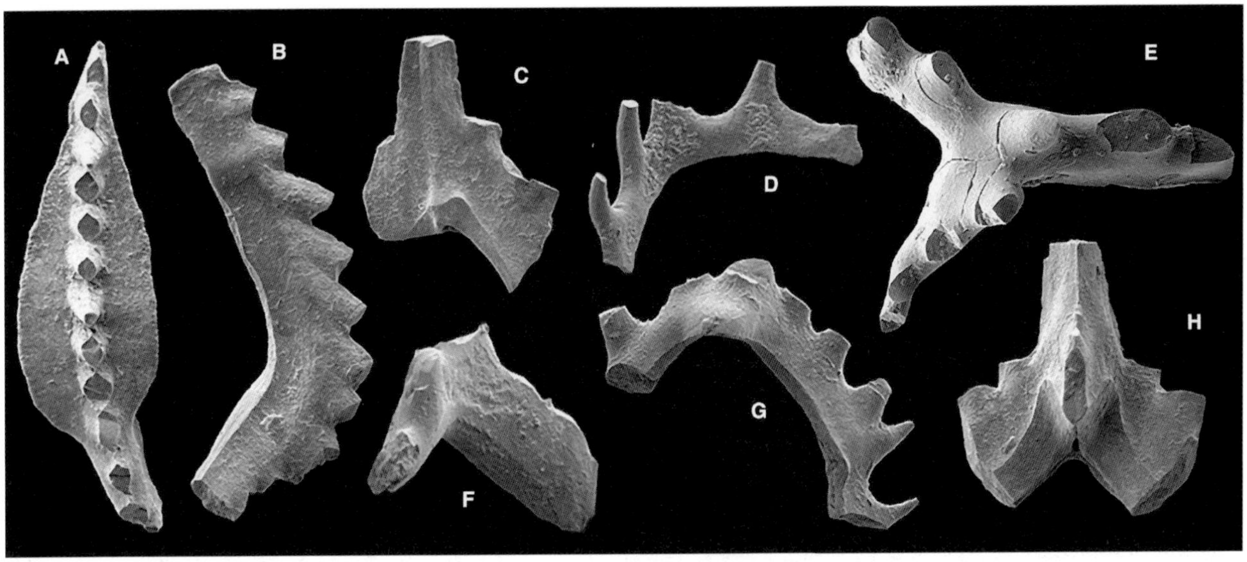

TEXT-FIG. 9. *Meekella meeki* (Paull). A, P_1, GSC 120154. B, P_2, GSC 120155. C, M, GSC 120156. D, S_4, GSC 120157. E, S_3, upper view, GSC 120158. F, S_1, GSC 120159. G, S_2, GSC 120160. H, S_0, posterior view, GSC 120161. A–C, sample W6 (GSC loc. C-301219); D–H, sample HS_1 (GSC loc. C-304705). Hotsprings, Idaho, USA. Smithian. All ×80.

antero-lateral process; a second process may be a very short adenticulate process but is generally developed only as an 'anticusp'. The S_0 element has antero-lateral processes branching from the cusp or from the denticle immediately anterior of it. Mature specimens of both S_3–S_4 and S_0 elements commonly have a posterior process that appears to be detached from the anterior part of the element at a point a few denticles posterior of the cusp.

Remarks. The Neogondolellinae are the conservative stock from which most other gondolelloids evolved. The apparatuses of its species are remarkably stable and in only a few included genera does it differ from the apparatus described by Orchard and Rieber (1999).

Genus NEOGONDOLELLA Bender and Stoppel, 1965
Text-figures 10–11

1965 *Neogondolella* Bender and Stoppel, p. 343.
1970 *Xaniognathus* Sweet, pp. 261–262.
1989 *Clarkina* Kozur, pp. 428–429.
1989 *Pridaella* Budurov and Sudar, pp. 250, 253.

Type species and holotype. Gondolella mombergensis Tatge, 1956, p. 132, pl. 6, fig. 2a–c.

Type stratum and locality. Upper Muschelkalk, Schmidtdiel Quarry, Momberg, near Marburg.

Original diagnosis and type species. The holotype of the type species is an unornamented segminiplanate P_1 element with a strong, partly fused carina of variable height ending in a commonly pronounced (sub)terminal cusp. *Neogondolella* was introduced for smooth forms that had formerly been included in *Gondolella* Stauffer and Plummer (Bender and Stoppel 1965, p. 343).

Multielement diagnosis. Species of *Neogondolella* bore an apparatus consisting of 15 elements (seven paired elements and an unpaired symmetrical S_0 element): P_1 is segminiplanate, P_2 is angulate, S_0 is alate, S_1 is breviform digyrate, S_2 is dolobrate, S_3 is bipennate with a bifid anterior process, S_4 is bipennate, and the M is breviform digyrate. The S_0 and S_3–S_4 elements commonly occur as disjunct anterior and posterior parts (Orchard and Rieber 1999).

Remarks. Orchard and Rieber (1999) gave a full description and discussion of this genus. One feature of interest is the change in the position of the antero-lateral processes relative to the cusp of the alate S_0 element: in all Lopingian and Scythian species, the processes branch from the cusp but within the Middle Triassic (middle Anisian) forms appear in which the processes diverge from the denticle anterior of the cusp. In collections of *N. mombergensis*, the type species, both morphotypes are present, but by the late Ladinian all S_0 elements have the lateral processes arising anterior of the cusp, as for example in both species illustrated herein. Examples of the older form were illustrated by Orchard and Rieber (1999, pls 1, 5). This appears to be a gradual evolutionary trend and is not judged to be a basis for further taxonomic differentiation. Furthermore, there appears to be no good reason to separate the youngest species (e.g. *N. inclinata*, Text-fig. 11) as a different genus, as has often been done (e.g. Kozur 1989).

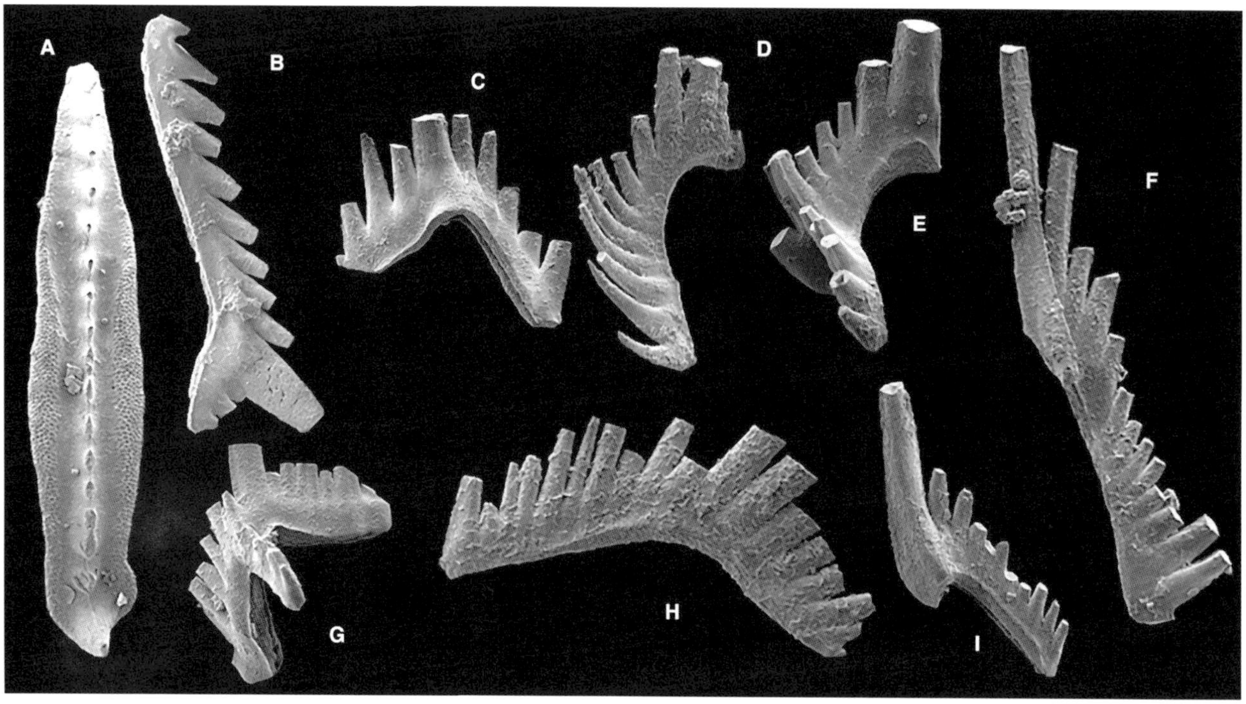

TEXT-FIG. 10. *Neogondolella* ex gr. *constricta* (Mosher and Clark). A, P_1, GSC 120070. B, P_2, GSC 120071. C, M, GSC 120072. D, S_4, GSC 120073. E, S_3, GSC 120074. F, S_1, GSC 120075. G, S_0, lateral view, GSC 120076. H, S_0, posterior process, GSC 120077. I, S_2, GSC 120078. All from sample HB534 (GSC loc. C-301228), Prida Formation, Fossil Hill, Humboldt Range, Nevada, USA. Ladinian. All ×80.

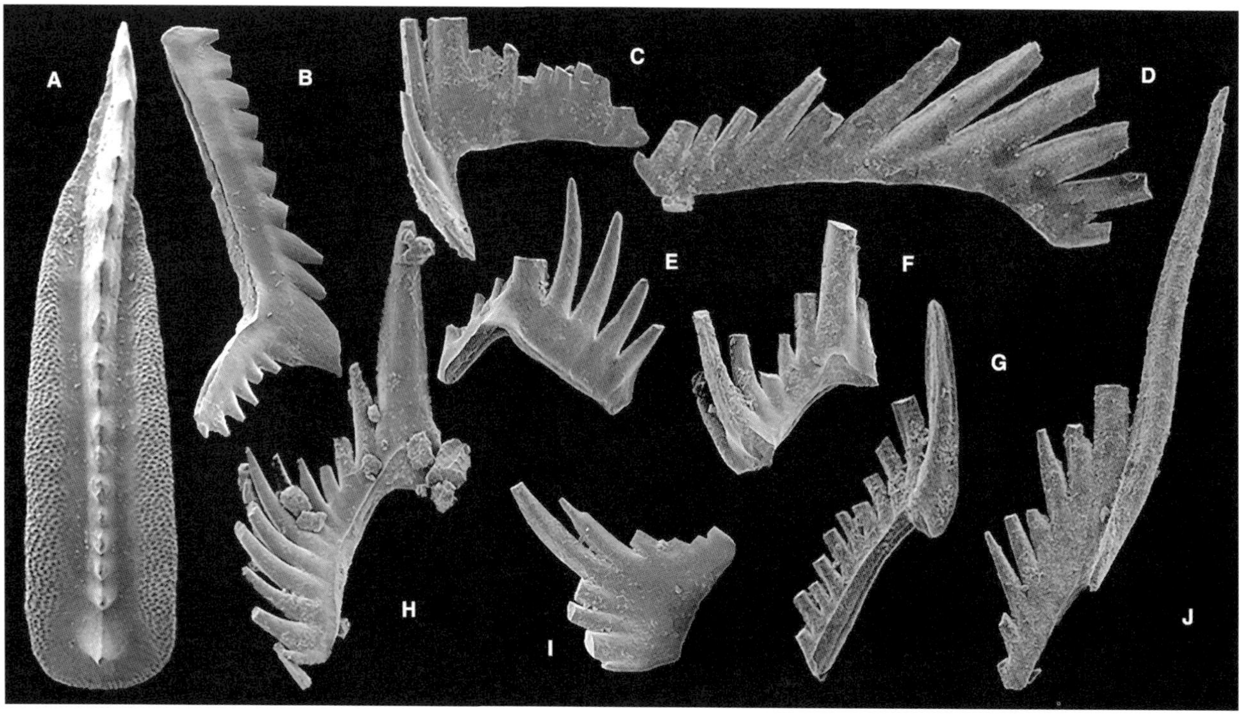

TEXT-FIG. 11. *Neogondolella inclinata* (Kovacs). A, P_1, GSC 120079. B, P_2, GSC 120080. C, S_0, lateral view, GSC 120081. D, S_0, posterior process, GSC 120082. E, M, GSC 120083. F, S_3, GSC 120084. G, S_2, GSC 120085. H, S_4, GSC 120086. I, S_1, distal part of long process, GSC 120087. J, S_1, GSC 120088. All Liard Formation, British Columbia; A–B, BHE-1 (GSC loc. C-305932), Brown Hill and E, G, I, GSP7A/8 (GSC locs. C-304346/C-304349), Glacier Spur, both Williston Lake; C, NWB-4 (GSC loc. C-305048), Besa River; D, F, sample 106B (GSC loc. 83825), Callazon Creek, British Columbia; H, sample GK-3–71H (GSC loc. 84212), Clearwater Creek; J, sample 220B (GSC loc. O-68226), Boiler Canyon, Liard River. Ladinian. All ×80.

Genus CHIOSELLA Kozur, 1989
Text-figure 12

Type species and holotype. Gondolella timorensis Nogami, 1968, pp. 127–128, pl. 10, fig. 17.

Type locality and stratum. Lacian, Landkreis Manatuto, Timor.

Original diagnosis. Based on segminate to segminiplanate P_1 elements with a very narrow or rudimentary platform developed, commonly asymmetrically, on each side of a high carina (Kozur 1989, pp. 415–416).

Multielement diagnosis. As for the subfamily, with the characteristic P_1 element. The apparatus is very similar to that of contemporaneous *Neogondolella* species, with the processes of the S_0 element meeting at the cusp.

Remarks. The type species, *C. timorensis*, has been assigned to both *Neospathodus* and *Neogondolella* in the past because the P_1 element had attributes of both. This species was long viewed as having evolved from *Neospathodus homeri* (see *Triassospathodus*) through platform gain (e.g. Bender 1970, text-fig. 8), but it apparently has a different apparatus from the latter. Rather, the genus may have developed from *Neogondolella* through platform loss. This has far-reaching implications for the proposed origins of *N. regale* and 'true' *Neogondolella*, which many workers (e.g. Kozur 1989, pp. 415–416) have regarded as originating in *Chiosella* by way of progressive platform gain that began from *Triassospathodus homeri*. In fact, *Neogondolella* was present in the Late Permian and ranges through the Lower Triassic so the separation of *Clarkina* cannot be justified on the basis of different origins.

Genus COLUMBITELLA gen. nov.
Text-figure 13

Derivation of name. After the ammonoid *Columbites* with which the type species commonly occurs.

Type species and holotype. Neogondolella elongata Sweet, 1970, pp. 241–243.

Type locality and stratum. Mittiwali Member, Mianwali Formation, Nammal, Salt Range, Pakistan.

Original diagnosis. A species composed of only segminiplanate elements with a short platform with parallel upturned margins, a free blade and a terminal cusp.

Multielement diagnosis. As for *Neogondolella* but with a digyrate S_2 element bearing one long and one short denticulated antero-lateral process. The antero-lateral processes of the alate S_0 element radiate from the cusp.

Remarks. The apparatus of this new genus differs from that of *Neogondolella* and all other members of the subfamily by having an S_2 element with two denticulate processes. In this respect they resemble the homologous element in the novispathodins although the longer process is shorter and straighter in *Columbitella*. In general, the elements of this genus are more robust than many other neogondolellins. The abrupt anterior termination of the platform, free blade and posterior cusp of the P_1 element is distinctive.

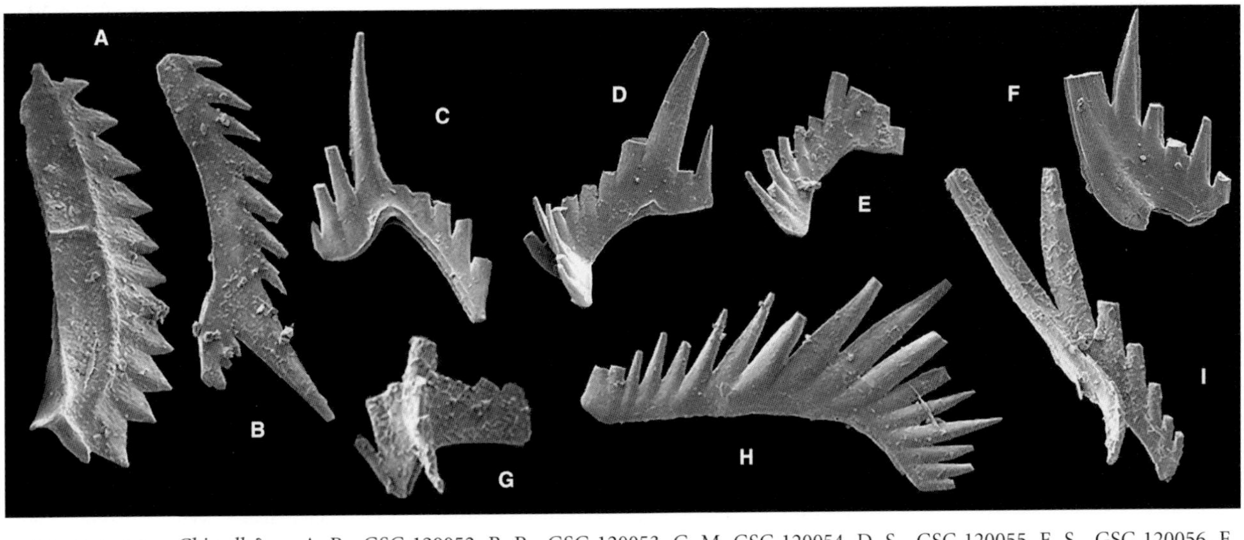

TEXT-FIG. 12. *Chiosella*? sp. A, P_1, GSC 120052. B, P_2, GSC 120053. C, M, GSC 120054. D, S_3, GSC 120055. E, S_4, GSC 120056. F, S_2, GSC 120057. G, S_0, lateral view, GSC 120058. H, S_0, posterior process, GSC 120059. I, S_1, GSC 120060. All from Desli Caira, Romania, sample D9039 (GSC loc. C-304291), except H, sample D9038 (GSC loc. C-304290). Anisian. All ×80.

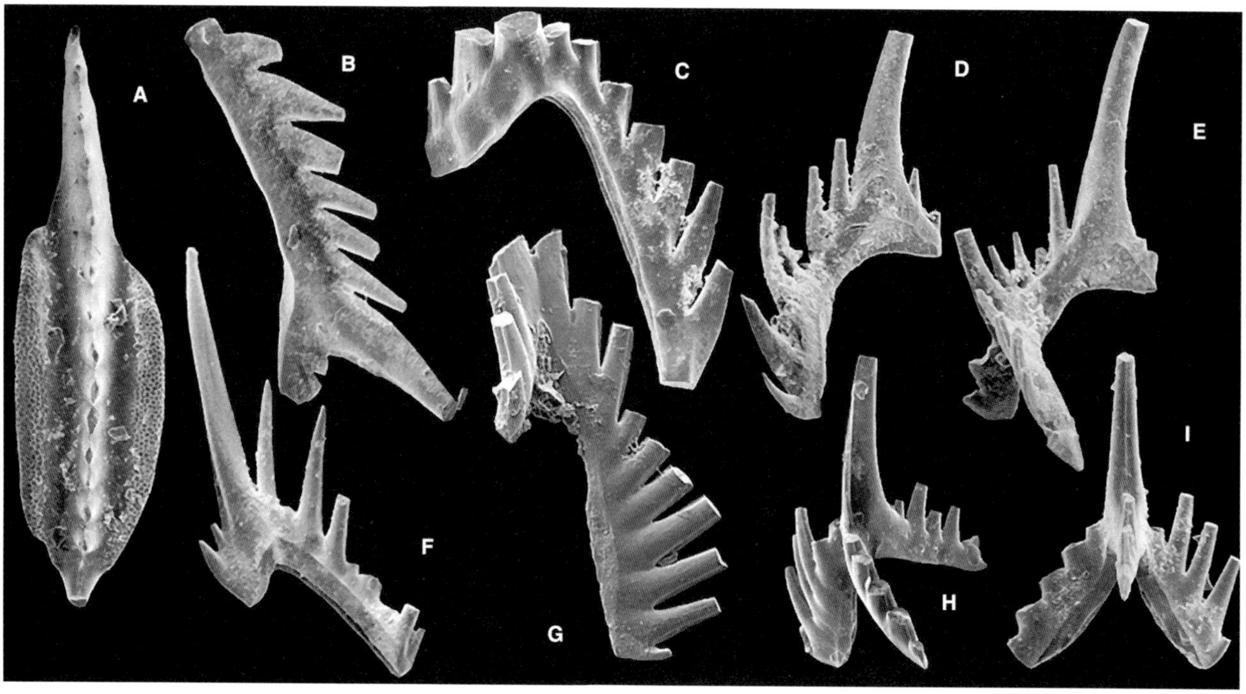

TEXT-FIG. 13. *Columbitella elongata* (Sweet). A, P_1, GSC 120061. B, P_2, GSC 120062. C, M, GSC 120063. D, S_4, GSC 120064. E, S_3, GSC 120065. F, S_2, GSC 120066. G, S_1, GSC 120067. H, S_0, lateral view, GSC 120068. I, S_0, posterior view, GSC 120069. All from sample W11 (GSC loc. C-301224), Thaynes Formation, Paris Canyon, Idaho, USA. Columbites beds, Spathian. All ×80.

Genus METAPOLYGNATHUS Hayashi, 1968

Type species and holotype. Metapolygnathus communisti Hayashi, 1968, p. 72, pl. 3, fig. 11a–c.

Type locality and stratum. Adoyama Formation, Ashio Mountains, central Japan.

Original diagnosis. A segminiplanate P_1 element with a subquadrate, anteriorly downturned platform margin, a smooth to weakly ornamented platform, a free blade and a carina that does not extend to the posterior end of the element. The lower side has a pit that is subcentral in position (Hayashi 1968, p. 72).

Multielement diagnosis. As for *Neogondolella* but with variably ornamented P_1 elements in which the downturned anterior platform and relatively high free blade distinguish it from most other neogondolellinids. The antero-lateral processes of the S_0 element branch from the denticle anterior of the cusp. The lower side of the P_1 element has a pit that ranges in position from subterminal to subcentral.

Remarks. The apparatus is identical to that of its forebear *Neogondolella inclinata*. The significance of the pit-position under the anterior half of the platform, as seen in the type species, has been given different weighting by various authors. Some (e.g. Kozur 1989) have adopted a narrow view of the genus and restricted it to forms with a marked anteriorly shifted pit. However, the apparatus of *Metapolygnathus* appears to be shared by all Carnian species with similar platform configuration and I regard the position of the pit, which shows an overall trend to anterior migration, as an intrageneric variable. The P_1 element of *Columbitella* is superficially similar to *Metapolygnathus* because both have a free blade, but the carina rarely extends to the posterior platform margin in *Metapolygnathus*; the S_2 and S_0 elements are also different as described above.

Genus NEOSPATHODUS Mosher, 1968
Text-figure 14

Type species and holotype. Spathognathodus cristagalli Huckriede, 1958, pp. 161–162, pl. 10, fig. 15.

Type locality and stratum. Lower 'Ceratitenschicten', Mittiwali near Chhidru, Salt Range, Pakistan.

Original diagnosis. The type species is a segminate P_1 element with a width:height:length ratio of 1:3:4, a posterior lower margin that is upturned beneath the posterior one-third of the element, and a short-terminal cusp (Mosher 1968, pp. 929–930). Mosher (1968) included forms with variable arching and denticulation. Sweet (1970) interpreted the genus as consisting of a single element type.

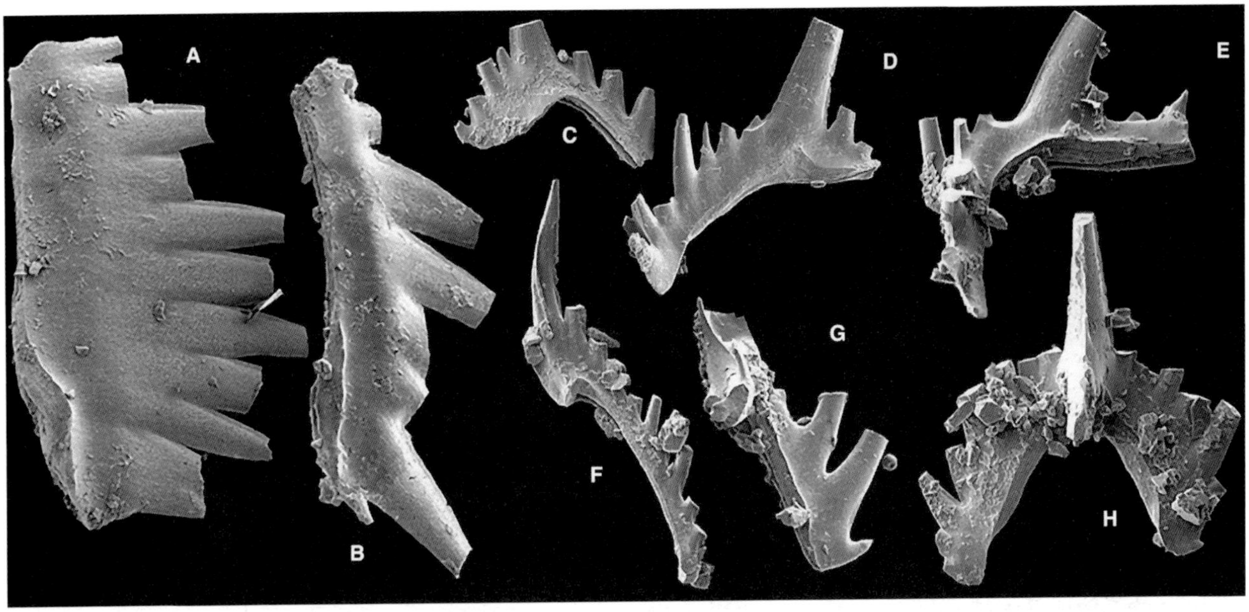

TEXT-FIG. 14. *Neospathodus* cf. *cristagalli* (Sweet). A, P_1, GSC 120089. B, P_2, GSC 120090. C, M, GSC 120091. D, S_4, GSC 120092. E, S_3, GSC 120093. F, S_2, GSC 120094. G, S_1, GSC 120095. H, S_0, posterior view, GSC 120096. All from sample CHOW3 (GSC loc. C-303938), Toad Formation, near headwaters of Chowade River, British Columbia. Dienerian. All ×80.

Multielement diagnosis. Apparatus as in *Neogondolella* but with the following differences: the segminate P_1 element generally lacks a platform although it might be vestigial; P_2 element has a very short (reduced) denticulate posterior process; the S_2 has a second, very short antero-lateral process that shows the beginning of denticulation; the S_3–S_4 and S_0 elements may have attached or disjunct posterior processes; the S_3 has a more posteriorly located bifurcation of the anterior process; and the lateral processes of the alate S_0 element arise from the denticle to the anterior of the cusp. Most of the elements show some inversion of the basal cavity.

Remarks. This genus evolved from *Neogondolella* in the Induan as shown by strong similarities in their apparatuses. However, *Neospathodus* shows the beginning of apparatus changes that led to other gondolelloid subfamilies, including species formerly included in *Neospathodus* but now placed in *Novispathodus*, and others here separated as Mullerinae. The beginning of the Triassic saw renewed innovation in apparatus morphology and derivatives from *Neogondolella* evolved rapidly. As interpreted here, *Neospathodus* is transitional between the Neogondolellinae and Novispathodinae, but the genus is retained in the present subfamily principally because of its bifid S_3 element and essentially neogondolellin S_2 element. The complex basal configuration of the P_1 element (see Sweet 1970, p. 248), and the position of the bifurcation in the anterior process of the S_3 element, shows a trend toward the Mullerinae.

Genus NORIGONDOLELLA Kozur, 1989
Text-figure 15

Type species and holotype. Paragondolella navicula steinbergensis Mosher, 1968, p. 939, pl. 117, figs 18–19.

Type locality and stratum. Hallstatter Kalk, Steinbergkogel, Hallstatt Salzberg, Austria.

Original diagnosis. Based on a segminiplanate P_1 element separated from *Neogondolella* by virtue of its appearance in the Norian without clear ancestry in the Middle Triassic stock (Kozur 1989, p. 421).

Multielement diagnosis. As in *Neogondolella* but differing in the following characters: both the P_2 and the M element have a cusp that is markedly broader at the base and denticles that are more discrete than is typical of the subfamily; the P_2 element has a higher anterior process and a relatively delicate posterior process which is strongly down-arched and distinctly lower than anterior process. The antero-lateral processes of the alate S_0 element branch from the denticle anterior to the cusp.

Remarks. The origins of *Norigondolella* remain obscure. Orchard (1991*a*, p. 179) speculated that it might have originated in *Metapolygnathus* ex gr. *communisti* Hayashi, 1968 at the base of the Norian but there are other poorly known precursor candidates in the Carnian.

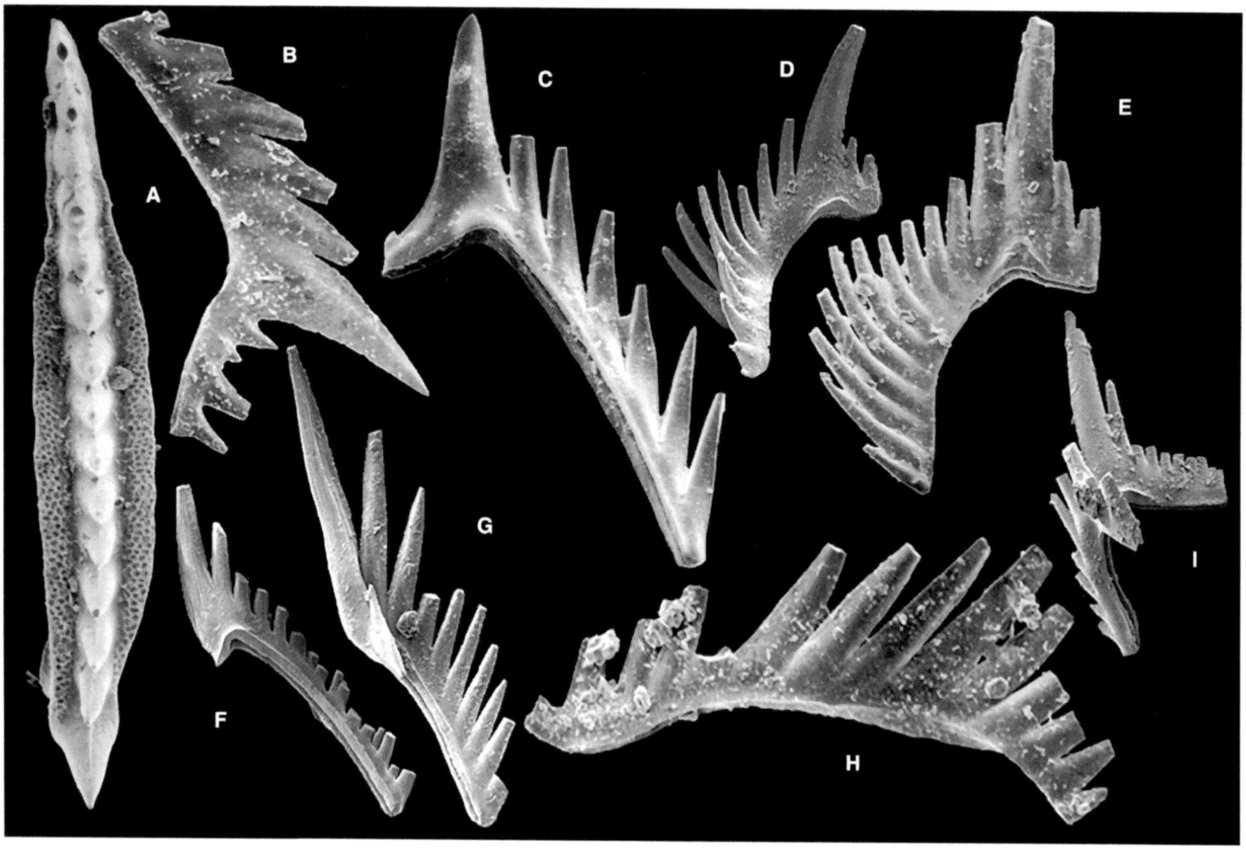

TEXT-FIG. 15. *Norigondolella steinbergensis* (Mosher). A, P_1, GSC 120097. B, P_2, GSC 120098. C, M, GSC 120099. D, S_3, GSC 120100. E, S_4, GSC 120101. F, S_2, GSC 120102. G, S_1, GSC 120103. H, S_0, posterior process, GSC 120104. I, S_0, lateral view, GSC 120105. All Pardonet Formation, British Columbia; A, E, sample 305a (GSC loc. C-101113), and B–C, H, sample 306a (GSC loc. C-101115), both Black Bear Ridge; D, G, I, sample PMB25a (GSC loc. C-303905), and F, sample PMB21 (GSC loc. C-303898), both Pink Mountain. Middle Norian. All ×80.

Subfamily NOVISPATHODINAE subfam. nov.

Characteristics. Species of this new subfamily generally have a 15-element apparatus characterized by rather variable P_1 and P_2 elements, non-bifid S_3 elements with attached posterior processes, and weakly digyrate S_2 elements with two denticulated antero-lateral processes, the longer one of which is commonly curved. The S_3 and S_4 elements differ from each other in the degree of downturning and flexure in the anterior process. The alate or modified alate S_0 elements commonly have antero-lateral processes that originate from a point that is one, two or rarely three denticles in front of a variably developed cusp.

Remarks. The Novispathodinae developed from the Neogondolellinae via *Neospathodus* through loss of a bifurcation in the S_4 element and by the growth of the anterior edge of the S_2 element into a short denticulated or adenticulate process.

Genus NOVISPATHODUS gen. nov.
Text-figure 16

Type species and holotype. *Neospathodus abruptus* Orchard, 1995, pp. 118–119, fig. 3.23–24.

Type locality and stratum. Mittiwali Member, Mianwali Formation, Chhidru B (east), Pakistan.

Multielement diagnosis. As for the subfamily, and with the following particulars: the P_1 element is essentially segminate but may be carminate with a few posterior denticles constituting an incipient posterior process, which is straight to downturned posteriorly; angulate P_2 elements may also be high-bladed with a posterior process half the length of the anterior one; short antero-lateral process of the S_2 element commonly comprises two or three denticles and may extend downward as a short process; bipennate S_3–S_4 elements commonly have a sinuous lower profile due to initial downturning of the posterior process immediately posterior of the cusp.

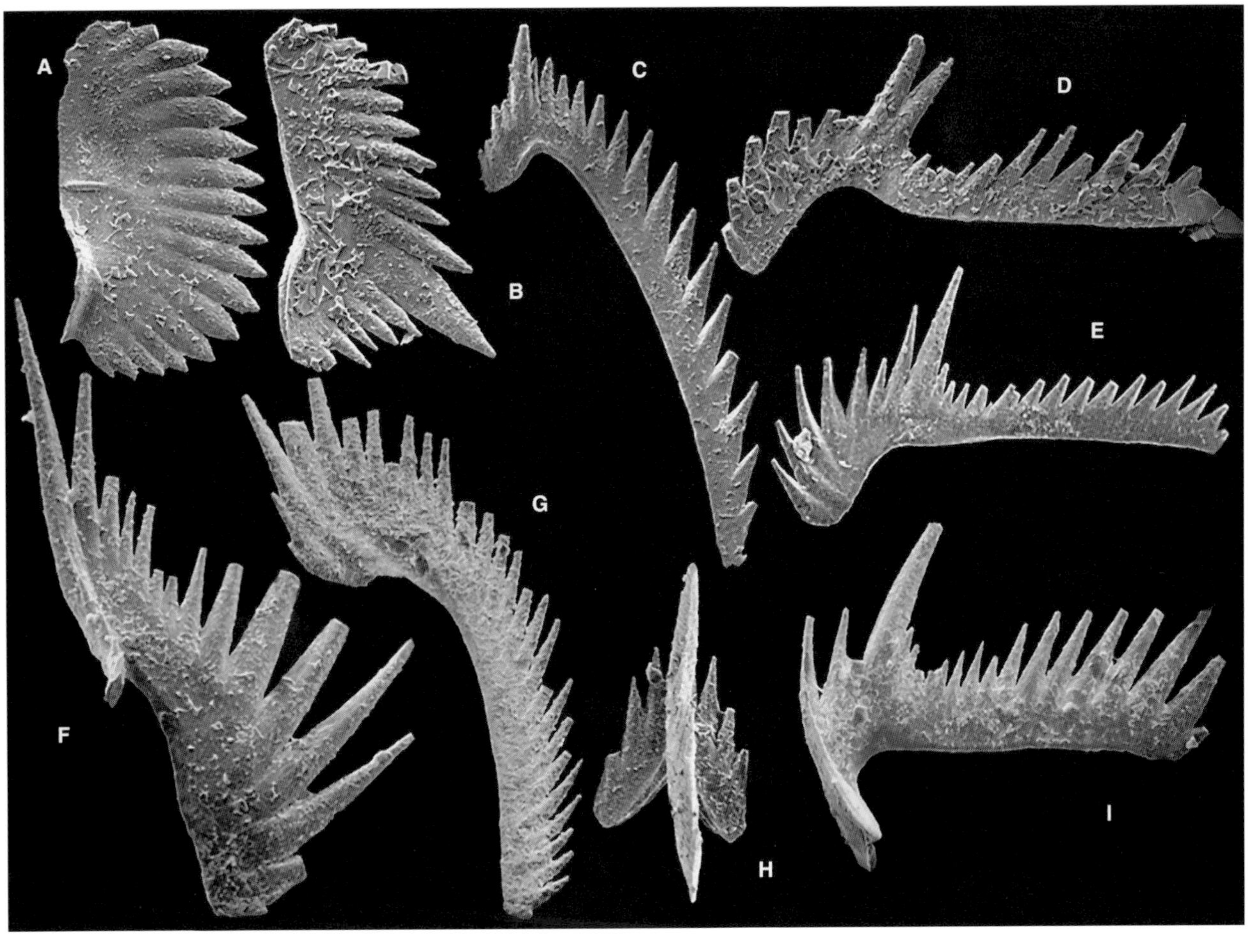

TEXT-FIG. 16. *Novispathodus abruptus* (Orchard). A, P_1, GSC 120106. B, P_2, GSC 120107. C, M, GSC 120108. D, S_3, GSC 120109. E, S_4, GSC 120110. F, S_1, GSC 120111x. G, S_2, GSC 120112. H, S_0, posterior view, GSC 120113. I, S_0, lateral view, GSC 120114. A–B, D–E, sample 103b (GSC loc. C-117673); C, F, H–I, sample 103a-1 (GSC loc. C-177651); G, sample 85314 (GSC loc. C-177663). All Jabal Safra, Oman. Spathian. All ×80.

Remarks. Constituent species vary in the degree of development of the cusp of the M, S_2–S_4 and S_0 elements. *Kashmirella* Budurov, Sudar and Gupta, 1988 may be a senior synonym but the apparatus of its type species (*K. alberti*) is unknown; as originally conceived, that genus included conodont genera from more than one subfamily.

Genus BUDUROVIGNATHUS Kozur, 1988
Text-figure 17

1973 *Carinella* Budurov, p. 799.

1988 *Sephardiella* March, Budurov, Hirsch and Marquez-Aliaga, p. 247.

Type species and holotype. Polygnathus mungoensis Diebel, 1956, pp. 431–432, pl. 4, fig. 1.

Type locality and stratum. Mungo-Kriede, Cameroon.

Original diagnosis. Based on a segminiplanate P_1 element with a free blade, straight to sinuous platform margins and carina, and bearing nodes on the anterior platform margins (Kozur 1988, p. 244).

Multielement diagnosis. Apparatus as for subfamily, with carminiplanate, smooth to ornamented P_1 elements, and characteristic P_2 elements that have a remarkably straight, unarched profile. S_2 element has one long denticulated process and a second one bearing two denticles. The larger process of the S_1 element is distinctively high in its proximal part.

Remarks. The generic diagnosis is based partly on that of Bagnoli *et al.* (1985), who described the apparatus of *Budurovignathus truempyi* as septimembrate (13-element). Although the elements of the apparatus are not common in my material, it seems probable that S_3 and S_4 elements with differing anterior curvature could be differentiated

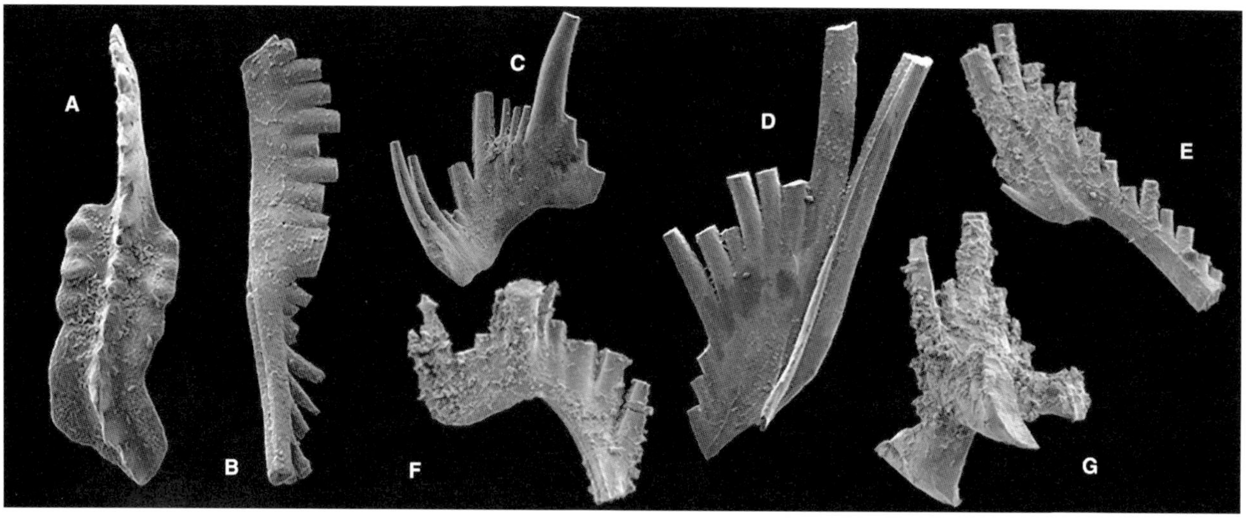

TEXT-FIG. 17. *Budurovignathus mungoensis* (Diebel). A, P_1, GSC 120131. B, P_2, GSC 120132. C, S_3, GSC 120133. D, S_1, GSC 120134. E, S_2, GSC 120135. F, M, GSC 120162. G, S_0, lateral view, GSC 120136. A, D, sample BHE-1 (GSC loc. C-305932), Brown Hill, and C, sample GSP_2 (GSC loc. C-304341), Glacier Spur, north-east British Columbia (Liard Formation); B, sample 155 (GSC loc. C-306576); E–G, sample SC-B2 (GSC loc. C-306610), New Pass, Nevada. Ladinian/Carnian boundary. All ×80.

as in other novispathodins. Mietto (1982) described fused pairs of P_1 elements of *Budurovignathus* but his accompanying reconstruction lacked P_2 elements.

Genus MOSHERELLA Kozur, 1972
Text-figure 18

Type species and holotype. Neospathodus newpassensis Mosher, 1968, p. 931, pl. 115, fig. 9.

Type locality and stratum. Middle member, Augusta Mountain Formation, New Pass Range, Nevada.

Original diagnosis. Based on a carminate P_1 element with a very short, downturned posterior process and subequal and confluent denticles and cusp (Kozur 1972, p. 14).

Multielement diagnosis. P_1 elements are carminate or segminate, the former with a short downturned posterior process; P_2 elements are angulate with a strongly reduced posterior process. Bipennate S_3–S_4 elements are similar but differ in the relative length of, and degree of, downturning and inturning of the anterior process. Both S_1 and S_2 are digyrate with a long denticulate antero-lateral process with distal denticles forming a high crest; the shorter process of the S_2 element is a downward-directed 'anticusp'. The modified alate S_0 element has two antero-lateral processes branching from the denticle in front of the cusp.

Remarks. The genus evolved from *Pseudofurnishius* close to the Ladinian/Carnian boundary by loss of the accessory denticles of the P_1 element; they are both characterized by a similar and very distinctive P_2 element (*Pollognathus sensu formo*). The S_2 element is very similar to that of *Triassospathodus*, but the other S elements most closely resemble those of *Novispathodus*. Like the Cornudininae, variation in the morphology of the anterior process in the S_3 and S_4 elements leads to potential differentiation of an additional element type.

PSEUDOFURNISHIUS van den Boogaard, 1966

Type species and holotype. Pseudofurnishius murcianus van den Boogaard, 1966, pp. 6–7, pl. 2, fig. 1.

Type locality and stratum. Sierra de Carrascoy, Murcia Province, south-east Spain.

Original diagnosis and holotype. Asymmetrical pectiniform element composed of a blade with a denticulate platform on the inner side and an inverted basal cavity and small pit on the underside (van den Boogaard 1966, pp. 5–6).

Multielement diagnosis. As in *Mosherella* but with lateral denticles developed on one side of the P_1 elements.

Remarks. Ramovs (1978) described numerous fused clusters of *Pseudofurnishius* from Slovenia, and also showed (Ramovs 1977) the individual elements reconstructed as a septimembrate (13-element) apparatus that he referred to a new family, the Pseudofurnishiidae. The so-called chirodelliform element ('Chi') of Ramovs (1977) is almost certainly a distal fragment of a hindeodelliform (S_3–S_4) element. Other illustrated elements resemble

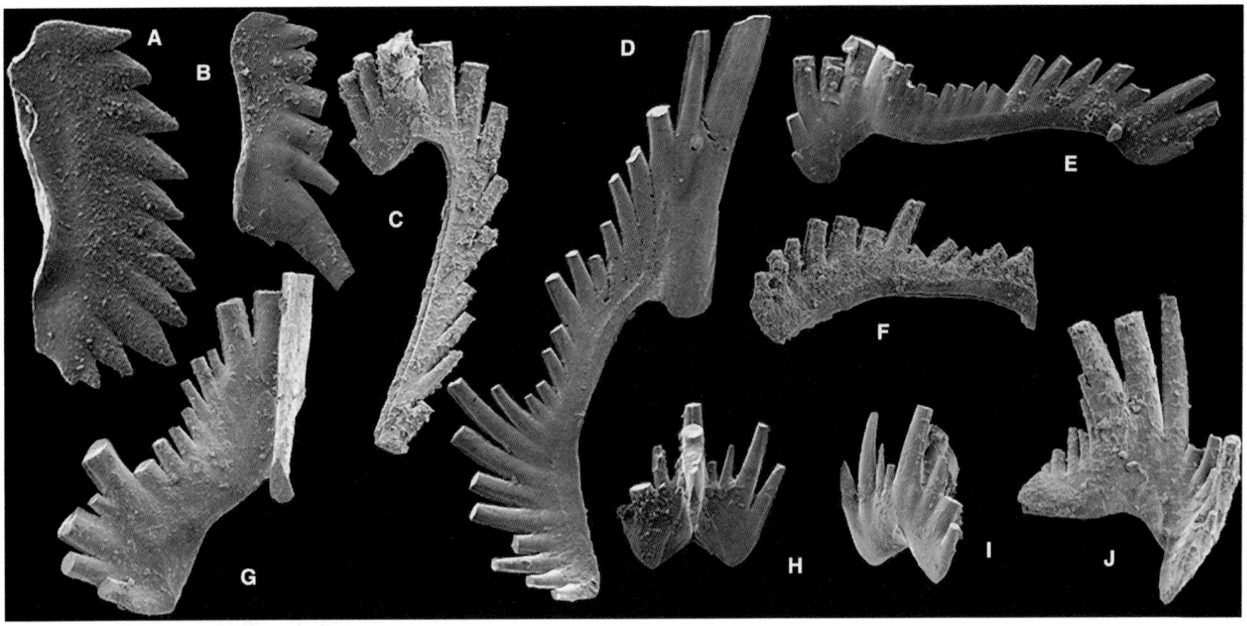

TEXT-FIG. 18. *Mosherella newpassensis* (Mosher). A, P_1, GSC 119963. B, P_2, GSC 119964. C, M, GSC 119965. D, S_2, GSC 119966. E, S_4, GSC 119967. F, S_4 variant, GSC 119968. G, S_1, GSC 119969. H, S_0, posterior view, GSC 119970. I, S_0, antero-lateral view, GSC 119971. J, S_3, GSC 119972. A–B, F, H, sample 105 (GSC loc. C-306570); C, sample SC-B6 (GSC 306614); D, I, sample 151 (GSC loc. C-306575); G, J, sample SC-A12 (GSC C-306606), all South Canyon, New Pass Range, Nevada, USA; E, sample 106B (GSC loc. 83825), Callazon Creek, British Columbia. Lower Carnian. All ×80.

those of *Mosherella*: non-bifid S_3 elements ('Hi') and an S_0 ('Hib') element with anterior processes branching from the denticle anterior of the cusp. The illustration of fused hindeodelliform (S_3–S_4) elements (Ramovs 1978, pl. 2, fig. 3c) shows that elements with both slightly to strongly inturned anterior processes occur. The M ('Pr') and S_1 ('En') elements are clear in Ramovs (1977) but no S_2 element was differentiated.

Genus TRIASSOSPATHODUS Kozur, 1998
Text-figure 19

Type species and holotype. Spathognathodus homeri Bender, 1970, pp. 528–529, pl. 5, fig. 16a–c.

Type locality and stratum. Marmarotrapeza Formation, Marathovuno, Chios, Greece.

Original diagnosis. The genus was introduced for a segminate (strictly carminate) P_1 element with small, subequal and confluent denticles and cusp, and a downturned basal posterior part (Kozur *et al.* 1998, pp. 2–3).

Multielement diagnosis. As for the subfamily, with these particular features: the S_2 is a distinctive arched element with one long posterior process, a large anterior cusp, and a short 'anticusp' with a straight to convex anterior margin. The two processes of the bipennate S_3–S_4 elements have a less arched lower margin than in other gondolelloids and differ in the relative length of the anterior and posterior processes and the orientation of the cusp, which is more inclined in the element assigned to S_3.

Remarks. This genus was separated from *Neospathodus* by Kozur *et al.* (1998) because of the downturning of the posterior lower margin of the P_1 element. However, the S_2–S_4 and S_0 elements of the apparatuses are also significantly different, as described above. The S_2 element is particularly distinctive: it resembles that of *Mosherella*. Kozur (1989, pp. 2–3) assigned many species to this genus but few, if any, appear to have the same apparatus.

Subfamily PARAGONDOLELLINAE subfam. nov.

Characteristics. Species of this subfamily have apparatuses characterized by S_0 elements in which the two antero-lateral processes branch far anterior of the cusp, from which they are separated by 4–5 denticles. The anterior process of the S_3 element may have an inwardly directed process at its distal end. The posterior processes of the S_3–S_4 elements are attached.

Remarks. Species of the Cornudininae have a similar S_0 element but the distinctive anterior processes of the S_3–S_4 elements and characteristic M elements serve to separate them from the present subfamily.

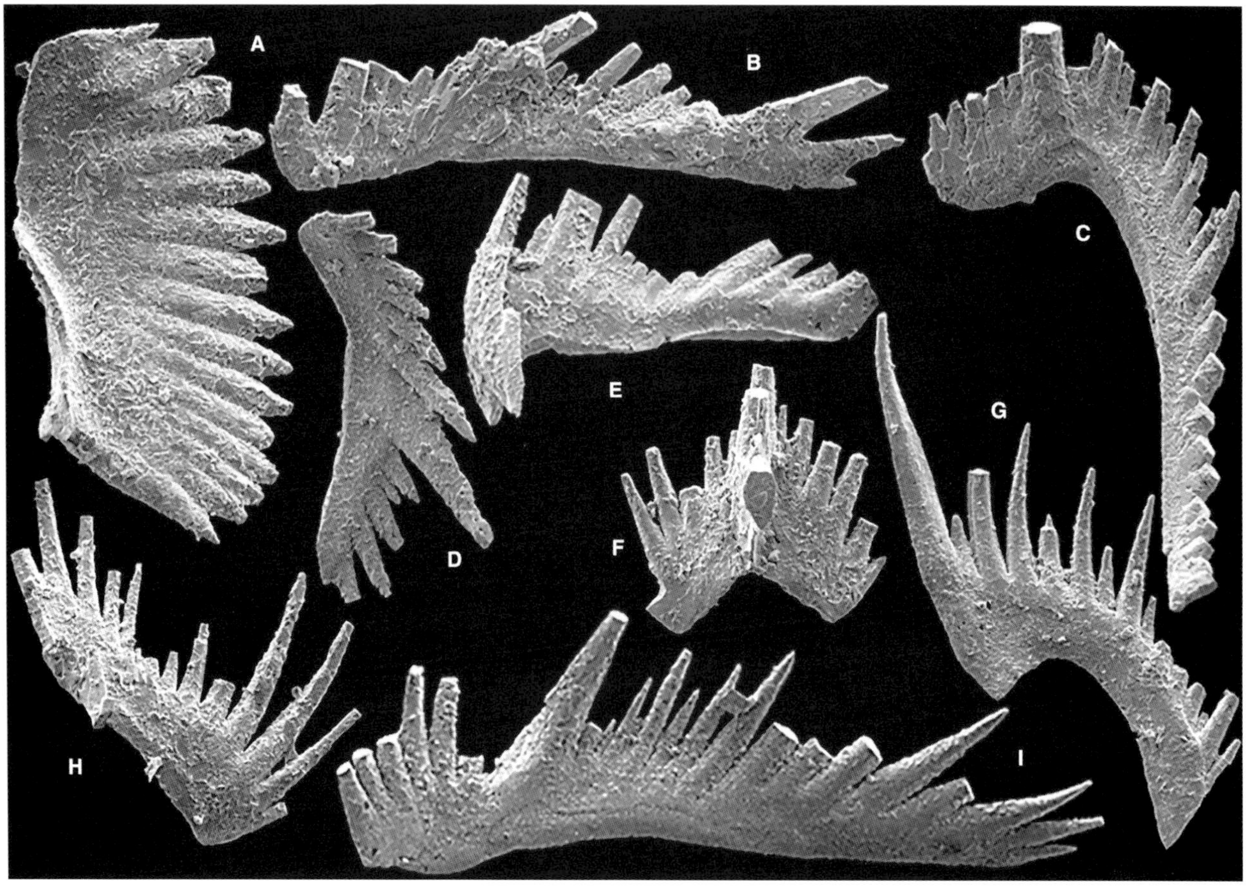

TEXT-FIG. 19. *Triassospathodus homeri* (Bender). A, P_1, GSC 120115. B, S_4, GSC 120116. C, M, GSC 120117. D, P_2, GSC 120118. E, S_0, lateral view, GSC 120119. F, S_0, posterior view, GSC 120120. G, S_2, GSC 120121x. H, S_1, GSC 120122. I, S_3, GSC 120123. A, C, F–G, sample 2 (GSC loc. C-306491); B, sample 3 (GSC loc. C-306492); D–E, H–I, sample 9 (GSC loc. C-306498). All Upper Guangdao section (UGD), Nanpanjiang Basin, South China. All ×80.

Genus PARAGONDOLELLA Mosher, 1968
Text-figure 20

Type species and holotype. Paragondolella excelsa Mosher, 1968, pp. 938–939, pl. 118, figs 7–8.

Type locality and stratum. Schreyeralmkalk, Feuerkogel, Austria.

Original diagnosis. The genus was based on an ovoid segminiplanate P_1 element with a distinctive growth pattern that emphasized a high crested blade at all stages (Mosher 1968, p. 938).

Multielement diagnosis. As for subfamily, and with the following features: P_1 elements have a wide, flat platform and a high carina; P_2 elements are long, arched units with a subcentral cusp; S_3 and S_4 elements have moderately down-arched anterior processes that include, in the S_3 element, an inwardly directed accessory process. The character of the S_2 element is uncertain.

Remarks. The present material is rather sparse, but the apparatus is clearly different from that of segminiplanate element-bearing species included in the Neogondolellinae. Several common species have formerly been assigned to this genus, e.g. the ubiquitous Upper Triassic species *inclinata* and *polygnathiformis*, but these species have a neogondolellin apparatus.

Genus TRAMMERELLA gen. nov.
Text-figure 21

Type species and holotype. Gondolella haslachensis trammeri Kozur, *in* Kozur and Mock 1972, p. 13, pl. 1, fig. 4a–b.

Type locality and stratum. Protrachyceras curionii bed, Koveskal, Balaton Highlands, Hungary.

Multielement diagnosis. A 15-element apparatus with a P_1 element that is segminate in early growth stages and segminiplanate in later growth stages, with an initially

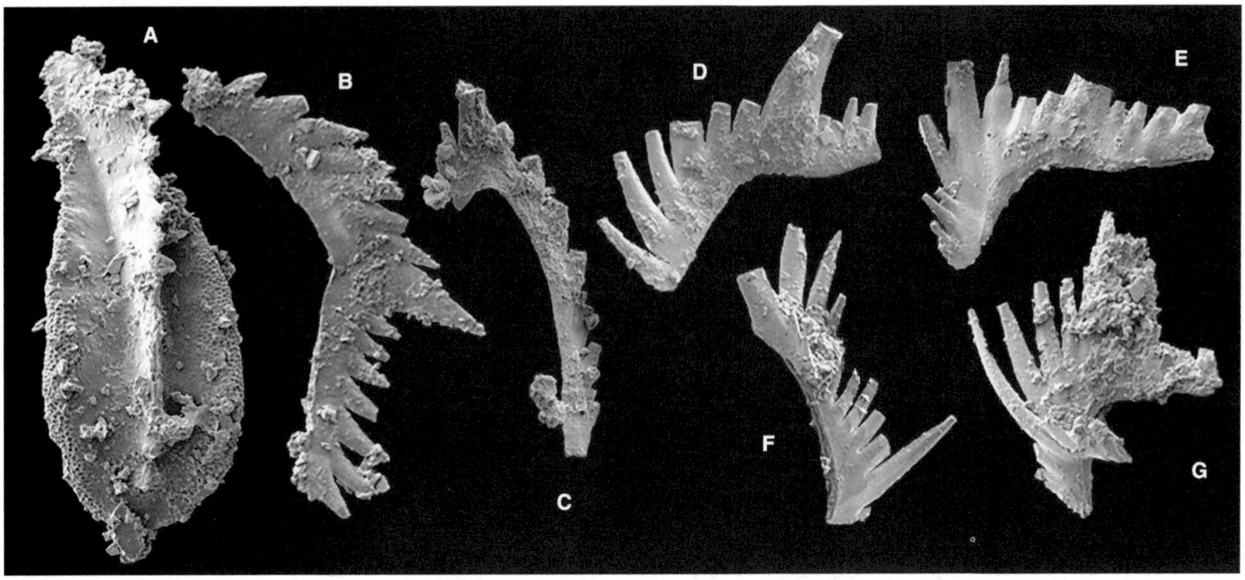

TEXT-FIG. 20. *Paragondolella* ex gr. *excelsa* Mosher. A, P_1, GSC 120124. B, P_2, GSC 120125. C, M, GSC 120126. D, S_4, GSC 120127. E, S_3, GSC 120128. F, S_1, GSC 120129. G, S_0, lateral view, GSC 120130. A, E, sample HB553 (GSC loc. C-300245); B–D, F, sample HB560 (GSC loc. C-300252); G, sample HB556 (GSC loc. C-300248). All from the Prida Formation at Fossil Hill, Humboldt Range, Nevada, USA. Anisian/Ladinian boundary. All ×80.

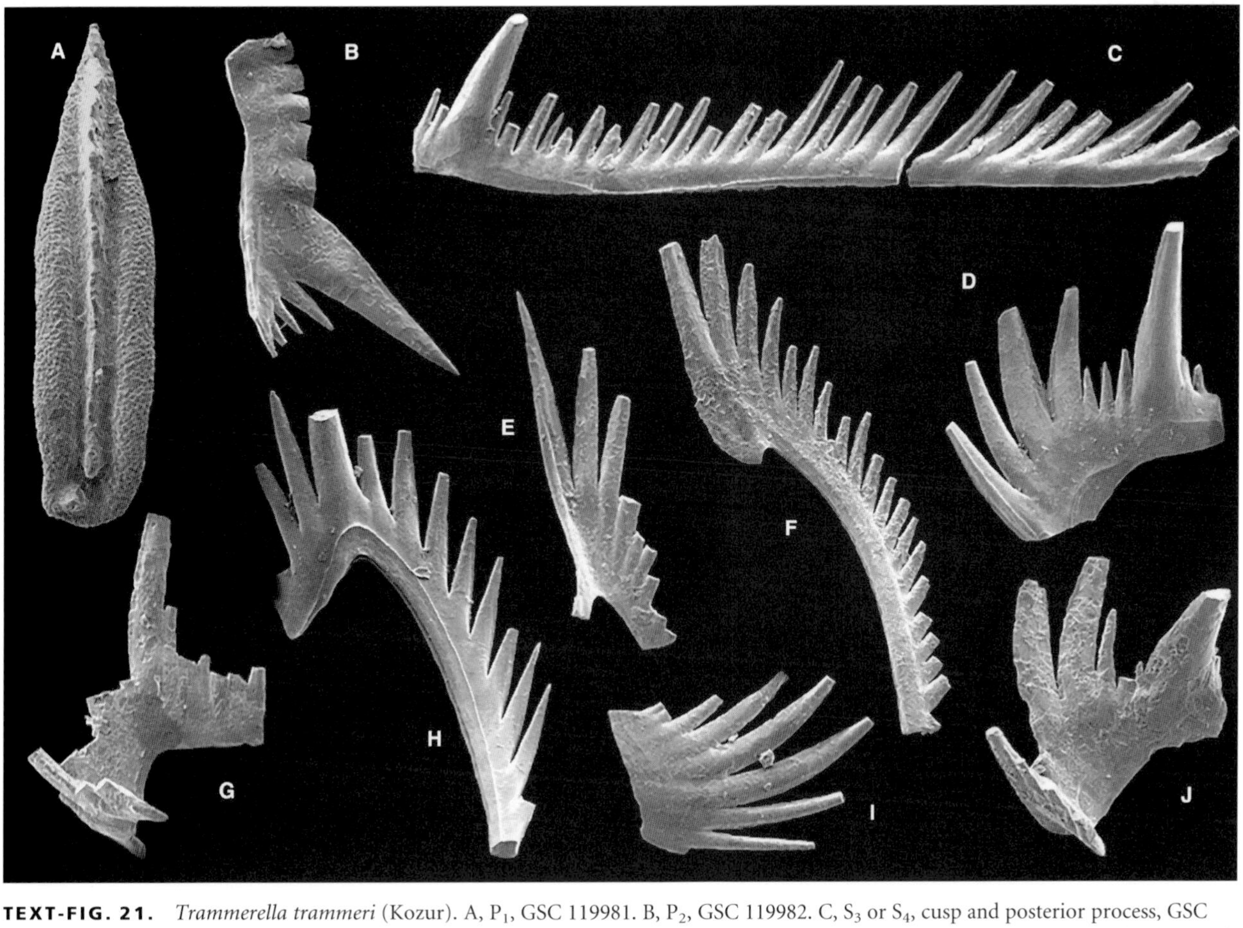

TEXT-FIG. 21. *Trammerella trammeri* (Kozur). A, P_1, GSC 119981. B, P_2, GSC 119982. C, S_3 or S_4, cusp and posterior process, GSC 119983. D, S_4, GSC 119984. E, S_1, GSC 119985. F, S_2, GSC 119986. G, S_0, lateral view, GSC 119987. H, M, GSC 119988. I, S_1, distal part of long process, GSC 119989. J, S_3, GSC 119990. All sample 6-5-Hun (GSC loc. C-301158), Felsôörs, Hungary. Ladinian. All ×80.

high carina that becomes progressively submerged with platform growth. P_2 element is relatively unarched, has a large cusp, and a relatively short, inwardly flexed posterior process bearing small denticles. S_3 elements have a short inner process projecting perpendicularly from the anterior end of the anterior process, whereas in the S_4 elements the anterior end of the process is variably inturned or may be costate; the anterior process of S_3–S_4 elements is characteristically large (high) with a strongly developed crest of long recurved denticles. M element is digyrate. S_2 elements are weakly digyrate with a single well-developed denticulate process. Modified alate S_0 element has multiple denticles between the cusp and the antero-lateral processes.

Remarks. The holotype is an early growth stage P_1 element that has not developed the diagnostic characteristics seen in larger specimens wherein the carina lies below the level of the platform margins. The genus is introduced primarily for the species *trammeri*, which is important in Middle Triassic Tethyan faunas. The S_2 element is close to that of the Neogondolellinae, which are easily separated on the basis of the branching of antero-lateral processes in both S_0 and S_3 elements.

Subfamily Uncertain

Genus ICRIOSPATHODUS Krahl, Kauffmann, Kozur, Richter, Foerster and Heinritzi, 1983

Text-figure 22

1987 *Spathoicriodus* Budurov, Sudar, and Gupta, pp. 175–176.

Type species and holotype. Neospathodus collinsoni Solien, 1979, pp. 302–303, pl. 3, fig. 13.

Type locality and stratum. Unit D, Thaynes Formation, near Salt Lake City, Utah, USA.

Original diagnosis. Based on a segminiscaphate P_1 element with two rows of nodes on the upper side, posterior nodes that may coalesce into incipient processes, beneath which there is a widely expanded basal cavity (Krahl *et al.* 1983).

Multielement diagnosis. Apparatus consists of 15 elements. The double-noded, 'icriode' P_1 element is distinctive. Moderately arched angulate P_2 elements have relatively high processes, the anterior one of which is longer; the cusp is slightly larger than the subequal and closely spaced anterior denticles, but much bigger than those of the posterior process. The M and S_2 elements are both digyrate but differ in the flexture of the two

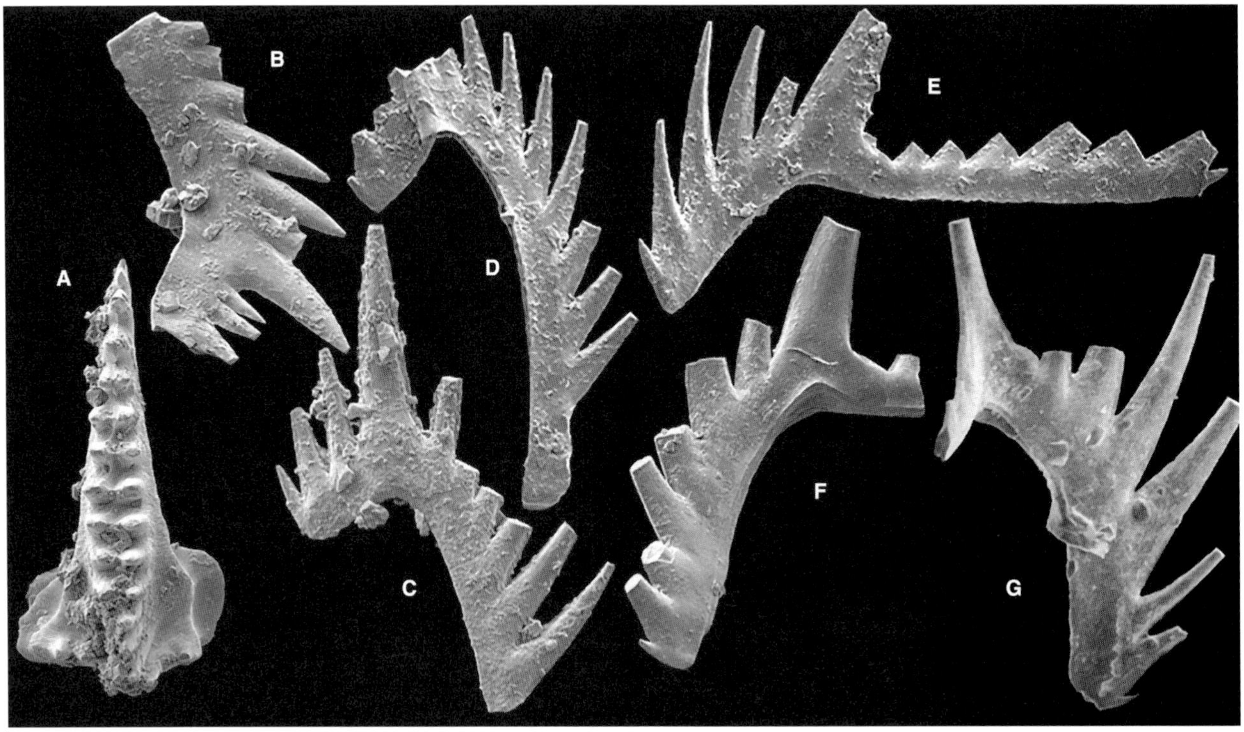

TEXT-FIG. 22. *Icriospathodus collinsoni* (Solien). A, P_1, GSC 120045. B, P_2, GSC 120046. C, S_2, GSC 120047. D, M, GSC 120048. E, S_4, GSC 120049. F, S_3, GSC 120050. G, S_1, GSC 120051. All Thaynes Formation, Idaho, USA. A–C, sample W13 (GSC loc. C-301226), Hot Springs; D–G, sample W11 (GSC loc. C-301224), Paris Canyon. Columbites beds, Spathian. All ×80.

unequal denticulated processes, which in both cases are well developed. The long anterior process of the S_1 element has distinctive symmetrical, crest-like denticulation with denticles increasing and then decreasing in height towards the cusp. Bipennate S_3 and S_4 elements differ in the relative length of the down-arched, non-bifid anterior processes. Nature of the S_0 element uncertain.

Remarks. The S_2 element has two distinct processes and hence resembles the homologous element in the Gladigondolellinae. The relatively discrete denticulation is also reminiscent of that subfamily, but there is no bifurcation of the S_3 element. The denticulated 'crest' of the larger process in the S_1 element is seen in other gondolelloids, but in other genera it occupies a distal position relative to the cusp.

Genus SCYTHOGONDOLELLA Kozur, 1989
Text-figure 23

Type species and holotype. Gondolella milleri Müller, 1956, p. 823, pl. 95, figs 4–6.

Type locality and stratum. Smithian ammonoid bed, Crittenden Springs, Elko Co., Nevada.

Original diagnosis. This genus was introduced for a P_1 element characterized by broad segminiplanate morphology with strong marginal nodes and denticles, and a variable free blade-carina-cusp configuration (see Kozur 1989, pp. 414, 429).

Multielement diagnosis. 15-element apparatus with a relatively short segminiplanate P_1 element that may be smooth to strongly ornamented. The two processes of the P_2 element make a distinct angle: the anterior process is half the length of the posterior one and commonly has an incipient platform flange developed. M, S_1 and S_2 elements digyrate with relatively little flexture in the S_2, which has a short but distinct second process bearing four denticles. Bipennate S_3 and S_4 elements have moderate but variable arching: the S_3 has a bifurcated anterior process branching from the third denticle anterior of the cusp. Alate S_0 element has antero-lateral processes branching from the cusp.

Remarks. The ramiforms of *Scythogondolella* closely resemble those of *Cratognathodus* but differ in the stronger downturning of the short anterior process in the S_2 element, which is also very similar to that of *Icriospathodus*. Another feature shared by *Cratognathodus* is the increase in size of posterior denticles on the posterior process of S_3–S_4 elements. However, the P_2 element differs markedly from many other gondolelloids in the relative length of the anter-

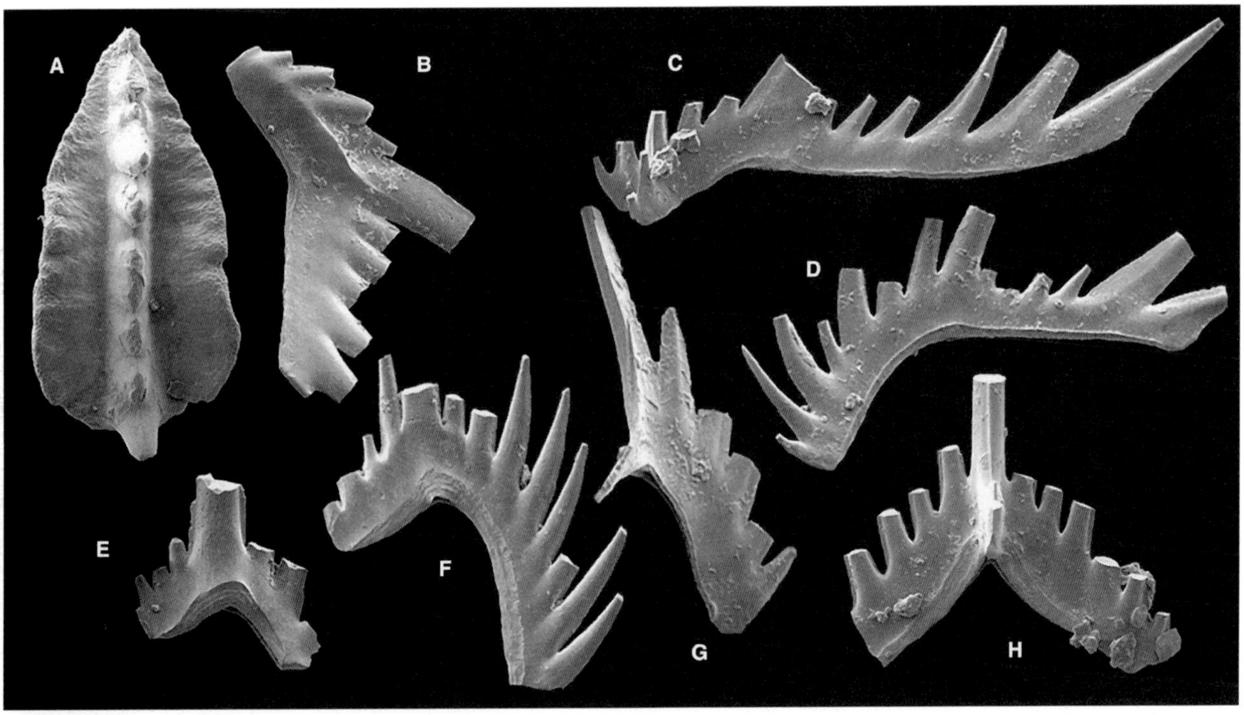

TEXT-FIG. 23. *Scythogondolella mosheri* (Kozur and Mostler). A, P_1, GSC 120030. B, P_2, GSC 120031. C, S_3, GSC 120032. D, S_4, GSC 120033. E, M, GSC 120034. F, S_2, GSC 120035. G, S_1, GSC 120036. H, S_0, posterior view, GSC 120037. A–B, E–F, H, sample 213c (GSC loc. O-99548), Toad Formation, Toad River, north-east British Columbia; C–D, G, sample W8 (GSC loc. C-301221), Georgetown, Idaho, USA. Tardus Zone, Smithian. All ×80.

ior process. Kozur (1989, p. 429) only tentatively included '*Gondolella*' *mosheri* Kozur and Mostler, 1976 in this genus because of its smooth platform, but this species has the same apparatus as *S. milleri* so I have no hesitation in including it here.

Genus WAPITIODUS gen. nov.
Text-figure 24

Derivation of name. Combination of the type locality at Wapiti Lake, and derivation from the Greek root for tooth.

Type species and holotype. Wapitiodus robustus sp. nov. (see below).

Multielement diagnosis. A 15-element apparatus in which the P_1 element is segminate in early growth stages but with growth it develops narrow, thick platform ledges that lack microreticulae. The other elements are characterized by discrete denticles. The S_3 element has a well-developed secondary anterior process branching from the second denticle in front of the large cusp. The S_2 elements are extensiform digyrate with two well-developed antero-lateral processes and a posteriorly extended basal cavity. Other components are angulate P_2, breviform digyrate M and S_1, bipennate S_4, and alate S_0 elements.

Remarks. The P_1 element and the S_3 elements are distinctive. Other elements, particularly the P_2 and S_2 elements,

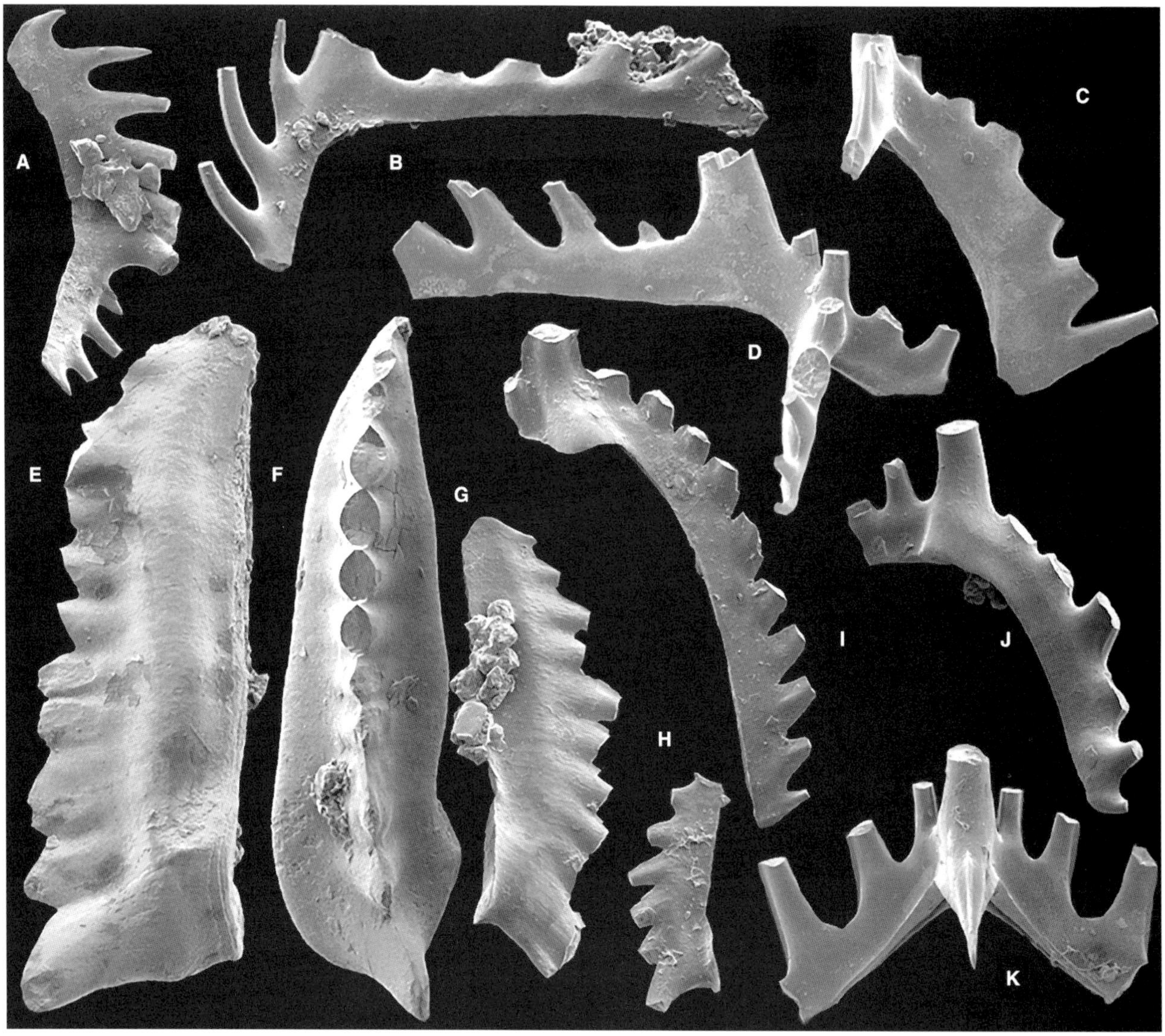

TEXT-FIG. 24. *Wapitiodus robustus* sp. nov. A, P_2, GSC 120038, holotype. B, S_4, GSC 120039. C, S_1, GSC 120040. D, S_3, GSC 120041. E, F, P_1, lateral and upper views, GSC 120042. G, P_1, GSC 120043. H, P_1, GSC 120044. I, M, GSC 120163. J, S_2, GSC 120164. K, S_0, posterior view, GSC 120165. All from Sulphur Mountain Formation, Wapiti Lake, British Columbia. A, J, sample ZC44-6 (GSC loc. C-416046); B, H–I, sample WapT35 (GSC loc. C-303562); C–G, K, sample ZC46-2 (GSC loc. C-416047). Smithian. All ×80.

closely resemble those of the Gladigondolellinae to which this group may be related. In general, *Wapitiodus* has similar denticulation to *Gladigondolella* and *Cratognathodus* but the M and S_3 elements are significantly different. Other related genera are presently combined as *Borinella* Budurov and Sudar, 1994. This name replaced *Kozurella* Budurov and Sudar, 1993 (= *Pseudogondolella sensu* Kozur, 1988, p. 244 = *Chengyuania* Kozur, 1994, pp. 529–530), the type species of which is *Gondolella nepalensis* Kozur and Mostler, 1976. However, Budurov and Sudar (1994) designated a different type species for *Borinella* (*Neogondolella buurensis* Dagys). Budurov and Sudar (1994) also assigned *G. nepalensis* to a new genus, *Kashmirella* (type species, *K. alberti*), but in doing so, they implicitly synonymized *Kashmirella* and *Kozurella*, with the latter as senior synonym. *Borinella* is thus the valid name for both *Kashmirella* and *Kozurella*, including the species *K. alberti*, *K. nepalensis* and *N. buurensis*. This was not the intent of Budurov and Sudar (1994), but separation of these taxa should be based on their apparatuses, which are unknown in detail. Limited material of '*Neogondolella*' *buurensis* available for study resembles elements of *Wapitiodus*.

Wapitiodus robustus sp. nov.
Text-figure 24

Derivation of name. Referring to the robust nature of the holotype.

Holotype. GSC 120038, from GSC loc. C-416047; Text-figure 24A.

Type locality and stratum. Sulphur Mountain Formation, AD Ridge, Wapiti Lake, British Columbia, Canada.

Diagnosis. As for genus.

Description. The segminate to segminiplanate P_1 element has 6–12 confluent, inclined denticles and a terminal cusp that projects posteriorly beyond the subcircular basal cup. In small, segminate specimens the denticles are more discrete but during growth a narrow platform flange appears, then thickens, and the denticles become increasingly fused. In large specimens, like the holotype, the platform ledges are thick and lack microreticulae. Other elements of *Wapitiodus* have discrete denticles. The angulate P_2 element is arched and posteriorly inturned with a medial, prominent cusp and 4–5 slender denticles on subequal processes. The digyrate M element has one long, downward-directed antero-lateral process and a second short one flexed forward; its basal cavity projects posteriorly. The depth of the larger of the two processes in the breviform digyrate S_1 elements increases distally, away from the cusp. The extensiform S_2 element has two well-developed antero-lateral processes and a posteriorly extended basal cavity. The bipennate S_3 element has a well-developed secondary anterior process branching from the second denticle in front of the large cusp. The bipennate S_4 element has an inturned and downturned anterior process about half the length of the posterior one. The two downturned anterior processes of the alate S_0 element form an obtuse angle when viewed from the posterior, where a third process is less well developed.

Acknowledgements. Many geologists have contributed well-dated carbonate samples that have yielded the conodont collections used in this work. Notable are Tim Tozer (Arctic, British Columbia, Oman); Hugo Bucher (Nevada); Wolfgang Weischat and Beth Carter (Idaho), Eugen Gradinaru (Romania) and Dan Lehrmann (China). Hillary Taylor is thanked for her help in microscopy, text-figure compilation and manuscript preparation. Peter Krauss was responsible for much of the laboratory preparation. I thank Mark Purnell, Bob Nicoll and Walt Sweet for their careful reviews and thoughtful comments.

REFERENCES

BAGNOLI, G., PERRI, M. C. and GANDIN, A. 1985. Ladinian conodont apparatuses from northwestern Sardinia, Italy. *Bollettino della Società Paleontologica Italiana*, **23**, 311–323.

BENDER, H. 1970. Zur gliederung der Mediterranen Trias II. Die Conodontenchronologie der Mediterranen Trias. *Annales Geologiques des Pays Helleniques*, **19**, 465–540.

—— and STOPPEL, D. 1965. Perm-Conodonten. *Geologisches Jahrbuch*, **82**, 331–364.

BOOGAARD, M. VAN DEN 1966. Post-Carboniferous conodonts from south-eastern Spain. *Proceedings of the Koninklijke Nederlandse Akademie Van Wetenschappen, Series B*, **69** (5), 1–19.

BUDUROV, K. 1972. *Ancyrogondolella triangularis* gen. et sp. n. (Conodonta). *Mitteilungen der Gesellschaft der Geologie- und Bergbaustudenten Österreich*, **21**, 853–860.

—— 1973. *Carinella* n. gen. und Revision der Gattung *Gladigondolella* (Conodonta). *Comptes Rendus de l'Académie Bulgare des Sciences*, **26**, 799–802.

—— BURYI, G. I. and SUDAR, M. N. 1988. *Smithodus* n. gen. (Conodonta) from the Smithian Stage of the Lower Triassic. *Mitteilungen der Gesellschaft der Geologie- und Bergbaustudenten, Österreich*, **34/35**, 295–299.

—— and SUDAR, M. N. 1989. New conodont taxa from the Middle Triassic. 250–254. *In*: *Geology of Himalayas–palaeontology, stratigraphy and structure. Contributions to Himalayan Geology*, **4**. Hindustan Publishing Corporation, India.

—— —— 1993. *Kozurella* gen. n. (Conodonta) from the Olenekian (Early Triassic). *Geologica Balcanica*, **23**, 24.

—— —— 1994. *Borinella* Budurov and Sudar, *nomen novum* for the Triassic conodont genus *Kozurella* Budurov and Sudar, 1993. *Geologica Balcanica*, **24** (3), 30.

—— —— and GUPTA, V. J. 1987. *Spathoicriodus*, a new Early Triassic conodont genera. *Bulletin of the Indian Geological Association*, **20**, 175–176.

—— —— —— 1988. *Kashmirella*, a new Early Triassic conodont genus. *Bulletin of the Indian Geological Association*, **21**, 107–112.

BURYI, G. 1979. *Lower Triassic conodonts of southern Primorye.* Institut Geologii i Geofiziki, Sibirskoye Otdeleniye, Akademiya Nauk SSSR, Moskva. [In Russian].

CLARK, D. L., SWEET, W. C., BERGSTRÖM, S. M., KLAPPER, G., AUSTIN, R. L., RHODES, F. H. T., MÜLLER, K. J., ZIEGLER, W., LINDSTRÖM, M., MILLER, J. F. and HARRIS, A. G. 1981. *Treatise on invertebrate paleontology. Part W Miscellanea, Supplement 2, Conodonta.* Geological Society of America, Boulder, and University of Kansas Press, Lawrence.

DIEBEL, K. 1956. Conodonten in der Oberkreide von Kamerun. *Geologie*, **5**, 424–450.

DZIK, J. 1976. Remarks on the evolution of Ordovician conodonts. *Acta Palaeontologica Polonica*, **21**, 395–455.

HAYASHI, S. 1968. The Permian conodonts in chert of the Adoyama Formation, Ashio Mountains, central Japan. *Earth Science*, **22**, 63–77.

HIRSCH, F. 1994. Triassic conodonts as ecological and eustatic sensors. 949–959. In EMBRY, A. F., BEAUCHAMP, B. and GLASS, D. J. (eds). *Pangea: global environments and resources.* Memoir of the Canadian Society of Petroleum Geologists, **17**, 982 pp.

HIRSCHMANN, C. 1959. Über Conodonten aus dem oberen Muschelkalk des Thüringer Beckens. *Freiberger Forschungshefte*, **76**, 35–86.

HUCKRIEDE, R. 1958. Die Conodonten der Mediterranen Trias und ihr stratigraphischer Wert. *Paläontologische Zeitschrift*, **32**, 141–175.

KOIKE, T. 1996. Skeletal apparatuses of Triassic conodonts of *Cornudina*. 113–120. *In* NODA, H. and SASHIDA, K. (eds). *Professor Hisayoshi Igo Commemorative Volume. Geology and Paleontology of Japan and Southeast Asia.* Gakujutu Tosho Publications, Tokyo.

—— 1999. Apparatus of a Triassic conodont species *Cratognathodus multihamatus* (Huckriede). *Paleontological Research*, **3**, 234–248.

KOZUR, H. 1972. Die Conodontengattung *Metapolygnathus* Hayashi 1968 und ihr stratigraphischer Wert. *Geologisch-Paläontologische Mitteilungen Innsbruck*, **2**, 1–37.

—— 1988. Division of the gondolellid platform conodonts. Abstract. *Proceedings of the Fifth European Conodont Symposium, Contributions 1.* Courier Forchungsinstitut Senckenberg, **102**, 244–245.

—— 1989. The taxonomy of the gondolellid conodonts in the Permian and Triassic. *Courier Forschungsinstitut Senckenberg*, **117**, 409–469.

—— 1994. *Chengyuania*, a new name for *Pseudogondolella* Kozur 1988 (Conodonta) [non *Pseudogondolella* Yang 1984 (hybodont fish teeth)]. *Paläontologische Zeitschrift*, **68**, 529–530.

—— and MOCK, R. 1972. Neue Conodonten aus der Trias der Slowakei und ihre stratigraphische Bedeutung. *Geologische und Paläontologische Mitteilungen, Innsbruck*, **2** (4), 1–20.

—— and MOSTLER, H. 1970. Neue Conodonten aus der Trias. *Berichte des Naturwissenschaftlich-Medizinischen Vereins in Innsbruck*, **58**, 429–464.

—— —— 1971. Probleme der Conodontenforschung in der Trias. *Geologische und Paläontologische Mitteilungen, Innsbruck*, **1** (4), 1–19.

—— —— 1976. Neue Conodonten aus dem Jungpaläozoikum und der Trias. *Geologische und Paläontologische Mitteilungen, Innsbruck*, **6**, 1–33.

—— —— and KRAINER, K. 1998. *Sweetospathodus* n. gen. and *Triassospathodus* n. gen., two important Lower Triassic conodont genera. *Geologia Croatica*, **51**, 1–5.

KRAHL, J., KAUFFMANN, G., KOZUR, H., RICHTER, D., FOERSTER, O. and HEINRITZI, R. 1983. Neue Daten zur Biostratigraphie und zur tektonischen Lagerung der Phyllit-Gruppe und der Trypali-Gruppe auf der Insel Kreta (Greichenland). *Geologische Rundschau*, **72**, 1147–1166.

LINDSTRÖM, M. 1970. A suprageneric taxonomy of the conodonts. *Lethaia*, **3**, 427–445.

MARCH, M., BUDUROV, K., HIRSCH, F. and MARQUEZ-ALIAGA, A. 1988. *Sephardiella* nov. gen. (Conodonta), emendation of *Carinella* (Budurov, 1973), Ladinian (Middle Triassic). Abstract. *Proceedings of the Fifth European Conodont Symposium, Contributions 1.* Courier Forchungsinstitut Senckenberg, **102**, 247.

MIETTO, P. 1982. A Ladinian conodont-cluster of *Metapolygnathus mungoensis* (Diebel) from Trento area (NE Italy). *Neues Jahrbuch für Geologie und Paläontologie, Monatshefte*, **1982**, 600–606.

MOSHER, L. C. 1968. Triassic conodonts from western North America and Europe and their correlation. *Journal of Paleontology*, **42**, 895–946.

MÜLLER, K. J. 1956. Triassic conodonts from Nevada. *Journal of Paleontology*, **30**, 818–830.

—— 1962. Zur systematischen Einteilung der Conodontophoridae. *Paläontologische Zeitschrift*, **36**, 109–117.

NOGAMI, Y. 1968. Trias-Conodonten von Timor, Malaysien und Japan (Palaeontological study of Portuguese Timor, 5). *Memoirs of the Faculty of Science, Kyoto University, Series of Geology and Mineralogy*, **34**, 115–136.

ORCHARD, M. J. 1983. *Epigondolella* populations and their phylogeny and zonation in the Norian (Upper Triassic). *Fossils and Strata*, **15**, 177–192.

—— 1991*a*. Late Triassic conodont biochronology and biostratigraphy of the Kunga Group, Queen Charlotte Islands, British Columbia. 173–193. *In* WOODSWORTH, G. W. (ed.). *Evolution and hydrocarbon potential of the Queen Charlotte Basin, British Columbia.* Geological Survey of Canada, Paper, **1990–10**, 569 pp.

—— 1991*b*. Upper Triassic conodont biochronology and new index species from the Canadian Cordillera. 299–335. *In* ORCHARD, M. J. and McCRACKEN, A. D. (eds). *Ordovician to Triassic conodont paleontology of the Canadian Cordillera.* Geological Survey of Canada, Bulletin, **417**, 335 pp.

—— 1995. Taxonomy and correlation of Lower Triassic (Spathian) segminate conodonts from Oman and revision of some species of *Neospathodus*. *Journal of Paleontology*, **69**, 110–122.

—— and RIEBER, H. 1999. Multielement *Neogondolella* (Conodonta, Upper Permian–Middle Triassic). 475–488. *In* SERPAGLI, E. (ed.). *Studies on conodonts. Proceedings of the*

Seventh European Conodont Symposium, Bologna-Modena, Italy, June 1998. Bollettino della Società Palaeontologica Italiana, **37**, 557 pp.

PAULL, R. K. 1983. Definition and stratigraphic significance of the Lower Triassic (Smithian) conodont *Gladigondolella meeki* n. sp. in western United States. *Journal of Paleontology*, **57**, 188–192.

PURNELL, M. A., DONOGHUE, P. C. J. and ALDRIDGE, R. J. 2000. Orientation and anatomical notation in conodonts. *Journal of Paleontology*, **74**, 113–122.

RAMOVS, A. 1977. Skelettapparat von *Pseudofurnishius murcianus* (Conodontophorida) im Mitteltrias (NW Jugoslawien). *Neues Jahrbuch für Geologie Paläontologie, Abhandlungen*, **153**, 361–399.

—— 1978. Mitteltriassische Conodonten-clusters in Slowenien, NW Jugoslawien. *Paläontologische Zeitschrift*, **52**, 129–137.

RIEBER, H. 1980. Ein Conodonten-cluster aus der Grenzbitumenzone (Mittlere Trias) des Monte San Giorgio (Kt. Tessin/Schweiz). *Annalen Naturhistorisches Museum, Wien*, **83**, 265–274.

SOLIEN, M. A. 1979. Conodont biostratigraphy of the Lower Triassic Thaynes Formation, Utah. *Journal of Paleontology*, **53**, 276–306.

SWEET, W. C. 1970. Uppermost Permian and Lower Triassic conodonts of the Salt Range and Trans-Indus Ranges, West Pakistan. 207–275. *In* KUMMELL, B. and TEICHERT, C. (eds). *Stratigraphic boundary problems, Permian and Triassic of West Pakistan.* University of Kansas Press, Department of Geology, Special Publications, **4**, 474 pp.

—— 1988. *The conodonta: morphology, taxonomy, paleoecology, and evolutionary history of a long-extinct animal phylum.* Oxford Monographs on Geology and Geophysics, **10**. Clarendon Press, Oxford, 212 pp.

TATGE, U. 1956. Conodonten aus dem germanischen Muschelkalk. *Paläontologische Zeitschrift*, **30**, 108–127, 129–147.

VON BITTER, P. H. and MERRILL G. K. 1998. Apparatus composition and structure of the Pennsylvanian conodont genus *Gondolella* based on assemblages from the Desmoinesian of northwestern Illinois, U.S.A. *Journal of Paleontology*, **72**, 112–132.

ZHANG SHUNXIN and YANG ZUNYI 1991. On multielement taxonomy of the Early Triassic conodonts. *Stratigraphy and Paleontology of China*, **1**, 17–47.

[Special Papers in Palaeontology, 73, 2005, pp. 103–116]

SILURIAN CONODONT BIOSTRATIGRAPHY AND PALAEOBIOLOGY IN STRATIGRAPHIC SEQUENCES

by JAMES E. BARRICK* *and* PEEP MÄNNIK†

*Department of Geosciences, Texas Tech University, Lubbock, TX 79409-1053, USA; e-mail: Jim.Barrick@ttu.edu
†Institute of Geology, Tallinn University of Technology, 10143 Tallinn, Estonia; e-mail: mannik@gi.ee

Abstract: Improvements leading to high-resolution correlation using Silurian conodonts must be made within the framework of stratigraphic sequences. Species ranges in shelf strata reflect both biological processes and preservation bias imposed by predictable patterns of deposition and erosion in response to eustatic sea-level fluctuations. Ranges used to erect biostratigraphic zones and on which evolutionary lineages and bioevents are based must be interpreted in the light of this bias to exclude occurrences that are explicable by sequence architecture. Stratigraphical disorder arising from advection by burrowers may obscure biological and geochemical events in condensed marine sections and decrease their time resolution. Chronostratigraphic boundaries placed in offshore, condensed sections that formed during sea-level highstands will be difficult to correlate accurately into shelf successions because of sequence architecture. Possible global climatic and oceanic events impose another level of discontinuity on the Silurian conodont record. Data supporting these events must be evaluated to remove the bias of sequence architecture before climatic, oceanic and palaeobiological models are devised. Stratigraphic sequences are the record of synchronous global events and can contribute as much to high-resolution correlation as biostratigraphy and short-lived geochemical and biotic events.

Key words: Conodonts, Silurian, stratigraphic sequences, GSSP, biostratigraphy.

THE evolutionary history, palaeoecology and palaeobiogeography of Silurian conodonts are poorly understood. The standard conodont zonation for the Silurian remains very much a work in progress, which has impeded our ability to arrange the conodont faunas in the detailed chronological framework required to reach the next levels of palaeobiological interpretation. Although we have recovered enough of the overall conodont succession to permit reasonably good correlations to be made at the stage level, we lack the biostratigraphical detail required for high-resolution stratigraphy. The global 'Standard Conodont Zonation' (Nowlan 1995) is a composite of zones based largely on successive first occurrences as interpreted from the stratigraphical record, determined by compositing data from different parts of the world. Some zonal indices with very short ranges are so rare that large sample sizes are required to obtain them (> 10 kg; Jeppsson 1997*a*). Longer ranging species occur sporadically through their ranges and first and last occurrences (FADs and LADs) are unreliable indicators of time. Lineages of some potentially biostratigraphically useful taxa are characterized by numerous gaps that omit evolutionary events, and lower confidence in lineage reconstruction.

Collation of the conodont zones with the boundaries on the standard Silurian Time Scale has also been problematic. Zonally diagnostic conodonts have not been obtained from all stratotype sections, many of which were selected only with regard to the graptolite succession (Aldridge and Schönlaub 1989). Interbedded graptolitic shales and conodont-bearing beds have not been discovered or studied for much of the Silurian, and correlation to stratotypes through the shelly faunas is even more difficult. Recent revisions of the zones for the Llandovery (Zhang and Barnes 2002*a*), the late Llandovery (Männik 1998), the late Llandovery–Wenlock (Jeppsson 1997*a*) and the Wenlock–Přídolí intervals (Corradini and Serpagli 1999) have improved the conodont zonation, but additional work must be completed.

The question is: What is the best strategy to employ in collecting and analysing new faunas and reinterpreting older collections in order for Silurian conodonts to become useful for high-resolution stratigraphy? The answer is not limited to Silurian conodonts, for it applies equally well to other Silurian faunal groups used for biostratigraphy, as well as to other intervals of geological time. This question might be framed in larger terms: What is the best approach to attain reliable high-resolution stratigraphy in the Palaeozoic? In this paper we restrict our discussion to Silurian conodonts, for this is the group we know best.

Are traditional approaches enough?

One possible solution to improving Silurian conodont biostratigraphy would be the discovery of rapidly evolving taxa that could serve as the basis for a more global biostratigraphical zonation. We would like to find taxa like *Palmatolepis*, *Siphonodella* or *Eoplacognathus*, forms in which rapid, obvious morphological features change through short increments of time. No Silurian genera have provided a long evolutionary succession of species that would allow the recognition of a series of zones based on successive FADs. Only a couple of short phylogenies are known, the best of which is the *Pterospathodus* succession in the Telychian (Männik 1998), but other potentially useful genera, such as *Kockelella* (Serpagli and Corradini 1999), have not provided us with material for a reliable detailed zonation.

For other intervals of time, the conodonts most useful for biostratigraphy are species that lived in offshore environments. During the Silurian, taxa characterized by apparatuses with coniform elements (e.g. *Dapsilodus*, *Decoriconus*, *Walliserodus* and *Belodella*), and to a lesser degree taxa such as *Aspelundia*, and some species of *Ozarkodina*, occupied the offshore position (Armstrong 1990; Aldridge and Jeppsson 1999; Zhang and Barnes 2002*b*). The common adherence to simple single character diagnoses for species of these genera will only continue the misconception that the species were too long ranging to be of interest (e.g. Zhang and Barnes 2002*a*). More intense study of these coniform-bearing taxa and taxa with carminate Pa elements, using features of the entire apparatus, may resolve more biostratigraphically useful species. However, the subtleties of morphological change in these taxa may preclude their routine use for zonation.

Without a long-lived, rapidly evolving series of morphologically complex species in offshore genera on which to base a zonation, we must critically evaluate the range data on a more empirical basis when attempting to composite ranges for zonations and evolutionary reconstructions. A serious problem, one that is rarely addressed explicitly, is whether the observed pattern is a result of biological changes (evolutionary and extinction events) or is an artefact of the structure of the stratigraphical record (MacLeod 1991; Brett 1995, 1998). Many species' FADs and LADs and other biotic events are 'stratal events', and few attain the supposed precision of interpreted 'phyletic events' (Johnson 1979). The combination of stratigraphical incompleteness, ecological incompleteness and biogeographical incompleteness, as discussed by Kemp (1999), diminishes the time value of observed biotic events, and hinders our attempts to interpret the biotic record directly from the observed stratigraphical record.

Completeness is a relative measure of the quality of the fossil record, and it is really the resolution, the size of the increments of time that are preserved in their entirety, that is significant (Kemp 1999). Estimates of the local 'completeness' of the palaeontological record using a database of Late Cambrian and Early Ordovician trilobites from the Arbuckle Group in southern Oklahoma (Stitt 1977) vary from 74 to 90 per cent at increments of 150,000 years, using a variety of statistical models and assumptions (Foote and Raup 1996; Foote 1997; Solow and Smith 1997). Schopf (1978) and Valentine (1989) had previously suggested that for well-skeletized taxa the completeness should be similar to the values (75–90%) derived from these models. Conodonts are a well-preserved group, even with the preservational limitations described in other papers in this volume. The conodont record preserved in the Arbuckle Group (Ethington and Dresbach 1990) should be at least as good as that of the co-occurring trilobite faunas. If this is the common situation, then conodont faunas should provide robust data for Palaeozoic biostratigraphy.

For well-preserved groups, the completeness of their fossil record largely reflects the loss of fossiliferous rock rather than the preservation of the skeletal remains (Schopf 1978; Valentine 1989; Foote and Raup 1996). The completeness of the fossil record of conodonts is largely a function of the size, spacing and number of unconformities in the stratigraphical succession. The incompleteness owing to the loss of fossiliferous rock, as well as the patterns of other stratal events that create ecological and biogeographical incompleteness are not random. They are predictable using the model of sequence stratigraphy (Brett 1995, 1998). Interpretations of stratigraphical architecture of marine strata show a multitude of possibilities for both stratigraphical and biotic continuity and discontinuity depending on eustatic events, subsidence rates, and rates of sediment production and supply. All of the possible combinations of these factors are incorporated within the basic model of stratigraphic sequences, which recognizes eustatic sea-level changes as the primary control on stratigraphical architecture (Nystuen 1998).

STRATIGRAPHIC SEQUENCES AND THE FOSSIL RECORD

The sequence stratigraphic model predicts the times and locations on the shelf to basin transect where sediment accumulation and preservation is most likely. This can be illustrated on a model chronostratigraphic chart, which displays graphically where during a eustatic rise and fall of sea level the stratigraphic record will be accumulating, and which parts of the cycle of sea-level rise and fall will be preserved (Text-fig. 1; Emery and Myers 1996). Different levels of sequences occur, but third-order sequences are the most commonly recognized sequences on a regional to global scale. Sloss (1996) used the data of Haq *et al.* (1987) to show that the vast majority of mid-Jurassic–Recent third-order sequences were 0·5–4·0 myr in duration, with modes at 1·0 and 1·5 myr. The durations of these third-order sequences are similar to those of many Palaeozoic conodont zones.

Nearshore, shallow shelf and cratonic successions are likely to be cut by unconformities of variable magnitudes (largely erosional), and only a small fraction of time of a sequence is preserved. Sloss (1996) has estimated that the proportion of time represented by the preserved sedimentary record on the craton is nearer 10 per cent than previous estimates of 50 per cent or more. In the offshore direction, stratigraphic completeness increases, as the magnitude and number of unconformities diminishes. Type 1 sequence boundaries are obvious regional unconformities, as when sea level falls below the edge of the shelf margin, erosion and incision of the shelf, as well as diagenetic alteration resulting from exposure, leaves unmistakable physical evidence. In Type 2 sequence boundaries, sea level does not fall below the shelf margin, and no erosional unconformity may form on the shelf margin. These boundaries must be located using regional stratigraphic studies that document a downward shift in coastal onlap. Each sequence comprises numerous parasequences, short-duration sedimentary cycles bounded by marine flooding surfaces at their bases and tops.

The chronostratigraphic chart (Text-fig. 1) illustrates the difficulty of obtaining complete representation of time across sequence boundaries, or within sequences. Note that the point designated as the offlap break (basically the shelf/slope break) is where deposition is essentially continuous, except for major falls in sea level (Type 1 sequence boundary). However, the offlap break varies in position across the shelf–basin transect, even shifting in position across Type 2 sequence boundaries. No one section or small area is likely to contain a continuous record of sedimentation through a single sequence, or any series of sequences.

Fossil ranges in stratigraphic sequences

Holland (1995, 1996, 2000) modelled the effects of sequence architecture on the distribution of species' FADs and LADs. His analysis goes beyond how unconformities affect ranges, as eustatic sea-level changes control the path of ecological shifts that marine faunas will track. Holland's overall conclusion is that the fossil record will appear to be highly episodic, even if the fossil group had a stable origination rate, extinction rate and ecological structure (Text-fig. 2). Events will tend to cluster at certain levels in the depositional sequence. FADs and LADs will cluster at sequence boundaries, at major flooding surfaces and in intervals of stratigraphical condensation. Biofacies and abundance shifts will occur at sequence

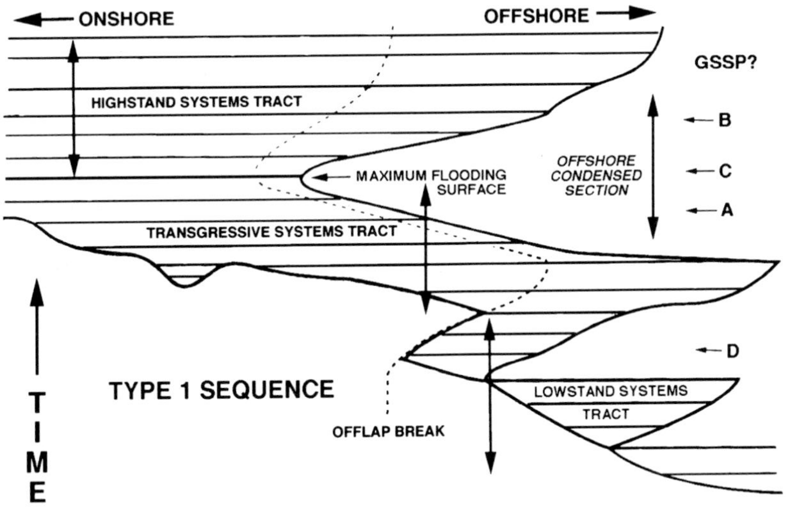

TEXT-FIG. 1. Chronostratigraphic representation of a Type 1 Stratigraphic Sequence. GSSP?, possible positions for placing Global Boundary Stratotype Section and Point: A, in an offshore condensed section below maximum flooding surface equivalent to transgressive systems tract on shelf; B, in an offshore condensed section above maximum flooding surface equivalent to highstand systems tract on shelf; C, in an offshore condensed section at a level coincident with the maximum flooding surface; D, in deposits of the lowstand system tract. See text for further discussion of GSSP placement. Modified from Emery and Myers (1996, fig. 5.2).

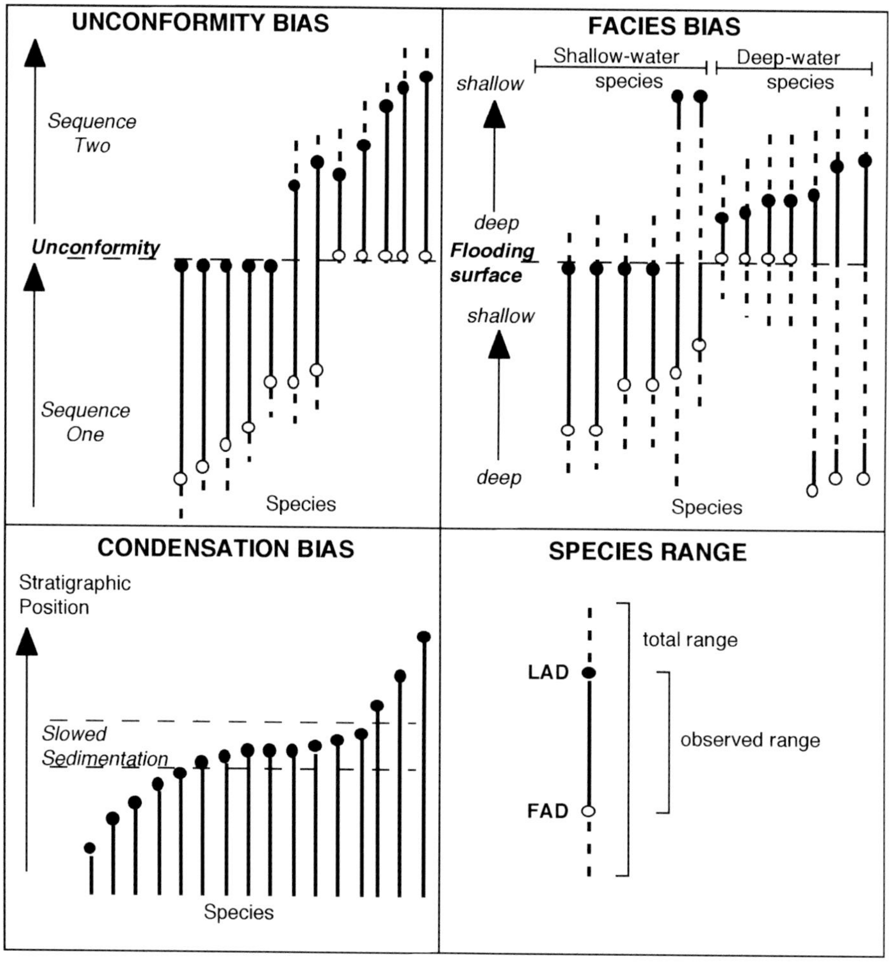

TEXT-FIG. 2. Biases recognized in sequence stratigraphic models of the fossil record. Unconformity bias causes first (FADs) and last occurrences (LADs) to cluster at unconformities marking sequences boundaries because a significant interval of time during which species originate and die is missing owning to non-deposition and erosion. Facies bias causes FADs and LADs to cluster at surfaces of abrupt facies changes, especially at flooding surfaces as shown here. Condensation bias causes FADs and LADs to appear clustered in intervals of reduced sedimentation rates. See text and Holland (1995, 2000) for further discussion of biases affecting ranges of fossil in stratigraphic sequences. Modified from Holland (2000, fig. 4).

boundaries and major flooding surfaces, especially with the shift to deeper water settings. Morphological changes, as used to delimit species in lineages, will appear episodic at the same levels. Clines will be broken up into steps and iterative evolution may be mimicked (Brett 1998). Holland (2000, fig. 6) showed that the median range offsets for LADs may exceed 1 myr across the shelf and shelf edge break, further distorting species ranges.

Sequence architecture will generate artificial LAD pulses that can be mistaken for extinction events (Holland 1996). When a data set of species with continuous extinctions through time was used, the sequence model consistently produced artificial LAD pulses at sequence boundaries. Species displayed a pulse in LADs in lowstand system tracts (LSTs). The flooding surfaces of transgressive system tracts (TSTs) were preceded by a pulse in LADs of shallow water stenotopes that became extinct before the return of shallow water biotopes in the highstand system tract (HST). These LAD pulses are artefacts of sequence architecture and cannot be distinguished from true extinction events. However, Holland (1996) pointed out that LAD pulses that occur during the HST and within the TST (not the maximum flooding surface: MFS) are likely to be biologically real extinctions, because artificial LAD pulses are not expected at these levels.

Third-order sequences may be subdivided into fourth- and fifth-order sequences, each of which has similar effects on species distributions. The rapid ecological shifts occurring at the flooding surfaces of parasequences may also create smaller and less apparent event horizons. Episodic control by parasequences will occur at a small time interval, on the order of tens of thousands of years (Holland 2000). The effects of smaller sequences and parasequences make resolving events over shorter intervals of time, such as attempts at high-resolution phenetics, interpretation of timing and patterns of speciation in lineages, and breaking down the components of palaeoecological, climatic and oceanic events more problematic (Brett 1998).

Condensed sections are not a solution

Conodont workers have emphasized the desirability of finding marine condensed sections, thin offshore sections that span long intervals of time. Klapper *et al.* (1994, p. 432) stated that 'It has been recognized by the Subcommission [on Devonian Stratigraphy] that, in general, Devonian sections in pelagic facies are more likely to be complete than those in the neritic facies.' This argument is supported by claims of good completeness and high resolution for near Recent deep-sea sediments (Anders *et al.* 1987). However, some caution must be shown in evaluating collections from condensed sections. Studies of deep-sea sediments show that resolution in slowly deposited sediments is problematic for two reasons: (1) advection by burrowing organisms and (2) erosional reworking.

Martin (1999) discussed mixing models developed for attenuation of geochemical signals and event horizons as a guide for advection of foraminiferal tests by burrowers in deep-sea sediments. In deep-sea cores, at rates of sedimentation up to 7 cm/1000 years, 3000-year events were severely attenuated by advection (Schiffelbein 1984). Microtektite layers in Pacific deep-sea cores were attenuated through a stratigraphical interval of 35–90 cm, or 33,000–320,000 years (60 cm/120,000 years average) (Glass 1969). Glass inferred from his studies of the microtektites that the last occurrences of events are more likely to be moved out of position than first occurrences. The mixing model of Berger and Heath (1968) showed that although the first occurrence of an event would be advected downward less than the thickness of the mixed layer, the last occurrence could be advected upward two to three times the thickness of the mixing zone. Boudreau (1997) found that the mean thickness of the mixing zone, depth of the zone of active bioturbation, is just less than 10 cm over a wide range of environments. Berger and Heath (1968) estimated that only 10^{-8} of material will be advected up through a sediment thickness of 1 m, but one would anticipate that significant advection could occur over a thickness of 30–40 cm (Martin 1993). This can create overlapping ranges of species occurring near each other in time, observations that have been used as evidence for 'phyletic events'. Based on similar concerns about advection by burrowers, Goldring (1995) believed that it is unwise to attach much biostratigraphical significance to sampling at less than 0·5-m intervals in Cretaceous chalks.

Advection strongly modifies the stratigraphical concentration of geochemical events, as well as the abundance of taxa through the species range and across event levels (Berger and Heath 1968; Hutson 1980; Martin 1993). Different deconvolution techniques have been proposed to locate better the position of geochemical events before advective processes spread them out stratigraphically (Martin 1999). Bard *et al.* (1987) applied a similar technique to the stratigraphical abundance of planktic foraminiferal species. An important observation of these models is that serious stratigraphical error may occur if the stratigraphic range of a species or event is similar in thickness to the mixing zone. This is most likely to affect the faunal record of short-lived oceanic events related to water-mass movements (Martin 1999).

Both Berger and Heath (1968) and Cutler and Flessa (1990) indicated that reworking of material by erosion is the dominant process mixing older material into younger pelagic sediments. Cutler and Flessa (1990) showed that unlike advection, physical reworking disorders sequences rapidly. Because periodic erosion or winnowing is one of the processes responsible for the thinness of condensed marine beds, reworking and time-averaging of conodonts may occur. Estimated intervals of time-averaging for 'biostratigraphically condensed assemblages' range from 100 kyr to over 10 myr. Sampling individual thin beds in condensed sections does not necessarily provide better resolution (Kidwell 1993). Although each individual bed may represent a single 'depositional event' of short duration, the bedding plane surfaces may bear assemblages that accumulated over significant intervals of time (Schindel 1980).

Detection of reworking is problematical, for the erosion and redeposition of fossils may not always affect the quality of preservation. Some authors consider preservational state to be a sensitive indicator of reworking (Fürsich and Aberhan 1990; Jeppsson 1998), but Martin (1999) urged a more cautious approach. An important factor in discoloration and deterioration of fossil debris is the length of time the debris spends exposed on the substrate surface (Flessa *et al.* 1993), which may or may not indicate reworking. One only needs to read the numerous papers discussing the details of species' FADs and LADs across the Cretaceous/Tertiary boundary to see that consensus about the presence or absence of reworked microfossils is difficult to attain (Ginsberg 1996).

The problems of advection and reworking in marine condensed intervals suggest that better resolution and fidelity of the conodont record, as well as that of other faunal and geochemical events, will occur where rates of rock accumulation are higher. Advection is an inefficient process for disrupting stratigraphical order on the scale of hundreds of thousands to millions of years in shelf environments (Cutler and Flessa 1990). Reworking is also less likely in shelf settings owing to the dominance of depositional events over minor erosional/winnowing events. The common-sense approach to overcome the effects of time averaging is to take samples so widely spaced that time averaging is not a factor (Kidwell and Flessa 1995). In the thicker shelf sections, a wider spacing of samples can avoid the effects of advection and reworking, but still collect the fauna at short time intervals. The Late Cambrian–Early Ordovician example discussed earlier as having excellent completeness and resolution was from a shallow-water shelf succession deposited in a rapidly subsiding basin (Foote and Raup 1996). Schindel (1980) provided estimates of the approximate scale of microstratigraphic sampling that would be required to resolve different short-lived biological processes in different depositional settings.

Sequence biostratigraphy and chronostratigraphy

The appearance of 'sequence biostratigraphy' (Loutit *et al.* 1991; Posamentier and Goodman 1992) in studies of Tertiary strata indicates how sequence architecture can be elevated to the primary role in subdividing the stratigraphical record, and biostratigraphy reduced to the secondary role of characterizing the stratigraphic sequences. Numerous recent studies show that biotic units (faunal units or zones) coincide with the boundaries of stratigraphic sequences in Palaeozoic successions: Middle and Upper Ordovician shelly faunas in eastern United States (Patzkowsky and Holland 1996); Upper Ordovician and Lower Silurian shelly faunas in western United States (Harris and Sheehan 1997); Middle and Upper Devonian shelly faunas in Midcontinent North America (Day 1996); Pennsylvanian fusulinids in North America (Ross and Ross 1995); and Pennsylvanian conodonts in Midcontinent United States (Heckel 1994). The correspondence between conodont ranges and Pennsylvanian cyclothems, which represent a mixture of fourth- and fifth-order sequences, is so strong that Ritter *et al.* (2002) used the phrase 'conodont sequence biostratigraphy' to describe how conodonts could be used to correlate sequences from the Midcontinent region into western USA.

Lane *et al.* (2000) proposed the term 'biothem' as the biostratigraphically based unit of sequence stratigraphy and discussed its practical application in hydrocarbon exploration. A biothem is essentially the biostratigraphical equivalent of a stratigraphic sequence, for like a sequence, it is bounded above and below by unconformities. In the absence of the stratigraphical and sedimentological data needed to resolve a stratigraphic sequence directly, the fossil content, if previously established for the sequence, can be used to identify the sequence and map its thickness and geographical distribution. Lane *et al.* (2000) considered biothems to be an alternative to traditional biostratigraphical zones, an alternative that fits in more directly with the sequence stratigraphic record.

With a few exceptions, such as the Devonian transgressive–regressive cycles (Johnson *et al.* 1985), Mississippian biothems (Lane *et al.* 2000) and Pennsylvanian cyclothems (Heckel 1994; Ritter *et al.* 2002), conodont faunas and zonations have not been placed in a stratigraphic sequence context. One reason for this is the emphasis on stratigraphical continuity among working groups attempting to place Global Boundary Stratotype Sections and Points (GSSPs) for chronostratigraphical boundaries. The international stratigraphical guides (Hedberg 1976; Salvador 1994) were developed before sequence stratigraphy was fully realized and the degree of stratigraphic discontinuity was not appreciated. The guides emphasize the importance of stratigraphical continuity in selecting GSSPs, and the avoidance of sections characterized by unconformities and major vertical lithofacies or biofacies changes. These criteria forced conodont workers to focus their work on condensed marine sections and to propose such sections as sites for GSSPs, as seen for Devonian series and stages. The result has been that many faunal ranges, faunal zones and GSSPs have been published without any reference to the stratigraphic sequences that characterize most marine strata and control the ranges of fossils in shelf successions. It is the same for the Silurian, where graptolites have served to characterize and correlate chronostratigraphical boundaries.

Integration of faunal ranges and zones developed from condensed sections, as well as GSSPs, into the succession of stratigraphic sequences is problematical. Most intervals of offshore marine condensation occur during the interval of time represented by the transgressive and the highstand systems tracts (Text-fig. 1). If a GSSP has been placed in such a condensed interval, then the point in time it represents will lie within a stratigraphic sequence, but an accurate high-resolution correlation may not be possible using an FAD. The FADs of species appearing offshore during the formation of the TST (Text-fig. 1, GSSP? = A) are often concentrated above the TST at the MFS (Holland 2000). The FADs of species appearing offshore during the formation of the HST (Text-fig. 1, GSSP? = B) may not occur in the shelf equivalent for ecological reasons, and the FADs on the shelf may be delayed by one sequence or more. Only a GSSP that lies at a time position equivalent to the MFS can be reliably correlated by an FAD from the condensed interval onto the shelf (Text-fig. 1, GSSP? = C). In each of these examples, the lower part of the shelf sequence would fall into the older chronostratigraphical unit, and the upper part into the younger chronostratigraphical unit.

For GSSPs that have not yet been defined, it might be better to place them between sequences (Boardman *et al.*

1990) and in the LST (Text-fig. 1, GSSP? = D). LSTs possess reasonable internal continuity, interrupted only by parasequences, but because they are thicker sections they may avoid the problems of advection and reworking that might time-average closely spaced FADs and LADs. However, care must be taken to avoid faunas that may be eroded from older shelf strata and redeposited in the lowstand deposits. If the GSSP is placed in an LST, then the sequence unconformity on the shelf will approximate the chronostratigraphical boundary, with the underlying sequence belonging entirely to the older chronostratigraphical unit, and the overlying one to the younger unit. The natural packaging of species ranges by the sequences, then, permits us to use 'sequence biostratigraphy' without requiring us to attempt an uncertain biostratigraphical subdivision of a sequence in order to locate the chronostratigraphical boundary within it.

SILURIAN CONODONTS AND SEQUENCES

The best approach to improve our understanding of Silurian conodont faunas and to employ Silurian conodonts as reliable biostratigraphical indices is to integrate conodont occurrences within the framework of stratigraphic sequences. However, Silurian stratigraphic sequences are poorly known because research on Silurian sections has been diverted by the construction of sea-level curves using palaeontological data, and sequences resolved on a stratigraphical basis are poorly constrained by biostratigraphy.

Silurian sea-level curves and Silurian sequences

Several sea-level curves for the Silurian have been published, most of which are based on analyses of faunal associations rather than sequence criteria. Because of the different methodology, these curves do not translate readily into stratigraphic sequences. Johnson (1996) developed a standard sea-level curve for the Silurian based primarily on water depths as inferred from benthic shelly assemblages. In a similar manner, Zhang and Barnes (2002*c*) employed multivariate analysis to identify recurrent conodont assemblages that were related to water depth, and used the distribution of these assemblages to resolve a sea-level curve for the Llandovery section on Anticosti Island. As Bolton (1990) and Aldridge *et al.* (1993) pointed out, certain faunal associations may be depth-related, but are unlikely to be depth-specific, as many other ecological parameters vary without changes in water depth on a shelf. Sequence stratigraphy shows that regressive sedimentary sections that shallow upward and contain shallowing-upward faunal assemblages are characteristic of shelf aggradation and progradation during a highstand in sea level, and do not indicate a fall in eustatic sea level. To overcome such problems, the position of the faunal change relative to the system tracts must be known in order to ascertain whether the change in faunal assemblage is the result of eustatic sea-level fluctuation or sedimentological processes within the system tracts, including parasequence formation. Johnson *et al.* (1998) focused on topographical relief at regional unconformities as a means to calibrate the magnitude of eustatic rises in sea level. By documenting the stratigraphical positions of these regional unconformities, it will be possible to integrate their sea-level curve with stratigraphic sequences.

Loydell (1998) recognized eustatic rises in Silurian sea level largely on the basis of appearances of graptolite-bearing dark grey shales in shelf sections. These graptolite-bearing shales probably correspond to MFSs associated with a rapid rise in eustatic sea level. The graptolitic shale horizons may be somewhat analogous to the core shales of Pennsylvanian cyclothems, which are used for physical correlation and serve as approximate time lines throughout their extent (Heckel 1994). The presence of a graptolite fauna in the shelf succession also permits biostratigraphical correlation to the graptolite zonation, with the limitations discussed above. A variant of sequence stratigraphy, the genetic stratigraphic sequence model (Galloway 1989), uses the sedimentary section bounded by MFSs as the basic stratigraphical unit. The genetic sequence model could be applied to the succession of graptolitic shales to define genetic sequences, but these units would not be identical with stratigraphic sequences, which are bounded by unconformable surfaces. Care would be needed to ascertain whether the occurrence of a thin graptolitic shale in the shelf section represents the flooding event of a third-, fourth- or fifth-order sequence event, or just the base of a parasequence. Recognition of flooding surfaces is the first step toward resolving sequences, but identifying sequence boundaries will require regional stratigraphical and sedimentological studies.

Some authors in North America have attempted to subdivide the Silurian succession directly into stratigraphic sequences. Ross and Ross (1995) compiled a series of transgressive–regressive events for North America, based on the areal extent of unconformity-bounded packages, and derived a sea-level curve. As Loydell (1998) pointed out, poor biostratigraphical control and miscorrelations of units render the timing of their sequences and eustatic events questionable. Brett *et al.* (1998) summarized nearly a decade of work on resolving the stratigraphical architecture of the Llandovery–Wenlock of the Appalachian Foreland Basin in terms of stratigraphic sequences. Based on this information, Ettensohn and

Brett (1998) interpreted the relative importance of eustatic vs. tectonic components of sea-level change in the Appalachian Basin. However, their efforts to collate their sequences with faunally derived eustatic sea-level curves suffered because of the coarse time resolution of brachiopods, ostracodes and conodonts, and absence of graptolites in their sequences. Regional sequence stratigraphic work by Shaver (1996) in the Midcontinent region and Harris and Sheehan (1997) in western United States suffer from the same problems of correlation that hamper comparisons of their regional stratigraphic sequences with the 'standard Silurian sea-level curve' of Johnson (1996).

Silurian conodont zonations and stratigraphic sequences

Silurian conodont zonations have been constructed without regard to stratigraphic sequences and with little reference to the sedimentology of the sections on which the zonations are based. The original zonation of Walliser (1964) was based largely on a single section at Cellon in the Carnic Alps of Austria, where a 'continuous' succession of strata from the upper Llandovery to the Lower Devonian appeared to be present. Not surprisingly, as data were gathered from other regions, it became apparent that the Cellon section is not continuous nor is the conodont fauna entirely typical. The stratigraphical peculiarities of the Cellon section and its fauna hindered the development of a Silurian conodont zonation, but the most damaging effect of reliance on the Cellon section occurred during the development of a composite standard using graphic correlation (Kleffner 1989, 1995). The Cellon section includes some strongly condensed intervals that may also contain hiatal surfaces in the Wenlock and lower Ludlow part of the section. One hiatal surface near the base of the Wenlock that superimposed the FAD of *Kockelella patula* directly above the LAD of *Pterospathodus amorphognathoides* was recognized through zonal work elsewhere (Barrick and Klapper 1976; Jeppsson 1997*a*). When Kleffner (1989) chose Cellon to serve as the composite reference standard (CS), variations in rates of rock accumulation introduced a strong time distortion into the CS, which shows an extremely 'short' Wenlock in terms of composite time units compared with the Ludlow and Přídolí. Not only does the Wenlock span a few composite time units in the CS, but FADs and LADs occurring during the Wenlock are separated by only a few tenths of a standard time unit. Resolution of the correct order of events is more difficult because of this close scaling of events. Recently, Kleffner (1995; work in progress) has attempted to 'expand' the Wenlock section by artificially adding rock thickness to the Wenlock part of the CS.

Because Kleffner (1989, 1995) integrated both conodont and graptolite ranges into a single composite standard, he needed to include sections like Cellon, where graptolitic shales are interbedded with conodont-bearing carbonates. Graptolites are the traditional biostratigraphical standard for the Silurian and most of the chronostratigraphical boundaries have been placed relative to graptolite ranges. Therefore, the conodont ranges must be calibrated against the graptolite zones. Graptolitic shales interbedded with thin carbonate beds are typical of condensed sections, where only a few centimetres may separate FADs and LADs that designate succeeding zones. The difficulties of collating conodont and graptolite ranges, as well the valuable information that can be obtained, can be seen in the results of Loydell *et al.* (1998) for Telychian faunas and Kríz *et al.* (1986) for Přídolí faunas.

The assumption of stratigraphical continuity misled Barrick and Klapper (1976) when they proposed an alterative Wenlock zonation based on thin carbonate sections of the Clarita Formation in southern Oklahoma. Although truncation of species ranges suggested that something unusual occurred between the *Kockelella amsdeni* and *K. stauros* zones, it was not until later that Barrick (1997) recognized that these range truncations were a result of a cryptic unconformity at a regional sequence boundary. With the aid of graphic correlation, a hiatus of some magnitude was identified, one that includes most of the early Homerian *Ozarkodina sagitta sagitta* Zone as well as the Mulde oceanic event of Jeppsson (1998). Preliminary evaluation of the conodont faunas from the overlying Ludlow–Přídolí Henryhouse Formation shows that the Ludlow part of the formation, the graptolite-bearing shale section, is bounded above and below by unconformities, and may also contain hiatal surfaces within it. The sedimentological and stratigraphical research required to place the southern Midcontinent Silurian succession into a sequence stratigraphic framework has only just started (Al-Shaieb and Puckette 2000). Additional progress will depend on recognition of sequences and the integration of conodont ranges with these sequences, using zonal biostratigraphy and graphic correlation.

A detailed zonation based on the evolutionary lineages of *Pterospathodus* was developed for Telychian strata by Männik (1998). At least two distinct *Pterospathodus* lineages, *P. amorphognathoides angulatus–P. a. lennarti–P. a. lithuanicus–P. a. amorphognathoides* and *P. celloni–P. pennatus procerus*, evolved separately. The former lineage evolved in carbonate–terrigenous open shelf environments and the latter in a deeper basinal environment characterized by the graptolite facies. Evolution was more rapid, and the morphological variation within each population greater, in open shelf environments. The appearance of *P. celloni* and *P. p. procerus* in open shelf environments occurred during times when the usual marine conditions were partially (*P. a. lithuanicus* time and the appearance of *P. celloni*) or completely (Ireviken oceanic event and the appearance of *P. p. procerus*) altered, causing temporary disappearances or final extinction of several lineages. The *Pterospathodus* succession has been recognized in more than 40 core sections from the eastern Baltic, and regions all over the world (Männik 1998). Studies are in progress to determine how this succession occurs relative to stratigraphic sequences in Estonia.

The revised conodont zonation of Corradini and Serpagli (1999), which includes late Llandovery–Přídolí zones, is the best fit of data from Sardinia with published ranges from other regions, and stands independently of other types of stratigraphical information. The basic framework of the zonations relied on the Cellon section and no direct reference was provided relative to sea-level events or stratigraphic sequences on Sardinia, or in other regions.

Zhang and Barnes (2002*a*) presented a conodont zonation for the entire Llandovery based on the Anticosti Island faunas. The Anticosti Island sections are relatively thick, and the carbonates and shales represent deposition in a carbonate ramp setting. Although Zhang and Barnes (2002*a*) stated the importance of sequence stratigraphy in testing patterns of conodont appearance and disappearance and cite studies on sequence stratigraphy on Anticosti Island (Sami and Desrochers 1992), they did not use this information in the biostratigraphical and ecological analyses of the conodont faunas (Zhang and Barnes 2002*a*, *b*). Instead, they focused on the construction of a sea-level curve based on the multivariate analysis of conodont occurrences and discussed how their sea-level curve compares with previously published sea-level curves for the Llandovery (Zhang and Barnes 2002*c*).

Jeppsson (1997*a*) has published parts of a revised zonation for the latest Llandovery through the Přídolí, based almost entirely on the carbonate–shale succession on Gotland. Many of his earlier ideas about conodont biostratigraphy were incorporated in the 'Conodont Global Zonation' chart that appears in Nowlan (1995), and more revisions continue to appear. The thick Silurian section on Gotland provides Jeppsson with the opportunity to study conodont faunas at a high degree of stratigraphical resolution wherever continuous stratigraphical sections occur. Using detailed sampling, he was able to utilize the high stratigraphical resolution to develop a detailed description of faunal change during the Ireviken oceanic event (Jeppsson 1997*b*). In many instances, extremely large samples (> 10 kg) were required to recover sufficient conodont elements to determine whether important, but rare, species were present or absent. Because of the low topographical relief on Gotland, long continuous sections are uncommon and smaller disjunct to isolated outcrops have to be placed in stratigraphical order using lithostratigraphy, facies analysis and often biostratigraphy. For this reason, the quality of the conodont zonation on Gotland varies from exceptional, where continuous data can be obtained, to marginal, where disjunct sections were composited. The high biostratigraphical resolution that can be attained on Gotland may not be transferable to other areas, where lower rates of deposition tend to merge or time-average events closely spaced in time or large sample sizes are not available.

Although much of the original sampling was conducted without reference to stratigraphic sequences, parts of the Gotland succession have now been interpreted in terms of stratigraphic sequences, and integrated with conodont occurrences (Calner 1999). As sequence stratigraphic work continues on Gotland, the conodont faunas can be organized by sequence, and then collections can be placed by position within the systems tracts. If sufficient conodont faunas can be obtained from all systems tracts, then the conodont ranges, as well as the numerous oceanic events and episodes that Jeppsson (1998) has inferred from Gotland conodont distributions, can be analysed within the context of stratigraphic sequences.

Silurian oceanic events and episodes

Chatterton *et al.* (1990) reported that FADs and LADs of Silurian trilobite and conodont taxa corresponded to certain positions in transgressive–regressive cycles in the deep shelf lithofacies in the MacKenzie Mountains in north-west Canada. They interpreted the recurring patterns of FADs and LADs to be a result of rapid eustatic changes in sea level, and inferred a cause and effect relationship between sea-level changes and FADs and LADs. Aldridge *et al.* (1993) argued that the correlation between sea-level variation and biotic events may not be causal, and pointed out that in some cases the faunal turnovers preceded the interpreted changes in sea level. Aldridge *et al.* (1993) proposed that the model of changes in global climate or global oceanic state developed by Jeppsson (1990) to explain similar patterns of FADs and LADs of Silurian conodonts in the Gotland sections would also apply to the Canadian sections.

In his most recent formulation of the oceanic episode/oceanic event model, Jeppsson (1998) proposed that the Silurian global ocean alternated between two oceanic states, primo and secundo episodes, the rapid transition between which generated events that are preserved in the conodont record. The model was initially developed to explain the stratigraphical recurrence of distinct conodont assemblages, anomalies in the conodont succession and variations in the frequency of certain lithotypes on Gotland. Primo episodes were times of low atmospheric CO_2, when the cold high latitudes generated cold dense water that produced well-ventilated and oxygenated oceans, characterized by common upwelling and high organic productivity. During primo episodes, sea level was low, oxygen-rich bottom sediments accumulated, and carbonate production was reduced, resulting in clastic-dominated shelf sediments. Secundo episodes were times with high atmospheric CO_2, when the high latitudes were warm, little cold water was produced and oceanic circulation was sluggish. Increased salinity in the mid-latitudes formed dense waters that produced a stratified, anoxic ocean. Without upwelling nutrients, plankton productivity was low and stable. During secundo episodes sea level was high, rapid carbonate deposition prevailed on shelf regions and black shales accumulated in deep waters. Oceanic events, rapid (> 100 kyr) transitions from one oceanic state to the other, were times of major faunal reorganization, and sudden brief declines in primary plankton production at primo–secundo events caused mass extinctions. The oceanic episode/event model argues that the episodic pattern of Silurian marine faunas globally, best expressed by conodonts on Gotland, is the result of these rapid shifts between the oceanic states. Others have extended the model to explain the distribution of Silurian reefs (Brunton *et al.* 1998) and graptolite occurrences in carbonate-dominated sections (Berry 1998) outside of the Baltic region.

An important aspect of the oceanic episode/event model is that major sea-level fluctuations are a consequence of changes in oceanic conditions, which are the result of climatic changes. Sea-level fluctuations are not considered to be a major control on sediment type and faunas, as the faunal and lithological changes typically precede sea-level changes (Jeppsson 1998). Loydell (1998) argued that the observed faunal (graptolites) and lithological changes in the Llandovery do not support Jeppsson's oceanic model, but that the latter are consistent with eustatic sea-level changes. The average duration of Silurian oceanic episodes is about 2 myr, which is comparable with the duration of third-order stratigraphic sequences through the Phanerozoic. It is possible that Jeppsson's oceanic episodes are the palaeontological equivalents of third-order sequences, or 'biothems' as used by Lane *et al.* (2000). The clustering of FADs and LADs, acme intervals, range gaps and recurring taxa that Jeppsson is able to document for the events in the thick Gotland sections, where a high degree of stratigraphical resolution is possible, may be as much a reflection of the control of sequence architecture over faunal distribution as described by Holland (2000) as real biotic change.

The Ireviken oceanic event (Jeppsson 1997*b*) represents such a major turnover in conodont faunas that is primarily a biotic event, although some of the details of the event may have been modified by sequence architecture. The late Wenlock Mulde oceanic event (Jeppsson 1998) is a significant biotic event as represented by extinctions in graptolites and a carbon isotope excursion. On Gotland, the corresponding conodont record is characterized by only a few extinction events in conodonts at Datum 1, and a couple of origination events well above Datum 2. The greatest changes in conodonts are shifts in relative abundance of a few species and gaps in the ranges of other species through the event. Calner (1999) placed the Mulde event on Gotland in a sequence stratigraphic context. Datum 1 coincides with the lower boundary of the Fröjel Formation (Slite siltstones), and lies near the MFS. Datum 2 lies at a sequence boundary. These are two levels at which Holland (2000) showed that FADs and LADs will be grouped by sequence architecture, even if no biotic event occurs. The Mulde event on Gotland can just as easily be interpreted as an artefact of stratigraphic sequences as an extinction event in conodonts. Calner (1999) pointed out that the model of Jeppsson (1990, 1998) also does not predict the sea-level changes documented on Gotland at the level of the Mulde event, which is a secundo–secundo event. The conodont record of most events that Jeppsson (1998) described for the Silurian, ten or more, is more like that of the Mulde event than the Ireviken event.

The rich dataset on conodont distribution on Gotland produced by Jeppsson provides an unparalleled opportunity to examine the Silurian conodont record in one small region at a high level of resolution. However, each proposed oceanic episode and oceanic event must first be evaluated in terms of its position relative to systems tracts of the stratigraphic sequences. Potential artefacts of sequence architecture must be evaluated cautiously, as they may not reflect any biotic event related to global oceanic or climate conditions. In the thick Gotland sections, short-lived eustatic events such as fourth- and fifth-order sequences and parasequences may be as good as or better than biostratigraphical zones, and these eustatic events can be used to correlate sections and test the synchroniety or ascertain the diachroneity of conodont ranges and abundance shifts. This sequence stratigraphic approach contrasts with how research on global events and event stratigraphy has been conducted. Kaljo *et al.* (1995) presented an immense compendium of information on Silurian bioevents and geochemical events and where possible discussed their distribution in time and interrelationships. Although some comparisons were made to interpreted sea-level curves, nowhere is any consideration given to the possibility that any of the events may be artefacts of sequence architecture.

CONCLUSIONS

The best way to improve Silurian conodont biostratigraphy, as well as to constrain models of global biotic and oceanic events, is to collect and interpret the data within the framework of sequence stratigraphy. The discontinuity of the stratigraphical record, as well as that of the fossil record that occurs in these strata, exists at several levels, from stratigraphic sequences millions of years in duration down to the small-scale ecological discontinuities of parasequences. Recognition of the discontinuous fabric of the stratigraphical record on shelf regions and appreciation of the limits of resolution in condensed offshore sections should temper any inclination to transform biostratigraphical datasets directly into biotic models and zonations that assume both continuity of record and high time resolution. The concepts of 'sequence biostratigraphy' and 'biothems', where the fossils are used to characterize and correlate sequences, may have as much relevance to the practical aspects of time-correlation as traditional zonal biostratigraphy. Both approaches will need to be fully integrated to achieve high-resolution correlations. Chronostratigraphical units and boundaries (GSSPs) will have to be characterized with respect to stratigraphic sequences as well as to biozonations.

We do not see that placing Silurian conodonts in their sequence stratigraphic context is an unusually difficult task. New data can certainly be collected with explicit reference to stratigraphic sequences that have previously been described, or the sequences can be described and interpreted as part of the basic stratigraphical work. Older data, for the most part, were collected with adequate stratigraphical documentation that when the sequence stratigraphy of the area is studied, the biostratigraphical data could easily be incorporated. Co-ordinated studies of Silurian conodont lineages, palaeoecology, palaeogeography, geochemical events, oceanic episodes and events, and sequence stratigraphy will yield the maximum results. We see no reason why this integration of research should not be in progress now.

The examples we discuss here are based on Silurian conodont faunas, as this is the group with which we have the greatest familiarity. We do not, however, believe that what we have described is unique either to the Silurian or to conodonts. It puzzles us that even though global eustacy and resulting stratigraphic sequences have been well established as the primary basis of the marine stratigraphical record, many palaeontologists and geochemists working on the Palaeozoic continue to collect and interpret their data without consideration of the sequence stratigraphic context.

Acknowledgements. We thank our many colleagues who over the years have shared their ideas and allowed us to view their conodont collections. In particular, we thank Lennart Jeppsson (Lund), who encouraged us in our work on Silurian conodonts, allowed us full access to his published and unpublished collections, has been willing discuss all aspects of conodonts, and tolerated us when we have disagreed with his conclusions. Charles Henderson (Calgary) shared his ideas on placing GSSPs in the lowstand system tracts. We owe our collaboration to the Twinning Program of the National Research Council (USA), which provided us with travel and living expenses to work with each other over a two-year period (2000–2001). The Twinning Program also allowed JB to visit L. Jeppsson during the grant period. PM's studies were partly supported by grants 4070 and 5406 from the Estonian Science Foundation. We thank Steve Leslie and Oliver Lehnert for their careful reviews and useful suggestions for improving the paper.

REFERENCES

ALDRIDGE, R. J. and JEPPSSON, L. 1999. Wenlock–Přídolí recurrent conodont associations. 37–41. *In* BOUCOT, A. J. and LAWSON, J. D. (eds). *Palaeocommunities: a case study from the Silurian and Early Devonian.* Cambridge University Press, Cambridge, 895 pp.

—— —— and DORNING, K. J. 1993. Early Silurian oceanic episodes and events. *Journal of the Geological Society, London,* **150**, 501–513.

—— and SCHÖNLAUB, H.-P. 1989. Conodonts. 264–267. *In* HOLLAND, C. H. and BASSETT, M. G. (eds). *A global standard for the Silurian System.* National Museum of Wales, Geological Series, **9**, 1–325.

AL-SHAIEB, Z. and PUCKETTE, J. 2000. Sequence stratigraphy of Hunton Group ramp facies, Arbuckle Mountains and Anadarko Basin, Oklahoma. 131–137. *In* JOHNSON, K. S. (ed.). *Platform carbonates in the southern Midcontinent, 1996 symposium.* Oklahoma Geological Survey, Circular, **101**, 1–359.

ANDERS, M. H., KREUGER, S. W. and SADLER, P. M. 1987. A new look at sedimentation rates and the completeness of the stratigraphic record. *Journal of Geology*, **95**, 1–14.

ARMSTRONG, H. A. 1990. Conodonts from the Upper Ordovician–Lower Silurian carbonate platform of North Greenland. *Bulletin, Grønlands Geologske Undersogelse*, **159**, 151 pp.

BARD, E., ARNOLD, M., DUPRAT, J., MOYES, J. and DUPLESSY, J. C. 1987. Reconstruction of the last deglaciation: deconvolved records of ^{18}O profiles, micropaleontological variations and accelerator mass spectrometric ^{14}C. *Climate Dynamics*, **1**, 101–112.

BARRICK, J. E. 1997. Wenlock (Silurian) depositional sequences, eustatic events, and biotic change on the southern shelf of North America. 47–65. *In* KLAPPER, G., MURPHY, M. A. and TALENT, J. A. (eds). *Silurian to Lower Carboniferous biostratigraphy, biofacies and biogeography, a volume in honor of J. Granville Johnson, his life and works.* Geological Society of America, Special Paper, **321**, 1–401.

—— and KLAPPER, G. 1976. Multielement Silurian (late Llandoverian – Wenlockian) conodonts of the Clarita Formation, Arbuckle Mountains, Oklahoma, and phylogeny of *Kockelella. Geologica et Palaeontologica*, **10**, 59–100.

BERGER, W. H. and HEATH, G. R. 1968. Vertical mixing in pelagic sediments. *Journal of Marine Research*, **26**, 134–143.

BERRY, W. B. N. 1998. Silurian oceanic episodes: the evidence from central Nevada. 259–264. *In* LANDING, E. and JOHNSON, M. E. (eds). *Silurian cycles: linkages of dynamic stratigraphy with atmospheric, oceanic, and tectonic changes. James Hall Centennial Volume.* New York State Museum, Bulletin, **491**, 1–327.

BOARDMAN, D. R., II, HECKEL, P. H., BARRICK, J. E., NESTELL, M. and PEPPERS, R. A. 1990. Middle–Upper Pennsylvanian chronostratigraphic boundary in the Midcontinent region of North America. 319–337. *In* BRENCKLE. P. L. and MANGER, W. L. (eds). *Intercontinental correlation and division of the Carboniferous System.* Courier Forschungsinstitut Senckenberg, **130**, 1–350.

BOLTON, T. E. 1990. Sedimentological data indicate greater range of water depths for *Costistricklandia lirata* in the southern Appalachians. *Palaios*, **5**, 371–374.

BOUDREAU, B. P. 1997. *Diagenetic models and their implementation: modeling transport and reaction in aquatic sediments.* Springer-Verlag, New York, 430 pp.

BRETT, C. E. 1995. Sequence stratigraphy, biostratigraphy, and taphonomy in shallow marine environments. *Palaios*, **10**, 597–616.

—— 1998. Sequence stratigraphy, paleoecology, and evolution: clues and responses to sea-level fluctuations. *Palaios*, **13**, 241–262.

—— BAARLI, B., GUDVEIG, B., CHOWNS, T., COTTER, E., DRIESE, S., GOODMAN, W. and JOHNSON, M. E. 1998. Early Silurian condensed intervals, ironstones, and sequence stratigraphy in the Appalachian foreland basin. 89–143. *In* LANDING, E. and JOHNSON, M. E. (eds). *Silurian cycles: linkages of dynamic stratigraphy with atmospheric, oceanic, and tectonic changes. James Hall Centennial Volume.* New York State Museum, Bulletin, **491**, 1–327.

BRUNTON, F. R., SMITH, L., DIXON, O. A., COPPER, P., NESTOR, H. and KERSHAW, S. 1998. Silurian reef episodes, changing seascapes, and paleobiogeography. 265–282. *In* LANDING, E. and JOHNSON, M. E. (eds). *Silurian cycles: linkages of dynamic stratigraphy with atmospheric, oceanic, and tectonic changes. James Hall Centennial Volume.* New York State Museum, Bulletin, **491**, 1–327.

CALNER, M. 1999. Stratigraphy, facies development, and depositional dynamics of the Late Wenlock Fröjel Formation, Gotland, Sweden. *GFF (Geologiska Föreningens i Stockholm Förhandlingar)*, **121**, 13–24.

CHATTERTON, B. D. E., EDGECOMBE, G. D. and TUFFNELL, P. A. 1990. Extinction and migration in Silurian trilobites and conodonts of northwestern Canada. *Journal of the Geological Society, London*, **147**, 703–715.

CORRADINI, C. and SERPAGLI, E. 1999. A Silurian conodont biozonation from late Llandovery to end Přídolí in Sardinia (Italy). 255–273. *In* SERPAGLI, E. (ed.). *Studies on conodonts – Proceedings of the Seventh European Conodont Symposium.* Bollettino della Società Paleontologica Italiana, **37**, 1–557.

CUTLER, A. H. and FLESSA, K. W. 1990. Fossils out of sequence: computer simulations and strategies for dealing with stratigraphic disorder. *Palaios*, **5**, 227–235.

DAY, J. 1996. Faunal signatures of Middle–Upper Devonian depositional sequences and sea level fluctuations in the Iowa Basin: U.S. Midcontinent. 277–300. *In* WITZKE, B. J., LUDVIGSEN, G. A. and DAY, J. E. (eds). *Paleozoic sequence stratigraphy: views from the North American craton.* Geological Society of America, Special Paper, **306**, 1–446.

EMERY, D. and MYERS, K. J. 1996. *Sequence stratigraphy.* Blackwell Scientific, Oxford, 297 pp.

ETHINGTON, R. L. and DRESBACH, R. I. 1990. Ordovician conodonts in the Arbuckle Group, southern Arbuckle Mountains, Oklahoma. 33–37. *In* RITTER, S. M. (ed.). *Early to Middle Paleozoic conodont biostratigraphy of the Arbuckle Mountains, southern Oklahoma.* Oklahoma Geological Survey, Guidebook, **27**, 1–114.

ETTENSOHN, F. R. and BRETT, C. E. 1998. Tectonic components in third-order Silurian cycles; examples from the Appalachian Basin and global implications. 145–162. *In* LANDING, E. and JOHNSON, M. E. (eds). *Silurian cycles: linkages of dynamic stratigraphy with atmospheric, oceanic, and tectonic changes. James Hall Centennial Volume.* New York State Museum, Bulletin, **491**, 1–327.

FLESSA, K. W., CUTLER, A. H. and MELDAHL, K. H. 1993. Time and taphonomy: quantitative estimates of time-averaging and stratigraphic disorder in a shallow marine habitat. *Paleobiology*, **19**, 266–286.

FOOTE, M. 1997. Estimating taxonomic durations and preservation probability. *Paleobiology*, **23**, 278–300.

—— and RAUP, D. M. 1996. Fossil preservation and the stratigraphic ranges of taxa. *Paleobiology*, **22**, 121–140.

FÜRSICH, F. T. and ABERHAN, M. 1990. Significance of time-averaging for paleocommunity analysis. *Lethaia*, **23**, 143–152.

GALLOWAY, W. E. 1989. Genetic stratigraphic sequences in basin analysis. I. Architecture and genesis of flooding-surface bounded depositional units. *American Association of Petroleum Geologists, Bulletin*, **73**, 125–142.

GINSBERG, R. N. 1996. The Cretaceous-Tertiary boundary: the El Kef blind test. *Marine Micropaleontology*, **29**, 65–103.

GLASS, B. P. 1969. Reworking of deep-sea sediments as indicated by the vertical dispersal of the Australasian and Ivory Coast microtektite layers. *Earth and Planetary Science Letters*, **6**, 409–415.

GOLDRING, R. 1995. Organisms and substrate: response and effect. 151–180. *In* BOSENCE, D. W. and ALLISON, P. A. (eds). *Marine palaeoenviromental analysis from fossils.* Geological Society, London, Special Publication, **83**, 1–272.

HAQ, B. U., HARDENBOL, J. and VAIL, P. R. 1987. Chronology of fluctuating sea levels since the Triassic (250 m.y.) to present. *Science*, **235**, 1156–1167.

HARRIS, M. T. and SHEEHAN, P. M. 1997. Carbonate sequences and fossil communities from the Upper Ordovician–Lower Silurian of the eastern Great Basin. *Brigham Young University Studies*, **42**, 105–128.

HECKEL, P. H. 1994. Evaluation of evidence for glacio-eustatic control over marine Pennsylvanian cyclothems in North America and consideration of possible tectonic effects. 65–87. *In* DENNISON, J. M. and ETTENSOHN, F. R. (eds). *Tectonic and eustatic controls on sedimentary cycles.*

SEPM, Concepts in Sedimentology and Paleontology, **4**, 1–264.

HEDBERG, H. D. (ed.) 1976. *International stratigraphic guide: a guide to stratigraphic classification, terminology, and procedure*. First edition. John Wiley and Sons, New York, 200 pp.

HOLLAND, S. M. 1995. Sequence stratigraphy, facies control, and their effects on the stratigraphic distribution of fossils. 1–13. *In* HAQ, B. U. (ed.). *Sequence stratigraphy and depositional response to eustatic, tectonic, and climatic forcing*. Kluwer Academic, Dordrecht, 381 pp.

—— 1996. Guidelines for interpreting the stratigraphic record of extinctions: distinguishing pattern from artifact. 174. *In* REPETSKI, J. E. (ed.). *Sixth North American Paleontological Convention, Abstracts of Papers*. Paleontological Society, Special Publication, **8**, 1–443.

—— 2000. The quality of the fossil record: a sequence stratigraphic perspective. 148–168. *In* ERWIN, D. H. and WING, S. L. (eds). *Deep time*. The Paleontological Society. Allen Press, Lawrence, Kansas, 371 pp.

HUTSON, W. H. 1980. Bioturbation of deep-sea sediments: oxygen isotopes and stratigraphic uncertainty. *Geology*, **8**, 127–130.

JEPPSSON, L. 1990. An oceanic model for lithological and faunal changes. *Journal of the Geological Society, London*, **147**, 663–674.

—— 1997*a*. A new latest Telychian, Sheinwoodian and early Homerian (Early Silurian) standard conodont biostratigraphy. *Transactions of the Royal Society of Edinburgh: Earth Sciences*, **88**, 91–114.

—— 1997*b*. The anatomy of the mid-Silurian Ireviken Event. 451–492. *In* BRETT, C. and BAIRD, C. G. (eds). *Paleontological events, horizons, ecological and evolutionary implications*. Columbia University Press, New York, 604 pp.

—— 1998. Silurian oceanic events: summary of general characteristics. 239–257. *In* LANDING, E. and JOHNSON, M. E. (eds). *Silurian cycles: linkages of dynamic stratigraphy with atmospheric, oceanic, and tectonic changes. James Hall Centennial Volume*. New York State Museum, Bulletin, **491**, 1–327.

JOHNSON, J. G. 1979. Intent and reality in biostratigraphic zonation. *Journal of Paleontology*, **53**, 931–942.

—— KLAPPER, G. and SANDBERG, C. A. 1985. Devonian eustatic fluctuations in Euramerica. *Geological Society of America, Bulletin*, **96**, 567–587.

JOHNSON, M. E. 1996. Stable cratonic sequences and a standard for Silurian eustacy. 203–212. *In* WITZKE, B. J., LUDVIGSEN, G. A. and DAY, J. E. (eds). *Paleozoic sequence stratigraphy: views from the North American craton*. Geological Society of America, Special Paper, **306**, 1–446.

—— RONG, J.-Y. and KERSHAW, S. 1998. Calibrating Silurian eustacy against the erosion and burial of coastal topography. 3–13. *In* LANDING, E. and JOHNSON, M. E. (eds). *Silurian cycles: linkages of dynamic stratigraphy with atmospheric, oceanic, and tectonic changes. James Hall Centennial Volume*. New York State Museum, Bulletin, **491**, 1–327.

KALJO, D., BOUCOT, A. J., CORFIELD, A. L. E., HÉRISSÉ, A., KOREN', T. N., KRÍŽ, J., MÄNNIK, P., MÄRSS, T., NESTOR, V., SHAVER, R. H., SIVETER, D. J. and VIIRA, V. 1995. Silurian bioevents. 173–224. *In* WALLISER. O. H. (ed.). *Global events and event stratigraphy in the Phanerozoic: results of international interdisciplinary cooperation in the IGCP Project 216 'Global Biological Events in Earth History'*. Springer-Verlag, Berlin, 333 pp.

KEMP, T. S. 1999. *Fossils and evolution*. Oxford University Press, Oxford, 284 pp.

KIDWELL, S. M. 1993. Taphonomic expressions of sedimentary hiatuses: field observations on bioclastic concentrations and sequence anatomy in low, moderate, and high subsidence settings. *Geologische Rundschau*, **82**, 189–202.

—— and FLESSA, K. W. 1995. The quality of the fossil record: populations, species, and communities. *Annual Review of Ecology and Systematics*, **26**, 269–299.

KLAPPER, G., FEIST, R., BECKER, R. T. and HOUSE, M. R. 1994. Definition of the Frasnian/Famennian Stage boundary. *Episodes*, **16**, 433–441.

KLEFFNER, M. A. 1989. A conodont-based Silurian chronostratigraphy. *Geological Society of America, Bulletin*, **101**, 904–912.

—— 1995. A conodont- and graptolite-based Silurian chronostratigraphy. 159–176. *In* MANN, K. O. and LANE, H. R. (eds). *Graphic correlation*. SEPM, Special Publication, **53**, 1–263.

KRÍŽ, J., JAEGER, H., PARIS, F. and SCHÖNLAUB, H. P. 1986. Prídolí – the fourth subdivision of the Silurian. *Jahrbuch der Geologischen Bundesanstalt*, **129**, 291–360.

LANE, H. R., FRYE, M. W. and COUPLES, G. C. 2000. Biothem: a new biostratigraphically based unit of sequence stratigraphy. 183–196. *In* JOHNSON, K. S. (ed.). *Platform carbonates in the southern Midcontinent, 1996 symposium*. Oklahoma Geological Survey, Circular, **101**, 1–359.

LOUTIT, T. S., HARDENBOL, J. and WRIGHT, R. C. 1991. Sequence biostratigraphy. *American Association of Petroleum Geologists, Bulletin*, **75**, 624.

LOYDELL, D. K. 1998. Early Silurian sea-level changes. *Geological Magazine*, **135**, 447–471.

—— KALJO, D. and MÄNNIK, P. 1998. Integrated biostratigraphy of the lower Silurian of the Ohesaare core, Saaremaa, Estonia. *Geological Magazine*, **135**, 769–783.

MACLEOD, N. 1991. Punctuated anagensis and the importance of stratigraphy to paleobiology. *Paleobiology*, **17**, 167–188.

MÄNNIK, P. 1998. Evolution and taxonomy of the Silurian conodont *Pterospathodus*. *Palaeontology*, **32**, 893–906.

MARTIN, R. E. 1993. Time and taphonomy: actualistic evidence for time-averaging of benthic foraminiferal assemblages. 34–56. *In* KIDWELL, S. M. and BEHRENSMEYER, A. K. (eds). *Taphonomic approaches to time resolution in fossil assemblages*. Paleontological Society, Short Course, **6**, 1–302.

—— 1999. *Taphonomy: a process approach*. Cambridge University Press, Cambridge, 508 pp.

NOWLAN, G. S. 1995. Left hand column for correlation charts. *Silurian Times*, **3**, 7–8.

NYSTUEN, J. P. 1998. History and development of sequence stratigraphy. 31–116. *In* GRADSTEIN, F. M., SANDVIK, K. O. and MILTON, N. J. (eds). *Sequence stratigraphy – concepts and applications*. Norwegian Petroleum Society, Special Publication, **8**, 1–437.

PATZKOWSKY, M. E. and HOLLAND, S. M. 1996. Extinction, invasion, and sequence stratigraphy: patterns of faunal change in the Middle and Upper Ordovician of the eastern United States. 131–142. *In* WITZKE, B. J., LUDVIGSEN, G. A. and DAY, J. E. (eds). *Paleozoic sequence stratigraphy: views from the North American craton.* Geological Society of America, Special Paper, **306**, 1–446.

POSAMENTIER, H. W. and GOODMAN, D. K. 1992. Biostratigraphy in a sequence stratigraphic framework. *Palynology*, **16**, 228–229.

RITTER, S. C., BARRICK, J. E. and SKINNER, M. R. 2002. Conodont sequence biostratigraphy of the Hermosa Group (Pennsylvanian) at Honaker Trail, Paradox Basin, Utah. *Journal of Paleontology*, **76**, 495–517.

ROSS, C. A. and ROSS, J. R. P. 1995. Foraminiferal zonations of late Paleozoic depositional sequences. *Marine Micropaleontology*, **26**, 469–478.

—— —— 1996. Silurian sea-level fluctuations. 187–192. *In* WITZKE, B. J., LUDVIGSEN, G. A. and DAY, J. E. (eds). *Paleozoic sequence stratigraphy: views from the North American craton.* Geological Society of America, Special Paper, **306**, 1–446.

SALVADOR, A. (ed.). 1994. *International stratigraphic guide: a guide to stratigraphic classification, terminology, and procedure.* Second edition. International Union of Geological Sciences and Geological Society of America, Boulder, 214 pp.

SAMI, T. and DESROCHERS, A. 1992. Episodic sedimentation on an early Silurian, storm-dominated carbonate ramp, Becscie and Merrimack formations, Anticosti Island, Canada. *Sedimentology*, **39**, 355–381.

SCHIFFELBEIN, P. 1984. Effect of benthic mixing on the information content of deep-sea stratigraphic signals. *Nature*, **311**, 651–653.

SCHINDEL, D. E. 1980. Microstratigraphic sampling and the limits of paleontologic resolution. *Paleobiology*, **6**, 408–426.

SCHOPF, J. M. 1978. Fossilization potential of an intertidal fauna, Friday Harbor, Washington. *Paleobiology*, **4**, 261–270.

SERPAGLI, E. and CORRADINI, C. 1999. Taxonomy and evolution of *Kockelella* (Conodonta) from the Silurian of Sardinia (Italy). 275–298. *In* SERPAGLI, E. (ed.). *Studies on conodonts – Proceedings of the Seventh European Conodont Symposium.* Bollettino della Società Paleontologica Italiana, **37**, 1–557.

SHAVER, R. H. 1996. Silurian sequence stratigraphy in the North American craton. 193–202. *In* WITZKE, B. J., LUDVIGSEN, G. A. and DAY, J. E. (eds). *Paleozoic sequence stratigraphy: views from the North American craton.* Geological Society of America, Special Paper, **306**, 1–446.

SLOSS, L. L. 1996. Sequence stratigraphy on the craton: caveat emptor. 419–424. *In* WITZKE, B. J., LUDVIGSEN, G. A. and DAY, J. E. (eds). *Paleozoic sequence stratigraphy: views from the North American craton.* Geological Society of America, Special Paper, **306**, 1–446.

SOLOW, A. R. and SMITH, W. 1997. On fossil preservation and the stratigraphic ranges of taxa. *Paleobiology*, **23**, 271–277.

STITT, J. H. 1977. Late Cambrian and earliest Ordovician trilobites, Wichita Mountains area, Oklahoma. *Oklahoma Geological Survey, Bulletin*, **124**, 1–79.

VALENTINE, J. W. 1989. How good is the fossil record? Clues from the Californian Pleistocene. *Paleobiology*, **15**, 83–94.

WALLISER, O. H. 1964. Conodonten des Silurs. *Abhandlungen des Hessischen Landesamts für Bodenforschung*, **41**, 1–106.

ZHANG, S. and BARNES, C. R. 2002*a*. A new Llandovery (Early Silurian) conodont biozonation and conodonts from the Becscie, Merrimack, and Gun River formations, Anticosti Island, Québec. *Paleontological Society, Memoir*, **57**, 1–46.

—— —— 2002*b*. Paleoecology of Llandovery conodonts, Anticosti Island, Québec. *Palaeogeography, Palaeoclimatology, Palaeoecology*, **180**, 33–35.

—— —— 2002*c*. Late Ordovician–Early Silurian (Ashgillian–Llandovery) sea level curve derived from conodont community analysis, Anticosti Island, Québec. *Palaeogeography, Palaeoclimatology, Palaeoecology*, **180**, 5–32.

[Special Papers in Palaeontology, 73, 2005, pp. 117–134]

CAMBRO-ORDOVICIAN SEA-LEVEL FLUCTUATIONS AND SEQUENCE BOUNDARIES: THE MISSING RECORD AND THE EVOLUTION OF NEW TAXA

by OLIVER LEHNERT*, JAMES F. MILLER†, STEPHEN A. LESLIE‡, JOHN E. REPETSKI§ *and* RAYMOND L. ETHINGTON¶

*Institut für Geologie und Mineralogie, Universität Erlangen, Schlossgarten 5, D-91054 Erlangen, Germany; e-mail: lehnert@geol.uni-erlangen.de; current address: Université des Sciences et Technologies de Lille (USTL) Paléontologie – Sciences de la Terre, CNRS (UMR 8014), Cite Scientifique SN5, F-59655 Villeneuve d'Ascq Cedex, France

†Department of Geography, Geology and Planning, Southwest Missouri State University, Springfield, MO 65804-0089, USA; e-mail: jfm845f@smsu.edu

‡Department of Earth Science, University of Arkansas at Little Rock, Little Rock, AR 72204, USA; e-mail: saleslie@ualr.edu

§US Geological Survey, 926A National Center, Reston, VA 20192, USA; e-mail: jrepetski@usgs.gov

¶Department of Geological Sciences, University of Missouri, Columbia, 101 Geosciences Building, Columbia, MO 65211, USA; e-mail: EthingtonR@missouri.edu

Abstract: The evolution of early Palaeozoic conodont faunas shows a clear connection to sea-level changes. One way that this connection manifests itself is that thick successions of carbonates are missing beneath major sequence boundaries due to karstification and erosion. From this observation arises the question of how many taxa have been lost from different conodont lineages in these incomplete successions. Although many taxa suffered extinction due to the environmental stresses associated with falling sea-levels, some must have survived in these extreme conditions. The number of taxa missing in the early Palaeozoic tropics always will be unclear, but it will be even more difficult to evaluate the missing record in detrital successions of higher latitudes. A common pattern in the evolution of Cambrian–Ordovician conodont lineages is appearances of new species at sea-level rises and disappearances at sea-level drops. This simple picture can be complicated by intervals that consistently have no representatives of a particular lineage, even after extensive sampling of the most complete sections. Presumably the lineages survived in undocumented refugia. In this paper, we give examples of evolution in Cambrian–Ordovician shallow-marine conodont faunas and highlight problems of undiscovered or truly missing segments of lineages.

Key words: conodont record, evolution, Cambrian, Ordovician, sea-level changes, sequence stratigraphy.

ALMOST all papers on conodont research document taxa that were actually found, but how many species of different lineages may be missing in incomplete successions is usually not discussed. We all recognize stratigraphic gaps in our sections, especially in the shallow-water environments of the vast Cambrian–Ordovician tropical carbonate platforms where major drops in sea-level exposed huge areas subaerially. Shallow-water taxa flourished on these platforms, and whole assemblages probably suffered extinction due to the environmental stresses associated with falling sea-levels, only to have part or all of their record obliterated. In some lineages, especially those adapted to shallow marine facies, taxa occur sporadically in time due to rarity and in space due to migration as facies shift during sea-level changes. Some taxa are probably buried in former continental depressions or grabens and in lowstand deposits along continental margins. In contrast to the record of shallow shelf areas, we can observe complete evolutionary lineages of taxa in deeper shelf and off-platform environments (e.g. the *Pygodus* lineage; see Bergström 1983).

Speciation in *Clavohamulus* Furnish demonstrates the missing record of faunas adapted to shallow water. *Clavohamulus* was widespread in the tropics during the early Ibexian and restricted to Laurentia during the late Ibexian. *Clavohamulus* migrated cratonward during transgressions, but its species may not be recorded from the preceding regresssive intervals because of limited deposition as well as subsequent erosion during exposure. Without records of their existence, some species may be doomed to be unnamed nodes on a cladogram. *Chosonodina herfurthi* Müller occurs in the Ibexian *Rossodus manitouensis* Zone followed by a long gap before *Chosonodina lunata* Harris and Harris appears in the early Whiterockian. Where did the genus survive in the meantime? How many species are missing? These questions can be asked for many genera.

Some taxa within evolutionary lineages may be hidden because of selective sampling. Few conodont workers have chosen to process dolomites and sandstones for conodonts, but these lithologies might fill gaps in the records of endemic shallow-water lineages. Sampling of such unusual facies or rocks that are difficult to process may be needed to recover such taxa and to reconstruct evolutionary lineages. Recently discovered species of *Cahabagnathus* from Nevada and Arkansas that are discussed below are good examples.

What about sampling cool- or cold-water areas? Taxa may be hidden in detrital successions in the peri-Gondwana and Gondwana regions. Thus far, few publications deal with conodonts from sandstones, shales and greywackes. Estimating how much information is missing from such successions is an uncertain undertaking. Another problem is that processing shales from tropical deep-water successions is often very difficult, so we tend to concentrate our efforts on the available interbedded calciturbidites, which will probably yield recycled material.

This paper is rather philosophical because it is not possible to quantify the number of taxa that were lost by erosion of huge areas exposed during sea-level lowstands and of regions that were uplifted by orogenic processes during plate collisions. It also is impossible to calculate accurately the number of taxa still hidden in 'unfavourable lithologies'. Recognition of these biases provides a useful framework for interpreting the samples we have collected, but also points to places where we might look to fill gaps that are usually ignored. Our main goal for this paper is to give examples, mainly from a Laurentian viewpoint, for the connection between sea-level changes, speciation events, extinction events and the missing record that is an inherent part of reconstructed evolutionary lineages of conodonts.

SEA-LEVEL CHANGES, EVOLUTION AND MASS EXTINCTIONS

Major sea-level changes in the Cambrian–Ordovician have been linked to climatic changes or fluctuations in organic production in the oceans (Saltzman *et al.* 2000) and therefore had a tremendous influence on most marine organisms. These changes are often expressed by shifts in stable isotopes (Ripperdan *et al.* 1992; Ripperdan and Miller 1995; Saltzman *et al.* 1995, 1998). Regressions that subaerially exposed huge cratonic regions and dramatically reduced the shelf areas of the Palaeozoic continents must have caused major damage to many established ecosystems.

A good starting point for discussing this topic is the biomere concept. Palmer (1965) defined a biomere as a sequence of rocks characterized by a fossil lineage that is enclosed between successions characterized by unrelated lineages. In re-evaluation of his idea, Palmer (1984) described in detail the rapid change from normal faunas existing for millions of years to taxa abundant in the short 'crisis interval' to, finally, a terminal genus that dominated the environment. Stitt (1975) and Taylor (1977) discussed a model of a well-stratified ocean in the late Cambrian and suggested that a rapid rise of the thermocline may have flooded the shelves with cool waters, thereby causing a faunal turnover in trilobites as well as other groups. Such changes in acrotretid brachiopods (Rowell and Brady 1976) and conodonts (Miller 1978) were documented at the base of the *Eurekia apopsis* Subzone of the *Saukia* (trilobite) Zone. Subsequent study of thin sections has shown hardgrounds and contrasting microfacies at biomere boundaries (Palmer 1984). Today, we know that much more time may be missing at planar hardgrounds that appear to be knife-sharp contacts in outcrops and thin-sections than is unrecorded in situations where we recognize immature karst features in a section. Irregular karst surfaces can be transformed into planar hardgrounds by marine abrasion during subsequent transgressions. Ethington *et al.* (1987) invoked the biomere concept to account for the rapid change in conodont faunas that appears to be universal across North America at the top of the *Rossodus manitouensis* Zone.

Miller (1984) compiled facies patterns, sea-level events, and the evolution of paraconodont and euconodont faunas during late Sunwaptan and early Ibexian (Skullrockian) times. Far more biostratigraphic data have been assembled subsequently, adding to our knowledge about evolution of early conodont lineages. Many stratigraphic studies published during that time have provided detailed information about global eustatic and regional sea-level fluctuations (Osleger and Read 1993; Miller *et al.* 1999, 2003). We conclude that a clear connection exists between sequence boundaries and the introduction of new faunal assemblages of trilobites, conodonts and other fossil groups.

Usually we see that changes in Upper Cambrian–Upper Ordovician conodont faunas are clearly related to regressions and transgressions on tropical platforms, which is well documented for Laurentia. This relationship is documented in the Cambrian–Ordovician Lawson Cove section in the Ibex area of Utah, where Miller *et al.* (1999) studied faunal changes and sea-level fluctuations based on 194 samples, approximately 44,000 identified elements and 92 conodont taxa. More recently, Miller *et al.* (2003) expanded this study using a somewhat larger data set; the range chart and sea-level curve for the lower and middle parts of the Lawson Cove section are displayed here in Text-figure 1. This figure demonstrates that faunal extinctions and the introductions of new conodont, trilobite and brachiopod assemblages clearly bear some relationship to sea-level fluctuations.

Expressions of global sea-level events like those of the Lange Ranch Eustatic Event (now identified as the Lange Ranch Lowstand; see LR LS in Text-fig. 1) have been found in several parts of the world (Miller 1992). The

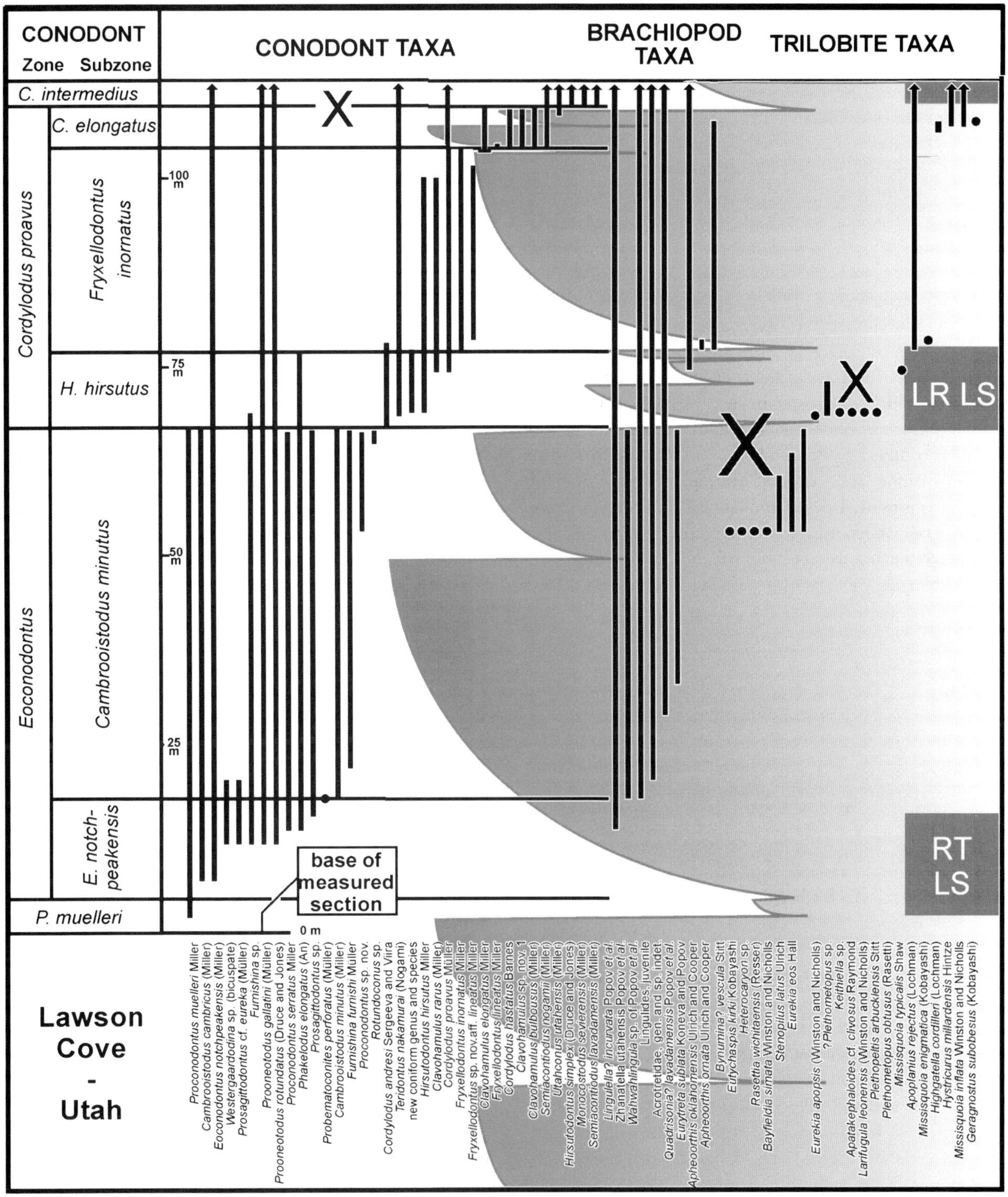

TEXT-FIG. 1. Chart showing the lower part of the Lawson Cove Section in Utah with faunal ranges and sea-level changes (modified from Miller *et al.* 2003). Extinctions are indicated by X and are related to major drops in sea-level. Whether trilobites or brachiopods were affected is shown by symbols. RT LS, Red Tops Lowstand; LR LS, Lange Ranch Lowstand.

obvious change in conodont faunas associated with the Lange Ranch Lowstand was used to define the base of the Ibexian Series and its Skullrockian Stage, which coincided with the lowest occurrence of *Cordylodus* Pander (Ross *et al.* 1997). This turnover at the base of the *Cordylodus proavus* Zone obviously affected conodonts, brachiopods and trilobites at Lawson Cove, and Miller *et al.* (1999) interpreted this event as the real start of the Great Ordovician Biodiversification even though the base of the Ordovician in terms of conodont zones has recently been defined at the base of the *Iapetognathus* Zone (Cooper and Nowlan 1999; Cooper *et al.* 2001) at a level just below the base of the Tremadoc Series as traditionally defined (Rushton 1982).

Conodont faunas from the uppermost Millardan Series are characterized by the dominance of *Proconodontus* Miller, *Eoconodontus* Miller and *Cambrooistodus* Miller. The abrupt extinction of *Proconodontus* and *Cambrooistodus* at the base of the *Cordylodus proavus* Zone indicates a change from a sea-level highstand to the subsequent Lange Ranch Lowstand. This event is indicated by the introduction of *Clavohamulus* Furnish, *Cordylodus* Pander, *Hirsutodontus* Miller and *Teridontus* Miller to the faunal succession.

This lowstand interval comprises only 11·6 m of strata in an area where the entire Cambrian and Ordovician succession is represented by 4500 m of carbonate rocks. The Lange Ranch Lowstand is complex and includes three small cycles whose rocks are termed packages by Miller *et al.* (1999). These three packages are delineated by four boundaries that mark increasingly regressive sea-level changes (Text-fig. 1).

The Lawson Cove section represents an excellent location to observe faunal changes without any of the 'stratigraphic artifacts' which usually occur at sequence boundaries (Holland 1996). It has a continuous faunal succession, which is exceptional when one considers how many decades the Cambrian-Ordovician Boundary Working Groups searched for a sequence without any stratigraphic gaps to serve as the boundary stratotype (Cooper *et al.* 2001). In most cases we have to deal with major lacunae at such important boundaries so that endemic species are absent from the rock record. The Lange Ranch section in central Texas is such an example. Miller *et al.* (2003) illustrated unconformity surfaces at some of the boundaries separating packages of the Lange Ranch Lowstand in that section and showed that the middle of the three packages is missing. As pointed out by Holland (1995), the study of just a few sections may not provide a reliable dataset and we might have 'artifacts of facies control and sequence architecture' in such regions. There, appearences and disappearences of taxa have not necessarily to be connected with extinctions or originations (e.g. Holland 1995).

Furthermore, tremendous problems accompany efforts to transfer such sequence-stratigraphic concepts to detrital successions in cold-water areas like northern Gondwana, resulting in difficulty in effecting biostratigraphic correlation of cold-water and warm-water strata. The problem is increased because detailed sequence-stratigraphic frameworks such as those shown in Text-figures 1–2 have not been developed in cold-water areas. If we could document such physical boundaries with better age control, we might be able to correlate across climatic barriers and determine how cold-water faunas reacted to falls and rises in sea-level. Unfortunately, the conodont record from detrital sequences in cold-water areas for the most part is too sparse for development of even a coarse sequence-stratigraphic correlation.

However, we can recognize significant extinction events during the early Palaeozoic on a global scale. Several Cambrian–Ordovician extinctions, including the end-Ordovician mass extinction, have been documented that were connected with sea-level falls (see Text-fig. 1). Some conodont groups (as well as groups of trilobites, graptolites, brachiopods and bryozoans) disappeared during the extreme sea-level fall and climatic change associated with the Hirnantian glaciation (Berry and Boucot 1973; Sheehan 1973). Armstrong (1996) described a complex model for the step-wise extinctions and recoveries of conodont faunas at the Ordovician–Silurian transition. This event may be exceptional due to a combination of factors that includes the only glaciation in Cambrian–Ordovician time. Zhang and Barnes (2002) recently documented in great detail Upper Ordovician/Lower Silurian sea-level fluctuations and associated changes in conodont communities and biofacies.

SEQUENCE BOUNDARIES AND EXTINCTIONS FROM THE RECORD

Following the appearance of euconodonts in late Cambrian time, conodont diversity reached its peak during the Ordovician. Many taxa flourished on the tropical platforms of Laurentia, Siberia and Baltica as well as in low-latitude portions of Gondwana and peri-Gondwana (e.g. China, Australia). Sketches of sequence-stratigraphic frameworks show that strata are cut out in cratonward directions beneath major sequence boundaries (e.g. Wilgus *et al.* 1988). Such diagrams imply a missing record, not only of strata but also of taxa that previously lived on the shallow shelf. Because of high faunal diversity during Ordovician times, it may be assumed that the number of unrecorded taxa is highest for this interval.

Sloss (1963) identified inter-regional unconformities that he believed to mark craton-wide interruptions in the continuity of sedimentation on Laurentia and to separate six packages of rock that he termed Sequences (Text-fig. 2). The internal architecture of the Sauk and Tippecanoe megasequences displays more time gaps in cratonic and wide shelf areas than does the record of sedimentation on tropical platforms. These unrecorded intervals include the time of early evolution of conodonts in the Cambrian, their peak of diversity in the Ordovician and

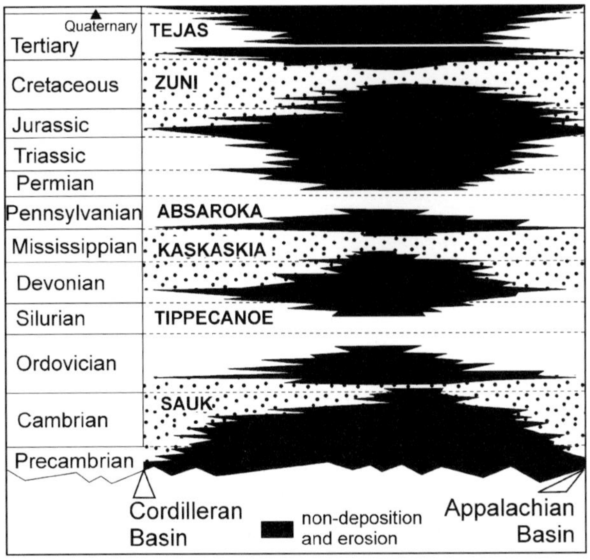

TEXT-FIG. 2. The six sequences of Sloss (1963) covering the North American craton.

their diversification after the major extinction connected with the Hirnantian glaciation. It follows that much palaeontological data were lost from areas that underwent karstification and erosion during certain periods of subaerial exposure. It was also pointed out by Holland (1995, 1996, 2001) that we do not see the real faunal turnovers and major extinction events in such areas on the platforms because the disappearence of species and the occurrence of new taxa will always cluster at these sequence boundaries and at flooding surfaces where their real ranges are not preserved. In contrast, during highstand and lowstand systems tracts low turnover rates may be observed (e.g. Holland 1996).

Many sequence-stratigraphic investigations of early Palaeozoic strata in North America reveal these features (Myrow *et al.* 1999, 2003; Cooper and Keller 2001) as do studies from other continents in tropical regions (Ryu *et al.* 2002). Although the Sloss sequences have been linked to Laurentian orogenies by Johnson (1971), they also are recognizable on some other continents (e.g. Russian Platform; Sloss 1972). Presumably these other continents also have missing stratigraphic records comparable with the situation in Laurentia.

Sweet (1995) created a conodont-based composite standard for the Ordovician System of North America, and he called attention to problems involved in finding places to bridge some of the time gaps so as to assemble a complete graphic record. These difficulties imply that such time intervals may be represented at only a few locations, and parts of the record may have been removed by erosion at most places. Some relatively continuous sections for the Cambrian and Ordovician have been described, but many sections in cratonic settings and in miogeoclinal environments display significant hiatuses in the stratigraphic record. Usually we select the most complete available section in order to establish biozonations, and we tend to ignore incomplete successions or those deposited in harsh environments that render them unfavourable for recovering faunal associations. However, we must consider strata deposited in such places as possible repositories of strange and hidden shallow-water taxa.

Clavohamulus: *a complex story with many gaps*

The evolution of *Clavohamulus* is interesting because of the unusual pattern whereby taxa spread out over the platform during transgressions. Usually we see open-marine taxa invading large shelf areas during a major sea-level rise (see Holland 1996). During such intervals the Ibexian species of *Clavohamulus* appeared as Lazarus taxa. They have no cratonic record for regressive phases, during which they were probably restricted to extremely shallow-water habitats in the inner shelf areas. Any shallow cratonic sediments that were deposited were removed later during subsequent regressions, thereby removing evidence of the existence of *Clavohamulus* at the time of their deposition.

For example, biostratigraphers realized that a widespread hiatus in Laurentia spans long time intervals across the Cambrian–Ordovician transition (Myrow *et al.* 1995, 1999, 2003; Runkel *et al.* 1999). However, complete, well-known sections spanning this time interval exist in the Southern Oklahoma Aulacogen and in miogeoclinal areas such as at Lawson Cove in Utah. In general, an interval from somewhat below the base of the Ibexian through the lower part of that series is missing as a major hiatus in many cratonic areas across Laurentia (e.g. Minnesota, Wisconsin, Colorado). The presence and magnitude of this large hiatus is not well understood by many stratigraphers in northern Europe who believe that Ordovician faunal development began with the Tremadocian transgression. The missing early representatives of the *Clavohamulus* lineage will probably never be recovered in cratonic areas because of this hiatus.

A monographic revision of *Clavohamulus* is in preparation by Lehnert, Ethington, Miller, Repetski and Sweet. The complex evolution of this important Ibexian index genus was first presented by Lehnert *et al.* (1998). *Clavohamulus rarus* (Miller) appears as the oldest species of *Clavohamulus* during the sea-level rise documented in the upper part of the lowermost Ibexian *Hirsutodontus hirsutus* Subzone of the *Cordylodus proavus* Zone. This horizon is at the transgressive base of the second package of the Lange Ranch Lowstand. The species continues through the rest of the Lange Ranch Lowstand and is rare through much of the next higher sequence (i.e. most of the *Fryxellodontus inornatus* Subzone).

Clavohamulus rarus was followed during the next transgressive phase by *Cl. elongatus* and *Cl. bulbousus* which are introduced at the base of the *C. elongatus* Subzone of the *Cordylodus proavus* Zone. Both species are abundant throughout their ranges, and they disappeared abruptly with the lowering of sea-level at the start of the Basal House Lowstand. An as yet unnamed species of *Clavohamulus* is known from the *C. elongatus* Subzone in Laurentia, as is *C. triangularis* Abaimova in Siberia. Another unnamed species of *Clavohamulus* (see Miller 1969, pl. 64, fig. 5 only) is introduced in the upper part of the *Hirsutodontus simplex* Subzone and continued into the *Clavohamulus hintzei* Subzone of the *Cordylodus intermedius* Zone. Coincident with this sea-level rise in the *Clavohamulus hintzei* Subzone are the appearances of *Clavohamulus hintzei* Miller in Laurentia and Australia and of another new species of the genus in China. Early to late forms of *C. hintzei* are shown in Text-figure 3 to indicate how rapid morphological change of elements of *Clavohamulus* might have been in response to sea-level change.

Clavohamulus disappeared with the *Acerocare* Regressive Event (Drum Mountain Lowstand), a major extinction level at the top of the *C. intermedius* Zone, and a subsequent long gap in the record of *Clavohamulus* spans the *C. lindstromi*, *Iapetognathus* and *C. angulatus* zones (Text-fig. 3). A record within this gap exists in peri-Laurentian slope deposits that received sediment eroded from an adjacent shallow platform, e.g. the mutual occurrence of *C. hintzei* and *C. elongatus* with *Cordylodus lindstromi* and *Iapetognathus* in the Green Point Formation of western Newfoundland (Barnes 1988). The order of first appearance of *C. elongatus* and *C. hintzei* in the data set for the Green Point Section (Barnes 1988; Cooper *et al.* 2001) is in reverse order to that found in well-controlled shelf

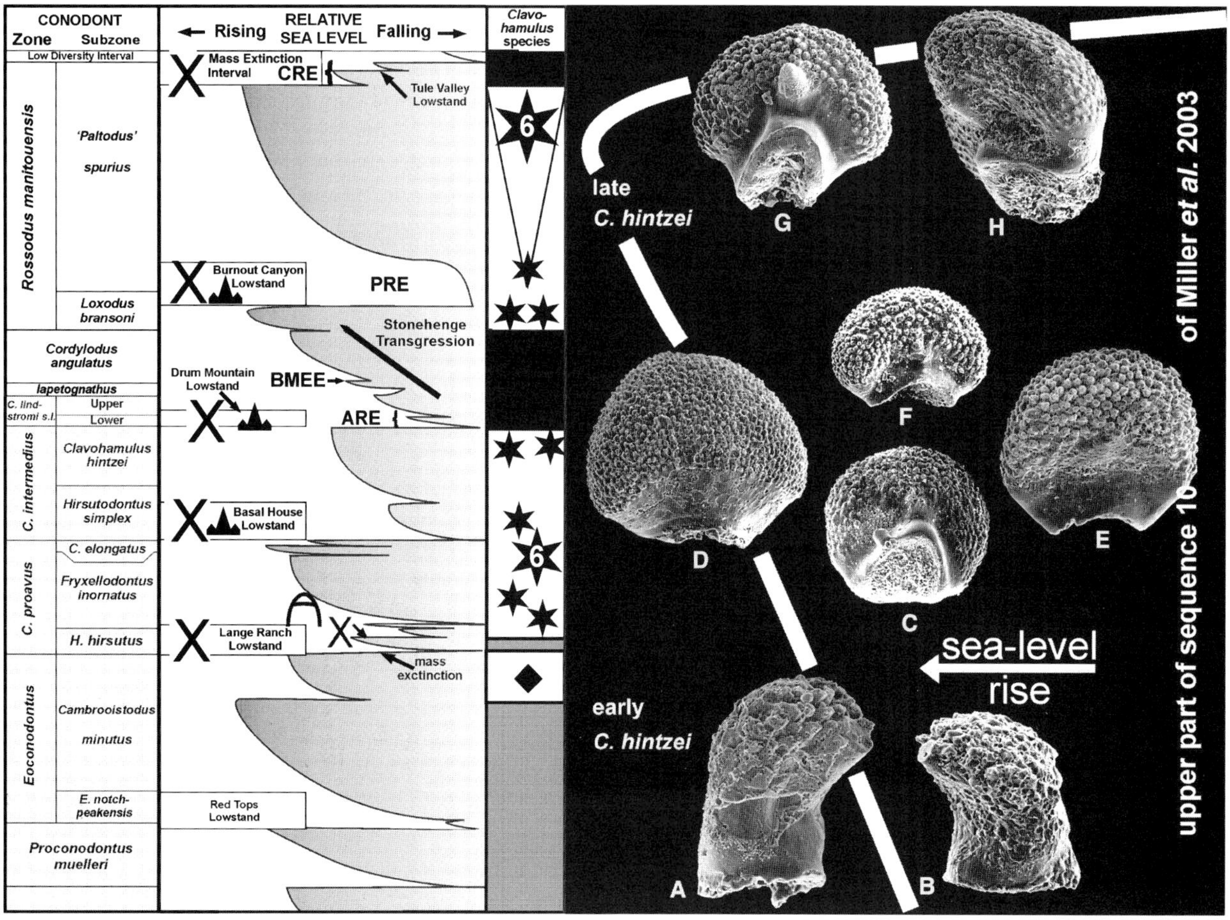

TEXT-FIG. 3. Left-hand side displays increase in taxa during sea-level rise, their decrease and absence during lowstands and, sometimes, during the early part of the following transgression (Stonehenge transgression). Grey shading indicates the Cambrian conodont zones. The number of *Clavohamulus* species related to sea-level change is documented by one star for a single species and by stars with numbers for multiple species; the sea-level curve is taken from Miller *et al.* (2003). Extinctions are marked by X. On the right is an example of plasticity among elements of one species, *Clavohamulus hintzei*, A–H (A, ×157; B, ×170; C, ×80; D, ×91; E, ×143; F, ×47; G, ×86; H, ×88), and its postulated relation to sea-level change.

sections, giving evidence of progressive down-section cannibalization of the shelf and redeposition of the species of *Clavohamulus* in reverse order in the accumulated debris.

Younger species of *Clavohamulus* are present in the lower part of the *Rossodus manitouensis* Zone. *Clavohamulus densus* and a species close to *C. hintzei* occur there. The latter species disappeared during the following Burnout Canyon Lowstand (*Peltocare* Regressive Event). With the next rise, *Clavohamulus sphaericus*, the reniform element of *C. reniformis* Ji and Barnes (an acontiodontiform element not belonging in the *Clavohamulus* apparatus was selected by Ji and Barnes 1994 as the holotype of *C. reniformis*), *Clavohamulus birdbeakensis* Seo and Ethington, and *Clavohamulus* aff. *C. birdbeakensis* appeared in ascending order during the ongoing transgression. At this time *Clavohamulus* became very plastic. *Clavohamulus densus* and other *Clavohamulus* species co-occurred during the rise of sea-level and are preserved in a green shale band in the Oneota Formation near Marquette, Iowa (undescribed collections of Ethington, Miller and Repetski). Geographically separate populations of these taxa have been observed. They became extinct together with many of the associated conodont species during the mass-extinction that accompanied the Tule Valley Lowstand (*Ceratopyge* Regressive Event) at the top of the *Rossodus manitouensis* Zone.

Clavohamulus has not been found in the Low Diversity Interval that follows the *Rossodus manitouensis* Zone, but four species are known to us from the interval spanning the *Macerodus dianae–Oepikodus communis* zones. No record of *Clavohamulus* has been found in uppermost Ibexian or younger Ordovician strata. The pattern of occurrences and gaps in the *Clavohamulus* lineage documented in the Ibex area of Utah has been confirmed in other parts of Laurentia, and to a lesser extent in Australia. Speciation in *Clavohamulus* must have occurred in unknown refugia during sea-level lowstands. The evolutionary patterns of *Clavohamulus* and the palaeogeographical distribution of its species are clearly connected with Ibexian sea-level changes. It is obvious that *Clavohamulus* reappeared in shallow-water environments during rising sea-level. This may be one of the rare cases which seems to contradict Holland's (1996) model that shallow-water taxa should disappear during the transgressive systems tract.

Leukorhinion *and successors: born in a flood and killed by a regression*

The evolutionary succession from *Leukorhinion ambonodes* Landing through species of two new genera provides a mechanism for subdivision of the extensive lower Ibexian *Rossodus manitouensis* Zone into four parts (Lehnert *et al.* 2000). Parts of the evolutionary succession are recorded in many shallow-water areas across Laurentia (western Newfoundland, Greenland, Appalachians, American Midcontinent and Utah). The ancestor of the lineage, *Leukorhinion* Landing, is monospecific, but in New Genus A we distinguish two species: New Genus A, new species 1 and New Genus A *reniformis* (Ji and Barnes 1994). New Genus B is monospecific, represented by New Genus B *manitouensis* (Seo and Ethington 1993). This well-defined lineage is restricted stratigraphically to the *Rossodus manitouensis* Zone and palaeogeographically to Laurentia.

The lowest occurrences of coniform *L. ambonodes* are associated with a sea-level rise at the base of the *Rossodus manitouensis* Zone in Texas and Utah. This species is followed by younger ones in which the lateral margins of the cusp are increasingly elaborated and eventually are transformed into symmetrical quasi-ramiform elements (Text-fig. 4). The lowest occurrence of the lineage is in the lower *R. manitouensis* Zone, but we know of no occurrences during the Burnout Canyon Lowstand (see Text-fig. 3). Most of the evolutionary succession occurs above this lowstand, within the transgressive to highstand parts of this zone. The lineage apparently became extinct during the stepwise regressions of the Tule Valley Lowstand (earliest *Ceratopyge* Regressive Event in Europe) at the top of the *R. manitouensis* Zone.

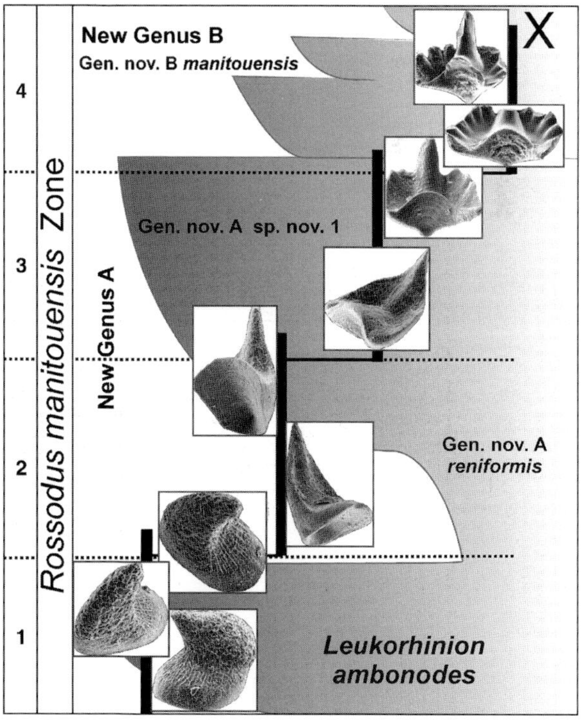

TEXT-FIG. 4. The phylogenetic lineage of *Leukorhinion ambonodes* Landing and its successors in the *Rossodus manitouensis* Zone (after Lehnert *et al.* 2000), which can be used to subdivide that zone in Laurentia.

Chosonodina: *more gaps than record*

The known stratigraphic record for some taxa shows large gaps between species of the same genus. For example, *Chosonodina* Müller comprises six species to date: *C. fisheri* Druce and Jones, *C. herfurthi* Müller, *C. tridentata* Zhang, *C. chirodina* Jiang, *C. lunata* Harris and Harris, and *C. rigbyi* Ethington and Clark. The first four of these species occur in low Ibexian (= low to middle Tremadocian) strata (Text-fig. 5), whereas *C. lunata* and *C. rigbyi* appear in the lower Whiterockian (= upper Arenigian).

Thus far *C. fisheri*, a species with three medial denticles with rounded tips between the marginal denticles, is known from Australia (*Cordylodus rotundatus-Cordylodus angulatus* Zone of Druce and Jones 1971) and north China (*Cordylodus rotundatus-'Acodus' oneotensis* Zone of An *et al.* 1983). The very similar *C. chirodina* Jiang (*in* An *et al.* 1983) with four instead of three rounded medial denticles is restricted to the *C. rotundatus-'A.' oneotensis* Zone of An *et al.* (1983). It is beyond the scope of this paper to discuss whether this taxon is a junior synonym of *C. fisheri*. The very distinct *C. tridentata* Zhang (*in* An *et al.* 1983) has only one sharply pointed medial denticle and is also known from north China in the *C. rotundatus-'A.' oneotensis* Zone of An *et al.* (1983). It has been reported by Chen and Gong (1986) from their *C. angulatus-C. herfurthi* Zone in the Hebei and Jilin provinces of China. Whether we treat the Chinese material as different form-taxa, interpret them as synonyms in the form sense or combine them as different elements of a multielement taxon, their stratigraphic range corresponds to the *C. angulatus* through lower *R. manitouensis* zones in terms of the North American zonation of Ross *et al.* (1997).

Chosonodina herfurthi has been reported widely, including occurrences in Korea (Müller 1964), southern Oklahoma (Mound 1968; Dresbach 1998), Colorado (Seo and Ethington 1993), Arkansas (Repetski and Ethington 1977), the Appalachian Mountains (e.g. Repetski and Perry 1980; Harris *et al.* 1994), Utah (Miller *et al.* 2003), Mexico (Stewart *et al.* 1999), Australia (Druce and Jones 1971) and New Zealand (Cooper and Druce 1975). In contrast, *Chosonodina lunata* is known thus far only from the Whiterockian uppermost part of the West Spring Creek Formation and from the Joins Formation, both in the Arbuckle Mountains of southern Oklahoma (Harris and Harris 1965; McHargue 1975). *Chosonodina rigbyi* has been found in the Kanosh Shale of western Utah (Ethington and Clark 1981), the Antelope Valley Limestone in central Nevada (Harris *et al.* 1979), the Knox and Beekmantown groups of the central Appalachian region (Harris and Repetski 1982), and the upper part of the Durness Group in north-west Scotland (undescribed collection of Repetski). Detailed studies of middle and upper Ibexian rocks and their equivalents in North America (Nowlan 1976; Ethington and Clark 1981; Stouge and Bagnoli 1988; Ji and Barnes 1994; Johnston and Barnes 2000), Europe (Lindström 1955; Löfgren 1978; Stouge and Bagnoli 1990; Dubinina 2000) and South America (Albanesi 1998) have not revealed any intermediate species to document a lineage connecting *C. herfurthi*, *C. fisheri*, *C. chirodina* and *C. tridentatus* with *C. lunata* and *C. rigbyi*.

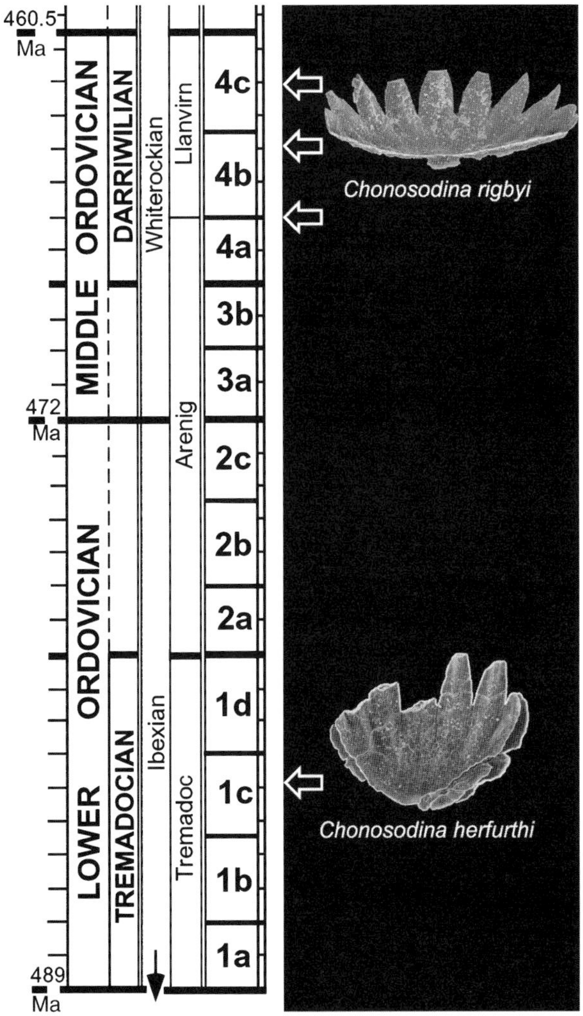

TEXT-FIG. 5. Stratigraphic occurrences of species of *Chosonodina* in Laurentia illustrating the stratigraphic gap between *C. herfurthi* and *C. rigbyi* (stratigraphic chart compiled by Webby *et al.* 2004).

A similar situation exists in the lineage of *Histiodella* Harris. The earliest species, *H. donnae* Repetski, has a rather short range within the lower Ibexian *Macerodus dianae* Zone, and it is widespread in Laurentian Ibexian successions. The next recorded species, *H. altifrons* Harris, does not appear until the earliest Whiterockian. The genus has no record for approximately the upper half of the Ibexian (lower half of the Arenigian) or four Laurentian conodont biozones (Repetski and Repetski 2002).

Interestingly, although the early species of both *Chosonodina* and *Histiodella* occupied niches in open-shelf, normal marine environments, the first successor species of these genera appear to have favoured much shallower, more inner shelf environments. These initial successor species appear to have been limited to Laurentia. Although *H. altifrons* occurs in shallow-water successions, later species of *Histiodella* lived in more outboard environments, even to the shelf break. If early species abandoned their open-shelf habitats and migrated to shallower niches, the gaps in the records of both genera are sufficiently large that adequate time is represented for near or complete elimination of the deposits of those habitats during the several periods of sedimentary offlap. In most of central Laurentia, where these shallow-water niches would have existed, much or all of the upper Ibexian record was eroded during the major lowstand that resulted in the Knox/Beekmantown unconformity.

Cahabagnathus*: a Middle Ordovician story*

A close connection of speciation in *Cahabagnathus* and Chazyan sea-level change was pointed out by Leslie and Lehnert (1999; Text-fig. 6 herein). In addition, the

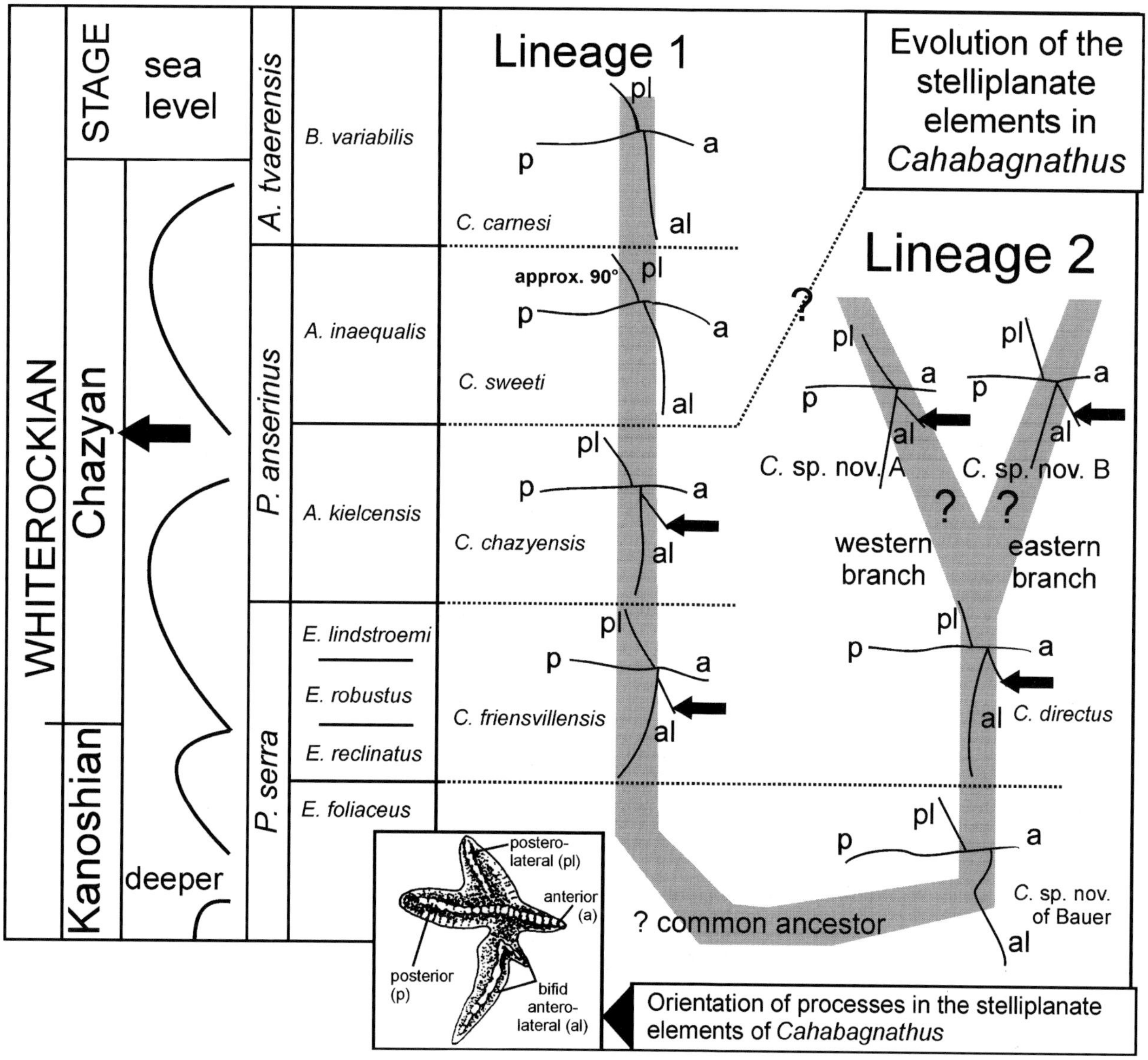

TEXT-FIG. 6. Biostratigraphy and phylogeny of the stelliplanate elements in two *Cahabagnathus* lineages, and their relationships to Chazyan sea-level changes (modified from Leslie and Lehnert 1999).

palaeogeographical distribution of *Cahabagnathus* species is clearly dependent on transgressions and regressions.

Bergström (1983) documented the evolutionary relationship in one lineage of *Cahabagnathus* (*C. friendsvillensis–C. chazyensis–C. sweeti–C. carnesi*; lineage 1, Text-fig. 6). Bauer (1987, 1990) subsequently suggested an expanded phylogeny for *Cahabagnathus* and the existence of two separate lineages based on the discovery of new taxa, *Cahabagnathus directus* Bauer and *Cahabagnathus* sp. nov. Leslie and Lehnert (1999) concurred with the concept of two distinct lineages, and discussed aspects of the palaeogeographical distribution, palaeoecology and evolution of *Cahabagnathus* triggered by Chazyan sea-level changes.

Leslie and Lehnert (1999) enlarged *Cahabagnathus* by reporting two new species from opposite sides of the North American transcontinental arch, where populations from lineage 2 diverged during a sea-level lowstand in the Chazyan *Pygodus anserinus* Zone. These two species are the youngest taxa recognized thus far in lineage 2. This lineage represents the proposed evolutionary relationship of *Cahabagnathus* sp. nov. Bauer (1990) and *C. directus* Bauer with *Cahabagnathus* sp. nov. A discovered by Lehnert in south-west Nevada and *Cahabagnathus* sp. nov. B found by Leslie in north-central Arkansas. The addition to lineage 2 of these taxa suggests that some gaps that we see on the platform may be filled by sampling lithologies traditionally considered to be 'unfavourable' for recovery of conodonts. The species of *Cahabagnathus* from Nevada and Arkansas were obtained from dolomitic sandstones and dolomites that were deposited in separate refugia where isolated populations diverged from within a common lineage.

A connection to the sea-level changes described by Ross and Ross (1995) can be detected in both *Cahabagnathus* lineages. *Cahabagnathus friendsvillensis* appeared during a first transgression in the late Whiterockian (Chazyan), and was followed by *C. chazyensis* during the first regression. Wide distribution of *C. sweeti* occurred during the second transgression, and restricted distribution of *C. carnesi* coincided with the second regression. In *Cahabagnathus* lineage 2, occurrences of *C. directus* represent the first Chazyan transgression. Descendants of *C. directus* may have been isolated from each other on the west and east sides of the arch during the first regression, thus giving rise to two new species, *Cahabagnathus* sp. nov. A and *C.* sp. nov. B.

THE UNDISCOVERED RECORD: FILLING THE GAPS

Where can we look for additional hidden taxa? Several possibilities come to mind: lowstand deposits along continental margins, fillings of karst features and pipes on carbonate platforms, clasts in a variety of settings. Some of the gaps that we see on the platform may be filled by sampling the unfavourable lithologies that we usually avoid such as dolomites, dolomitic sandstones, sandstones, shales and greywackes.

Clast research

Erosion of strata over wide areas may result in complete elimination of any palaeontological record. Fortunately, in many cases, fragments of eroded material were deposited in new settings and geographical regions. Research on such clast material can provide valuable information regarding the palaeobiogeographical distribution of known taxa. Such studies might also lead to discovery of new forms. Olistoliths or clasts in debris flows that were shed into continental slope environments may preserve information about platform faunas that we cannot find in autochthonous successions.

For example, new ostracod taxa were described from such clasts in the Argentine Precordillera by Schallreuter (1999). *Rhodesognathus elegans*, which dominates the conodont fauna in one clast from these slope deposits (undescribed collection of Lehnert), demonstrates a relatively warm period during the Caradocian that has not been recognized in autochthonous sections. Pohler and James (1989) reconstructed the former inner-shelf palaeoenvironments of Newfoundland by using information from clasts preserved in the slope deposits of the Cow Head Group. Another example of the reconstruction of an early Palaeozoic succession is the research on clasts from the resurge breccia of the Tvären Bay crater in Sweden by Ormö (1994). Armstrong *et al.* (1999) and Armstrong and Owen (2000) described Middle Ordovician conodont faunas from clasts in the Old Red Sandstone of the Midland Valley in Scotland. They reconstructed a cryptic arc terrane as the source area of the clasts and discussed its palaeogeographical position. Similarly, Ordovician conodonts commonly occur in clasts of shallow-water limestones in the siliceous shales of the thick Pennsylvanian flysch sequence of the Arkoma Basin in Arkansas and Oklahoma (undescribed collections of Ethington).

Many new taxa of ostracods and other fossils groups have been reported from glacial erratics in northern Germany (see Schallreuter 1998 for references). As Wolska (1961) and Dzik (1976) did in northern Poland, we could look for new conodont taxa in glacial erratics. For example, the Tertiary basin on north-east Ellesmere Island was filled mainly by detritus of early Palaeozoic carbonates derived from the Arctic Platform. Research was initiated by the Canadian Geological Survey in co-operation with the Geologische Bundesanstalt (Hannover, Germany) to determine the source of the clasts and the geodynamic history (polyphase movements and erosion) of the basins that formed during the opening of Baffin Bay and the Labrador Sea (see Lee *et al.* in press). In addition to well-known conodont species, a few forms were found such as a new genus that is similar to *Rhipidognathus* (Lee *et al.* in press). A major problem with describing new taxa from clasts is that the available material is usually limited in quantity and out of stratigraphic context. Nevertheless, conodont taxa that are not available from other settings may be found there.

Cold-water clastics

Conodont elements recovered from detrital deposits of cold-water areas (sandstones, shales) are often fragmented, and their number is usually too limited to permit reconstruction of apparatuses. For example, Pierre Breuer (University of Liége, Belgium) recently discovered conodonts in the uppermost Tremadocian (?) greywackes in the Salm Group of Belgium (Breuer 2002; Vanguestaine *et al.* in press). The assemblage is low in diversity and includes only non-geniculate coniform elements.

Even when suberectiform elements are found that resemble those of some *Drepanoistodus*, the characteristic geniculate elements of this apparatus are missing. Most of the drepanodontiform elements may belong to *Drepanodus* apparatuses and resemble those of *Drepanodus arcuatus* Pander or other species of *Drepanodus*. There are few elements that are comparable with pipaform Pb and graciliform M elements because the material is often fragmented at the base and hard to interpret. Possible reasons for the absence of oistodiform M elements typical for species of *Drepanoistodus* are: (1) such geniculate elements are absent because the collection is too small; (2) some of the poorly preserved pipaform elements present belong to an early *Drepanoistodus*? apparatus but cannot be distinguished from those poorly preserved ones of *Drepanodus arcuatus*; (3) this is a uniquely cold-water taxon with drepanodontiform and suberectiform elements similar to those of species of *Drepanoistodus* but lacking other elements characteristic of the apparatus. Many such situations might remain unresolved because we may never have abundant, well-preserved material from cold-water clastic deposits. The few other known assemblages from northern peri-Gondwana and Gondwana, e.g. *Acontiodus franconicus* and associated taxa from the Frankenwald area (Sannemann 1955) and that reported from the lower Loire Valley (Lindström 1976), share these problems of uncertainty regarding assignment to a known apparatus.

The estimation of the amount of material and number of taxa reworked into detrital successions is another difficulty. One of us (Lehnert) has recently begun to process material from tuffaceous layers in the Arenigian of the lower Klabava Formation in Bohemia (in co-operation with Olda Fatka and Petr Kraft, Charles University, Prague). It is not yet clear how much reworked and fragmented Upper Cambrian and Tremadocian material is involved.

Stratigraphic gaps in deep-water clastics, an example from western Texas

Leslie *et al.* (2002) presented data from cores from south-west Texas that demonstrate how our understanding of distributions of species in space and time are affected by limited access to deeply buried sections and by potential problems with interpreting biostratigraphic significance of conodonts in carbonate beds in deep-water clastic successions. The Brown-Basset well in Terrell County, Texas, penetrated Upper Ordovician shallow-water rocks with *Noixodontus*, which indicates a Hirnantian (late Cincinnatian, late Ashgillian) age. A middle–late Mohawkian (middle Caradocian) conodont fauna containing *Aphelognathus* cf. *A. gigas*, *Belodina compressa*, *Drepanoistodus suberectus*, *Erismodus radicans*, *Panderodus gracilis*, *Phragmodus undatus*, *Plectodina aculeata*, *Oistodus* sp., *Curtognathus* spp. and *Scyphiodus primus* occurs below this Hirnantian interval. The presence of *Scyphiodus primus* is of particular interest because it greatly expands the known geographical range of this characteristic species of the western Midcontinent Fauna, indicating how tied we often are to surface and shallow-core data when interpreting the geographical spread of species.

Deep-water shales exposed in the Marathon fold and thrust belt of south-west Texas are in part coeval with the cored interval and are famous for their graptolite faunas (e.g. Berry 1960). Bergström (1978) described the conodont biostratigraphy of the Middle and Late Ordovician deep-water succession in the Woods Hollow Shale and Maravillas Formation, and Goldman *et al.* (1995) revised the graptolite biostratigraphy of part of this interval. The top of the Woods Hollow Shale is in the upper part of the *Pygodus anserinus* conodont Zone and the base of the Maravillas Formation is in the *Amorphognathus ordovicicus* conodont Zone (Bergström 1978; Goldman *et al.* 1995). This indicates that the unconformity between these formations encompasses the *A. tvaerensis* and *A. superbus* conodont zones (Text-fig. 7). Graptolites place the top of the Woods Hollow Shale in the lower *Climacograptus bicornis* graptolite Zone (= lower–middle *D. foliaceous* Zone) and the base of the Maravillas Formation in the *Dicellograptus gravis* graptolite Zone (= upper *A. manitoulinensis* Zone). This suggests that correlatives of the North American *C. lanceolatus*, *O. ruedemanni*, *C. spinferus* and *G. pygmaeus* graptolite zones are not present in the Marathon region.

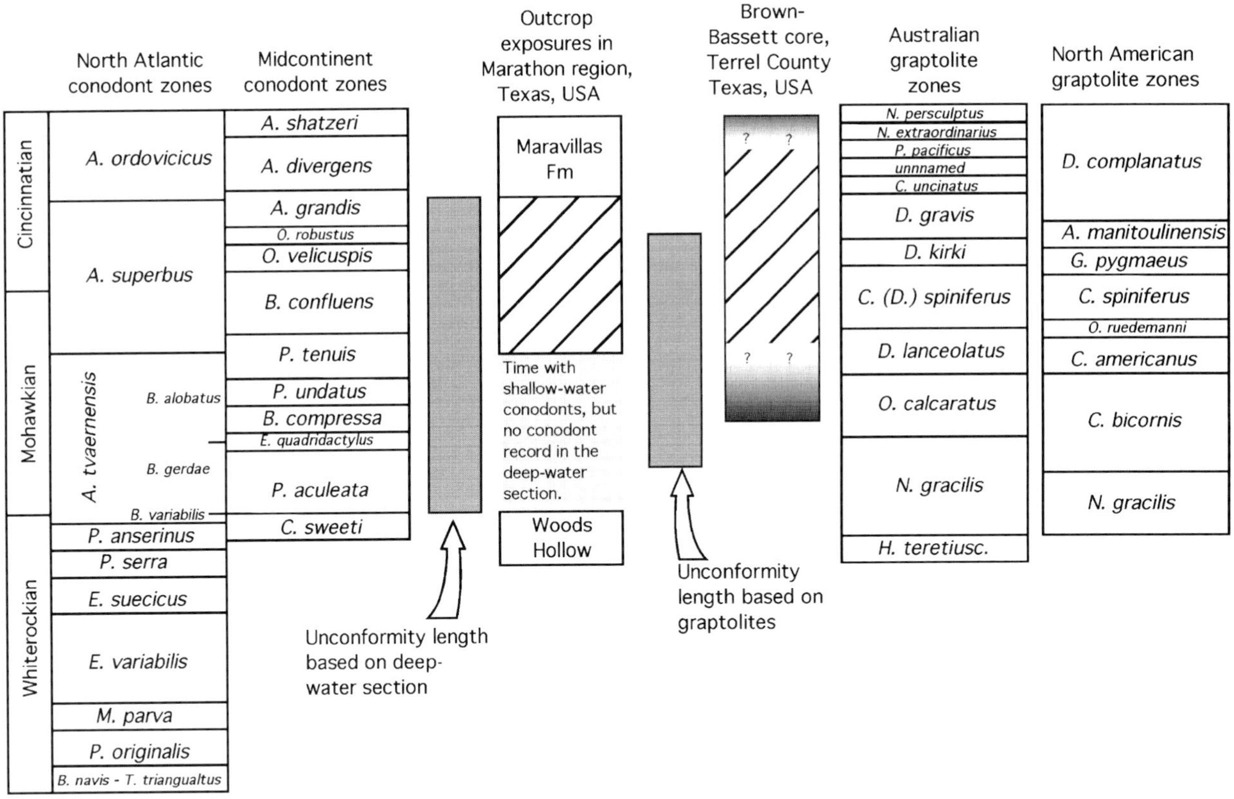

TEXT-FIG. 7. Stratigraphic chart with the range of faunas in the Marathon area, west Texas (from Leslie *et al.* 2002).

The shallow-water conodont fauna recovered from the relatively nearby (*c.* 150 km) Brown-Bassett core is difficult to place within the temporal context of the deeper-water conodont fauna recovered from outcrops; none of the key species are known from the Marathon exposure. Apparently, part of the section that is missing in the deeper rocks (the Mohawkian part) is preserved on the platform (Ross *et al.* 1982). Using the graptolite ages together with the conodonts from the Brown-Bassett core, Leslie *et al.* (2002) suggested two alternative explanations: (1) that the upper Woods Hollow Shale is Mohawkian and that the conodonts recovered from carbonate beds in the Woods Hollow may be a Lazarus fauna that was deposited in carbonate turbidites as sea-level fell and incision of the platform-slope carbonates began, or (2) tectonic subsidence occurred on the platform at the same time as shallowing in the deep-water interval in an aulacogen or foreland basin.

Unexpected discoveries: an example from Laurentia

The faunas described by Ethington *et al.* (1986) from subsurface rocks of eastern Indiana demonstrate the possibility of recovery of strange and previously unknown taxa. They were adapted to harsh environments during the early flooding of huge karst regions with gentle to moderate relief that led to areas of very shallow water and restricted circulation. Taxa of these unusual faunas are now known elsewhere only from the western flank of the Illinois Basin, where a rather thin succession of rocks coeval with those of the Ancell Group of southern Indiana crop out. Elements of this fauna are found in the uppermost beds of the Dutchtown Formation and in beds transitional from the Dutchtown into the basal part of the overlying Joachim Dolomite in at least five sites in Scott, Cape Girardeau and Perry counties, south-east Missouri (Repetski 1973, pp. 118–119; undescribed collections of Repetski, Ethington and Tom Shaw). It now appears that the uppermost Dutchtown Formation of the south-east Missouri outcrop belt probably correlates best with the lower part of the Dutchtown in south-east Indiana. The valleys became extremely shallow-water estuaries with restricted circulation during the early transgression of the Ancell sea into the southern part of the Illinois Basin from the Arkoma Basin to the south-west and the Appalachian Basin to the south-east (Ethington *et al.* 1986). Strange hyaline taxa occupied these habitats under these unfavourable conditions and dominated the conodont faunas for a short time. These specially adapted conodonts are known only from a few drill holes and a few localities along the edge of one basin. In general, such endemics may have been restricted to short periods and small areas.

Ethington *et al.* (1986) described four new genera, two of them in open nomenclature, and several new species from the Indiana subsurface. These are the coniform elements of *Lumidens vitreus* and the spatula-like elements of *Scapulidens* as well as the superficially *Belodina*- or *Pseudobelodina*-like New Genus B (Text-fig. 8A), and the *Icriodus*-like specimens of New Genus A (Text-fig. 8D). Other new species of Ethington *et al.* (1986) include *Leptochirognathus resimus*, *Stereoconus crepidiformis* and *Stereoconus* aff. *S. plenus*. They also reported the presence of *Prionognathus ordovicicus*, which occurs elsewhere in the Harding Sandstone in Colorado (Branson and Mehl 1933). The stellate plates that they documented (Pl. 2, fig. 27; Text-fig. 8C herein) resemble the less robust *Cambropustula kinnekullensis* described only from the Upper Cambrian of Sweden by Müller and Hinz (1991). Several species were reported in open nomenclature (*Coleodus*? spp. A–C and *Erismodus* sp.). Albid taxa (*Dapsilodus* sp., *Panderodus* sp., *Phragmodus* sp. and *Plectodina* aff. *P. joachimensis*) are very rare and occur only at a few

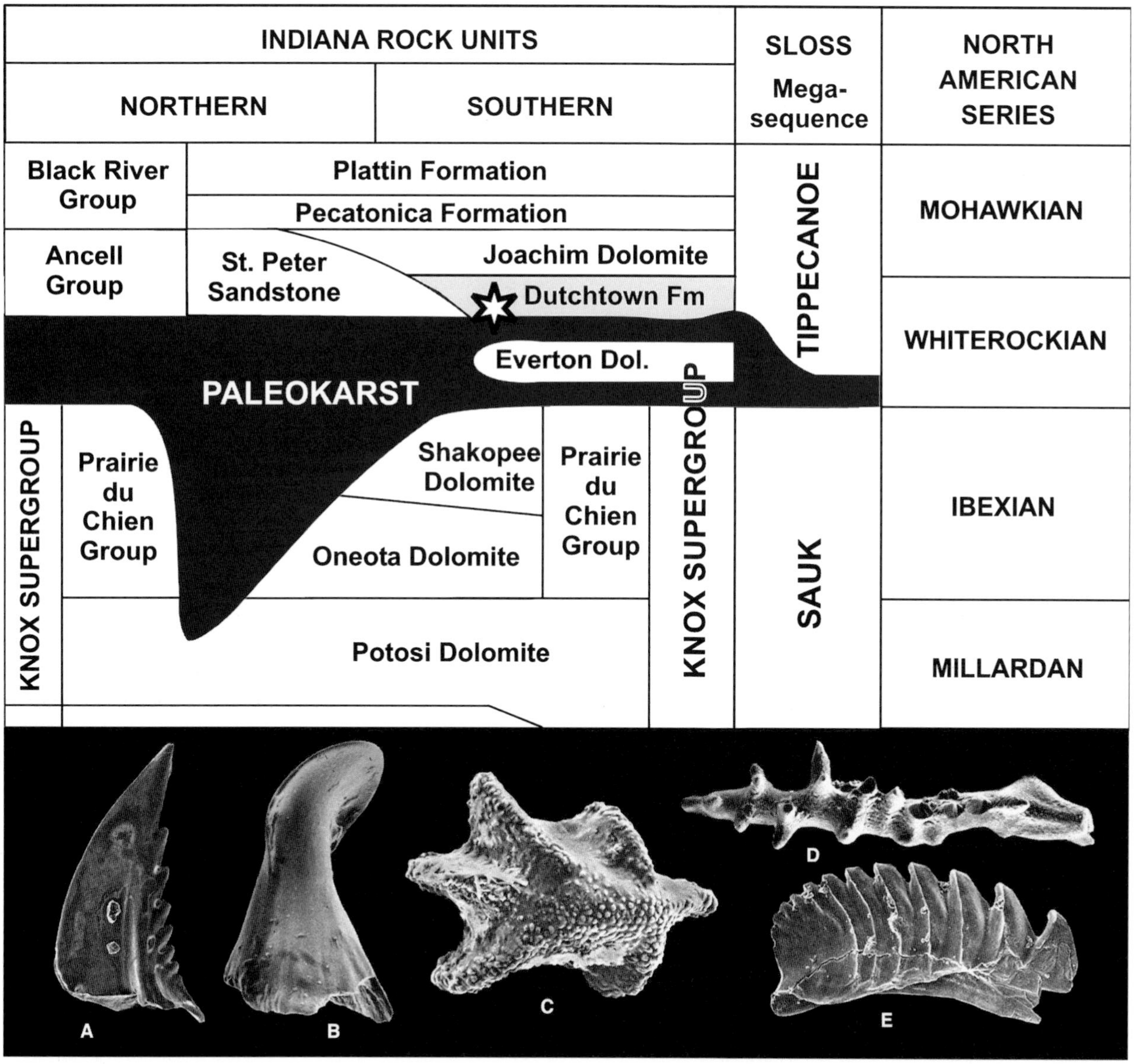

TEXT-FIG. 8. Stratigraphic chart showing the lithological units in the Sauk and Tippecanoe sequences of Indiana. The position of the enigmatic shallow-water fauna described by Ethington *et al.* (1986) from the lower part of the Dutchtown Formation (grey) is indicated by a star; A–E display some representatives of the Dutchtown fauna. A, New Genus B of Ethington *et al.* 1986; ×27. B, *Scapulidens primus*; ×40. C, 'stellate plate'; ×91. D, New Genus A of Ethington *et al.* 1986; ×75. E, *Coleodus*? sp. B; ×35.

levels. The Missouri outcrops in this interval contain *Scapulidens primus* (Text-fig. 8B), *Prionognathus ordovicicus*, *Coleodus* aff. *C. delicatus* Branson and Mehl, and species of *Erismodus* and *Curtognathus*. As with the samples from the subsurface of Indiana, albid taxa are extremely rare in these Missouri samples.

Intervals of major regressions and karstification that exposed conodonts to unusual ecological conditions may have occurred frequently for early Palaeozoic tropical continents. Regions where these conditions existed would be good candidates for hosting missing segments of lineages that have been defined incompletely on the basis of well-documented successions of rocks that accumulated under conditions more favourable to conodonts.

CONCLUSIONS

Regressions associated with sequence boundaries caused extensive erosion and karstification of large tropical platforms during early Palaeozoic times and probably erased the record of many endemic conodonts. Conodont provincialism increased during the Early Ordovician as diversity increased, and these trends are associated with times of significant sea-level fluctuations when conodont (and trilobite) assemblages underwent wholesale turnovers. Elimination of taxa from the fossil record may have been more frequent than during any post-Ordovician interval. Based on the examples from evolutionary lineages that are well-documented but still have stratigraphic gaps, many hidden taxa await discovery in unfavourable lithologies and in remote areas. These faunas, which are critical for establishing conodont phylogeny, may be found in rocks that are likely to yield only small and sparse faunas.

Our conclusions include five main points.

1. A close relationship exists between Cambrian–Ordovician sea-level changes, stratigraphic sequences and taxonomic ranges; documenting all three of these factors is critical if one wishes to understand evolutionary lineages. The effect is greatest in cratonic areas, and different scales of sequence boundaries result in different scales of missing rock and faunal records.
2. Even when studying rather complete stratigraphic successions (e.g. the Ibex Area in western Utah), precise sampling reveals gaps in some evolutionary lineages (e.g. *Clavohamulus*); such gaps commonly follow extinctions that, in turn, are associated with times of sea-level reduction.
3. Migrations from outside areas may occur during sea-level falls, or new taxa may appear in the record for the first time.
4. More common than migrations during sea-level falls are appearances of new taxa within evolutionary lineages during sea-level rises (e.g. species of *Clavohamulus*).
5. Refugia where speciation occurred may be recorded in lithologies that are unusual, difficult to process, or that occur in remote areas that have not received thorough study.

Acknowledgements. The paper was improved significantly as a result of thorough reviews by Paul Smith (University of Birmingham), John F. Taylor (Indiana University of Pennsylvania), C. Blaine Cecil (USGS), and one anonymous reviewer. Studies on Cambro-Ordovician conodonts by Lehnert were supported by several research grants from the German Science Foundation (DFG). The recent research on Laurentian material was funded by a habilitation grant of the DFG (Le 847-1–Le 847-4). The publication was compiled during a stay at the Charles University in Prague (Czech Republic) with a Feodor Lynen Fellowship of the Alexander von Humboldt Foundation (Bonn, Germany). Lehnert thanks both organizations for supporting the work on early Palaeozoic conodonts from different palaeogeographical areas. Miller's research was supported by the National Science Foundation and by Southwest Missouri State University Faculty Research Grants. Ethington's work was supported by grants from the National Science Foundation and from the Petroleum Research Fund of the American Chemical Society.

REFERENCES

ALBANESI, G. 1998. Taxonomia de conodontes de las secuencias Ordovicicas del Cerro Potrerillo, Precordillera Central de San Juan, R. Argentina. *Actas de la Academia Nacional de Ciencias* (Córdoba, República Argentina), **12**, 101–227.

AN TAIXIANG, ZHANG FANG, XIANG WEIDA, ZHANG YOUQIU, XU WENHAO, ZHANG HUIJUAN, JIANG DEBIAO, YANG CHANGSHENG, LIN LIANDI, CUI ZHANTANG and YANG XINCHANG 1983. *The conodonts of North China and the adjacent regions*. Science Press, Beijing, 233 pp., 33 pls. [In Chinese, English abstract].

ARMSTRONG, H. A. 1996. Biotic recovery after mass extinction: the role of climate and ocean-state in the post-glacial (Late Ordovician–Early Silurian) recovery of the conodonts. 105–117. *In* HART, M. B. (ed.). *Biotic recovery from mass extinction events*. Geological Society, London, Special Publication, **102**, 394 pp.

—— and OWEN, A. W. 2000. Age and provenance of limestone clasts in Lower Old Red Sandstone conglomerates: implications for the strike-slip accretion of the Midland Valley Terrane. 459–471. *In* FRIEND, P. F. and WILLIAMS, B. P. (eds). *New perspectives on the Old Red Sandstone*. Geological Society, London, Special Publication, **180**, 400 pp.

—— —— and CLARKSON, E. N. K. 1999. Ordovician limestone clasts in the Lower Old Red Sandstone, Pentland Hills, southern Midland Valley Terrane. *Scottish Journal of Geology*, **35**, 1–5.

BARNES, C. R. 1988. The proposed Cambrian-Ordovician global boundary stratotype and point (GSSP) in western Newfoundland, Canada. *Geological Magazine*, **125**, 381–414.

BAUER, J. A. 1987. Conodont biostratigraphy, correlation, and depositional environments of Middle Ordovician rocks in Oklahoma. Unpublished PhD dissertation, Ohio State University, Columbus, 346 pp.

—— 1990. Stratigraphy and conodont biostratigraphy of the upper Simpson Group, Arbuckle Mountains, Oklahoma. 39–46. *In* RITTER, S. M. (ed.). *Early to Middle Paleozoic conodont biostratigraphy of the Arbuckle Mountains, southern Oklahoma*. Oklahoma Geological Survey, Guidebook, **27**, 114 pp.

BERGSTRÖM, S. M. 1978. Middle and Upper Ordovician conodont and graptolite biostratigraphy of the Marathon, Texas graptolite zone reference standard. *Palaeontology*, **21**, 723–754.

—— 1983. Biogeography, evolutionary relationships, and biostratigraphic significance of Ordovician platform conodonts. *Fossils and Strata*, **15**, 35–58.

BERRY, W. B. N. 1960. Graptolite faunas of the Marathon region, West Texas. *University of Texas, Publication*, **6005**, 179 pp.

—— and BOUCOT, A. J. 1973. Glacio-eustatic control of Late Ordovician – Early Silurian platform sedimentation and faunal changes. *Geological Society of America, Bulletin*, **84**, 275–284.

BRANSON, E. B. and MEHL, M. G. 1933. Conodonts from the Harding Sandstone of Colorado. *University of Missouri Studies*, **8**, 19–38.

BREUER, P. 2002. L'assemblage d'acritarches messaoudensis-trifidium et de la faune franconicus à conodontes dans les formations de Jalhay et d'Ottre, Ordovicien Inférieur du Synclinal de Chevron (Massif de Stavelot, Belgique): biostratigraphie, paléobiogéographie et paléogeographie. Unpublished MSc thesis, University of Liège, Belgium, 61 pp., 21 pls.

CHEN JUN-YUAN and GONG WEI-LI 1986. Conodonts. 93–223, pls. 17–54. *In* CHEN JUN-YUAN (ed.). *Aspects of Cambrian-Ordovician boundary in Dayangcha, China*. China Prospect Publishing House, Beijing, 410 pp., 98 pls.

COOPER, J. D. and KELLER, M. 2001. Paleokarst in the Ordovician of the southern Great Basin, USA, and implications for sea-level history. *Sedimentology*, **48**, 855–873.

COOPER, R. A. and DRUCE, E. D. 1975. Lower Ordovician sequence and conodonts, Mount Patriarch, north-west Nelson, New Zealand. *New Zealand Journal of Geology and Geophysics*, **18**, 551–582.

—— and NOWLAN, G. S. 1999. Proposed Global Stratotype Section and Point for the base of the Ordovician System. *Ordovician News*, **16**, 61–89.

—— —— and WILLIAMS, S. H. 2001. Global stratotype section and point for base of the Ordovician System. *Episodes*, **24**, 19–28.

DRESBACH, R. I. 1998. Early Ordovician conodonts and biostratigraphy of the Arbuckle Group in Oklahoma. Unpublished PhD dissertation, University of Missouri, Columbia, 275 pp.

DRUCE, E. C. and JONES, P. J. 1971. Cambro-Ordovician conodonts from the Burke River structural belt, Queensland. *Australian Bureau of Mineral Resources, Geology and Geophysics, Bulletin*, **110**, 163 pp.

DUBININA, S. V. 2000. Conodonts and zonal stratigraphy of the Cambrian-Ordovician boundary deposits. *Geological Institute of the Russian Academy of Sciences, Transactions*, **517**, 237 pp.

DZIK, J. 1976. Remarks on the evolution of Ordovician conodonts. *Acta Palaeontologica Polonica*, **21**, 395–455.

ETHINGTON, R. L. and CLARK, D. L. 1981. Lower and Middle Ordovician conodonts from the Ibex area, western Millard County, Utah. *Brigham Young University, Geology Studies*, **28**, 155 pp. [issued 1982].

—— DROSTE, J. B. and REXROAD, C. B. 1986. Conodonts from subsurface Champlainian (Ordovician) rocks of eastern Indiana. *Indiana Geological Survey, Special Report*, **37**, 32 pp.

—— ENGEL, K. M. and ELLIOTT, K. L. 1987. An abrupt change in conodont faunas in the Lower Ordovician of the Midcontinent Province. 111–127. *In* ALDRIDGE, R. J. (ed.). *Palaeobiology of conodonts*. Ellis Horwood, Chichester, 493 pp.

GOLDMAN, D., MITCHELL, C. E. and BERGSTRÖM, S. M. 1995. Revision of the graptolite fauna of the Marathon Limestone and the age of Fauna 13. *Lethaia*, **28**, 115–128.

HARRIS, A. G. and REPETSKI, J. E. 1982. Conodonts across the Lower-Middle Ordovician boundary – U.S. Appalachian basin: Maryland to New York. *In* JEPPSSON, L. and LÖFGREN, A. (eds). *Third European Conodont Symposium (ECOS III), Abstracts*. Publications from the Institutes of Mineralogy, Palaeontology and Quaternary Geology, University of Lund, Sweden, **238**, 13.

—— BERGSTRÖM, S. M., ETHINGTON, R. L. and ROSS, R. J. Jr 1979. Aspects of Middle and Upper Ordovician conodont biostratigraphy of carbonate facies in Nevada and southeast California and comparison with some Appalachian successions. *Brigham Young University, Geology Studies*, **26** (3), 7–44.

—— STAMM, N. R., WEARY, D. J., REPETSKI, J. E., STAMM, R. G. and PARKER, R. A. 1994. Conodont Color Alteration Index (CAI) Map and conodont-based age determinations for the Winchester 30′ × 60′ Quadrangle and adjacent area, Virginia, West Virginia, and Maryland. *US Geological Survey, Miscellaneous Field Studies Map*, **MF-2239**, 40 pp.

HARRIS, R. W. and HARRIS, B. 1965. Some West Spring Creek (Ordovician, Arbuckle) conodonts from Oklahoma. *Oklahoma Geology Notes*, **25**, 34–47.

HOLLAND, S. M. 1995. The stratigraphic distribution of fossils. *Paleobiology*, **21**, 92–109.

—— 1996. Recognizing artifactually generated coordinated stasis: implications of numerical models and strategies for field tests. *Palaeogeography, Palaeoceanography, Palaeoclimatology*, **127**, 147–156.

—— 2001. Fossils in sequence stratigraphy. 548–553. *In* BRIGGS, D. E. G. and CROWTHER, P. R. (eds). *Palaeobiology II*. Blackwell, Oxford, xv + 583 pp.

JI ZAILIANG and BARNES, C. R. 1994. Lower Ordovician conodonts of the St. George Group, Port au Port Peninsula, western Newfoundland, Canada. *Palaeontographica Canadiana*, **11**, 149 pp.

JOHNSON, J. G. 1971. Timing and coordination of orogenic, epirogenic, and eustatic events. *Geological Society of America, Bulletin*, **82**, 3263–3298.

JOHNSTON, D. I. and BARNES, C. R. 2000. Early and Middle Ordovician (Arenig) conodonts from St. Paul's Inlet and Martin Point, Cow Head Group, western Newfoundland, Canada. *Geologica et Palaeontologica*, **34**, 11–87.

LEE, C., LEHNERT, O. and NOWLAN, G. S. *in press*. Cretaceous and Tertiary sediments. *In* MAYR, U. (ed.). *Geology of NE Ellesmere Island, Nunavut, Canada*. Geological Survey of Canada, Bulletin.

LEHNERT, O., ETHINGTON, R. L., MILLER, J. F., REPETSKI, J. E. and SWEET, W. C. 1998. The complex evolution of the Ibexian (Early Ordovician) conodont genus *Clavohamulus* Furnish. *Geological Society of America, Abstracts with Programs*, **30**, 32.

—— REPETSKI, J. E., MILLER, J. F., ETHINGTON, R. L. and DRESBACH, R. 2000. Subdividing the Lower Ibexian *Rossodus manitouensis* conodont zone using *Leukorhinion* and its successor taxa. *Geological Society of America, Abstracts with Programs*, **32**, A–285.

LESLIE, S. A. and LEHNERT, O. 1999. New insight into the phylogeny and paleogeography of *Cahabagnathus* (Conodonta). *Acta Universitatia Carolinae, Geologica*, **43**, 443–446.

—— BARRICK, J. E., MOSLEY, J. and BERGSTRÖM, S. M. 2002. Conodonts from a deep core in the Upper Ordovician platform rocks of West Texas. Eighth International Conodont Symposium in Europe, ECOS VIII, abstracts. *Strata, Série 1*, **12**, 96.

LINDSTRÖM, M. 1955. Conodonts from the lowermost Ordovician strata of south-central Sweden. *Geologiska Föreningens i Stockholm Förhandlingar*, **76**, 517–604.

—— 1976. Conodont palaeogeography of the Ordovician. 501–522. *In* BASSETT, M. G. (ed.). *The Ordovician System. Proceedings of a Palaeontological Association Symposium, Birmingham, September 1974*. University of Wales Press and National Museum of Wales, Cardiff, 696 pp.

LÖFGREN, A. 1978. Arenigian and Llanvirnian conodonts from Jämtland, northern Sweden. *Fossils and Strata*, **13**, 1–129.

McHARGUE, T. R. 1975. Conodonts of the Joins Formation (Ordovician), Arbuckle Mountains, Oklahoma. MS thesis, University of Missouri, Columbia, 151 pp.

MILLER, J. F. 1969. Conodont fauna of the Notch Peak Limestone (Cambro-Ordovician), House Range, Utah. *Journal of Paleontology*, **43**, 413–439.

—— 1978. Upper Cambrian and lowest Ordovician conodont faunas of the House Range, Utah. *Southwest Missouri State University, Department of Geography and Geology, Geoscience Series*, **5**, 1–33.

—— 1984. Cambrian and earliest Ordovician conodont evolution, biofacies, and provincialism. 43–68. *In* CLARK, D. L. (ed.). *Conodont biofacies and provincialism*. Geological Society of America, Special Paper, **196**, v + 340 pp.

—— 1992. The Lange Ranch eustatic event: a regressive-transgressive couplet near the base of the Ordovician System. 395–407. *In* WEBBY, B. D. and LAURIE, J. R. (eds). *Global perspectives on Ordovician geology*. Balkema, Rotterdam, 524 pp.

—— EVANS, K. R., HOLMER, L., LOCH, J. D., ETHINGTON, R. L., STITT, J. H., HOLMER, L. and POPOV, L. E. 2003. Stratigraphy of the Sauk III Interval (Cambrian–Ordovician) in the Ibex area, western Millard County, Utah. *Brigham Young University, Geology Studies*, **47**, 23–118, plus CD-ROM with 23 tables and 17 section descriptions.

—— LOCH, J. D., STITT, J. H., ETHINGTON, R. L., POPOV, L. E., EVANS, K. R. and HOLMER, L. 1999. Origins of the Great Ordovician biodiversification: the record at Lawson Cove, Ibex area, Utah, USA. *Acta Universitatis Carolinae, Geologica*, **43**, 459–462.

MOUND, M. C. 1968. Conodonts and biostratigraphy of the lower Arbuckle Group (Ordovician), Arbuckle Mountains, Oklahoma. *Micropaleontology*, **14**, 393–434.

MÜLLER, K. J. 1964. Conodonten aus dem unteren Ordovizium von Südkorea. *Neues Jahrbuch für Geologie und Paläontologie, Abhandlungen*, **119**, 93–102.

—— and HINZ, I. 1991. Upper Cambrian conodonts from Sweden. *Fossils and Strata*, **28**, 53 pp.

MYROW, P. M., ETHINGTON, R. L. and MILLER, J. F. 1995. Cambro-Ordovician proximal shelf deposits of Colorado. 375–379. *In* COOPER, J., DROSER, M. L. and FINNEY, S. C. (eds). *Ordovician odyssey*. Pacific Section, SEPM, **77**, 498 pp.

—— TAYLOR, J. F., MILLER, J. F., ETHINGTON, R. L., BRACHLE, T. and OWEN, M. R. 1999. Stratigraphic synthesis of the Cambrian–Ordovician rocks of Colorado. *Acta Universitatis Carolinae, Geologica*, **43**, 9–11.

—— —— —— —— RIPPERDAN, R. L. and ALLEN, J. 2003. Fallen arches: dispelling myths concerning Cambrian and Ordovician paleogeography of the Rocky Mountain region. *Geological Society of America, Bulletin*, **115**, 695–713.

NOWLAN, G. S. 1976. Late Cambrian to Late Ordovician conodont evolution and biostratigraphy of the Franklinian Miogeosyncline, eastern Canadian Arctic Islands. Unpublished PhD dissertation, University of Waterloo, Ontario, 591 pp.

ORMÖ, J. 1994. The pre-impact stratigraphy of the Tvären Bay impact structure, SE Sweden. *Geologiska Föreningens i Stockholm Förhandlingar*, **116**, 139–144.

OSLEGER, D. A. and READ, J. F. 1993. Comparative analysis of methods used to define eustatic variations in outcrop: Late Cambrian interbasinal sequence development. *American Journal of Science*, **293**, 157–216.

PALMER, A. R. 1965. Trilobites of the Late Cambrian pterocephaliid biomere in the Great Basin, United States. *US Geological Survey, Professional Paper*, **493**, 105 pp.

—— 1984. The biomere problem: evolution of an idea. *Journal of Paleontology*, **58**, 599–611.

POHLER, M. L. and JAMES, J. P. 1989. Reconstruction of a Lower/Middle Ordovician carbonate shelfmargin: Cow Head Group, western Newfoundland. *Facies*, **21**, 189–262.

REPETSKI, J. E. 1973. The conodont fauna of the Dutchtown Formation (Middle Ordovician) of southeast Missouri. Unpublished MA thesis, University of Missouri, Columbia, 182 pp.

—— and ETHINGTON, R. L. 1977. Conodonts from graptolite facies in the Ouachita Mountains, Arkansas and Oklahoma. *Arkansas Geological Commission, Symposium on the Geology of the Ouachita Mountains*, **1**, 91–106.

—— and PERRY, W. J. Jr 1980. Conodonts from structural windows through the Bane Dome, Giles Conty, Virginia. *Virginia Division of Mineral Resources, Publication*, **27**, 12–22.

—— and REPETSKI, R. 2002. The apparatus of *Histiodella donnae* and its phylogenetic relationships. *Strata, Série 1*, **12**, 101.

RIPPERDAN, R. L. and MILLER, J. F. 1995. Carbon isotope ratios from the Cambrian-Ordovician boundary section at Lawson Cove, Ibex area, Utah. 129–132. *In* COOPER, J., DROSER, M. L. and FINNEY, S. C. (eds). *Ordovician odyssey*. Pacific Section, SEPM, **77**, 498 pp.

—— MAGARITZ, M., NICOLL, R. S. and SHERGOLD, J. H. 1992. Simultaneous changes in carbon isotopes, sea level, and conodont biozones within the Cambrian-Ordovician boundary interval at Black Mountain, Australia. *Geology*, **20**, 1039–1042.

ROSS, C. A. and ROSS, J. R. P. 1995. North American Ordovician depositional sequences and correlations. 309–314. *In* COOPER, J., DROSER, M. L. and FINNEY, S. C. (eds). *Ordovician odyssey*. Pacific Section, SEPM, **77**, 498 pp.

ROSS, R. J. Jr, AMSDEN, T. W., BERGSTROM, D., BERGSTRÖM, S. M., CARTER, C., CHURKIN, M., CRESSMAN, E. A., DERBY, J. R., DUTRO, J. T. Jr, ETHINGTON, R. L., FINNEY, S. C., FISHER, D. W., FISHER, J. H., HARRIS, A. G., HINTZE, L. F., KETNER, K. B., KOLATA, D. L., LANDING, E., NEUMAN, R. B., SWEET, W. C., POJETA, J. Jr, POTTER, A. W., RADER, E. K., REPETSKI, J. E., SHAVER, R. H., THOMPSON, T. L. and WEBERS, G. F. 1982. The Ordovician System in the United States: correlation chart and explanatory notes. *International Union of Geological Sciences, Publication*, **12**, 73 pp.

—— HINTZE, L. F., ETHINGTON, R. L., MILLER, J. F., TAYLOR, M. E. and REPETSKI, J. E. 1997. The Ibexian, lowermost series in the North American Ordovician. *United States Geological Survey, Professional Paper*, **1579-A**, 50 pp.

ROWELL, A. J. and BRADY, M. J. 1976. Anatomy of a biomere boundary (the Cambrian pterocephaliid/ptychaspid boundary) in the Great Basin, U.S.A. *International Geological Congress Abstracts*, **25** (1), 280–281.

RUNKEL, A., MILLER, J. F., McKAY, R., SHAW, T. and BASSETT, D. J. 1999. Cambrian-Ordovician boundary strata in the central midcontinent of North America. *Acta Universitatia Carolinae, Geologica*, **43**, 17–20.

RUSHTON, A. W. A. 1982. The biostratigraphy and correlation of the Merioneth–Tremadoc Series boundary in North Wales. 41–59. *In* BASSETT, M. E. and DEAN, W. T. (eds). *The Cambrian-Ordovician boundary: sections, fossil distributions, and correlations*. National Museum of Wales, Cardiff, Geological Series, **3**, 227 pp.

RYU IN-CHANG, DOH, SEONG-JAE and CHOI, SEONGYU 2002. Carbonate breccias in the Lower–Middle Ordovician Maggol Limestone, Taebacksan Basin, South Korea: implications for global correlation and regional tectonism. *Facies*, **46**, 35–56.

SALTZMAN, M. R., BRASIER, M. D., RIPPERDAN, R. L., ERGALIEV, G. K., LOHMANN, K. C., ROBISON, R. A., CHANG, W. T., PENG, S. and RUNNEGAR, B. 2000. A global carbon isotope excursion during the Late Cambrian: relation to trilobite extinctions, organic-matter burial and sea level. *Palaeogeography, Palaeoceanography, Palaeoclimatology*, **162**, 211–224.

—— DAVIDSON, J. P., HOLDEN, P., RUNNEGAR, B. and LOHMANN, K. C. 1995. Sea-level-driven changes in ocean chemistry at an Upper Cambrian extinction horizon. *Geology*, **23**, 893–896.

—— RUNNEGAR, B. and LOHMANN, K. C. 1998. Carbon isotope stratigraphy of Upper Cambrian (Steptoean Stage) sequences of the eastern Great Basin: record of a global oceanic event. *Geological Society of America, Bulletin*, **110**, 285–297.

SANNEMANN, D. 1955. Ordovicium und Oberdevon der bayerischen Fazies des Frankenwaldes nach Conodontenfunden. *Neues Jahrbuch für Geologie und Paläontologie, Abhandlungen*, **102**, 1–36.

SCHALLREUTER, R. 1998. Klastenforschung unter besonderer Berücksichtigung der Geschiebeforschung. *Archiv für Geschiebekunde*, **2**, 267–322.

—— 1999. Eine neue Ostrakodenfauna aus dem Ordoviz Argentiniens. 55–71. *In* REICH, M. (ed.). *Festschrift zum 65. Geburtstag von Ekkehard Herrig*. Greifswalder Geowissenschaftliche Beiträge, **6**, 535 pp.

SEO, K.-S. and ETHINGTON, R. L. 1993. Conodonts from the Manitou Formation, Colorado, U.S.A. *Journal of the Paleontological Society of Korea*, **9**, 77–92.

SHEEHAN, P. M. 1973. The relationship of the end-Ordovician glaciation to the Ordovician–Silurian changeover in North American brachiopod faunas. *Lethaia*, **6**, 147–154.

SLOSS, L. L. 1963. Sequences in the cratonic interior of North America. *Geological Society of America, Bulletin*, **74**, 93–114.

—— 1972. Synchrony of Phanerozoic sedimentary-tectonic events of the North American craton and the Russian platform. *24th International Geological Congress, Montreal, Section*, **6**, 24–32.

STEWART, J. C., POOLE, F. G., HARRIS, A. G., REPETSKI, J. E., WARDLAW, B. R., MAMET, B. L. and MORALES-RAMIREZ, J. M. 1999. Neoproterozoic (?) to Pennsylvanian inner-shelf, miogeoclinal strata in Sierra Agua Verde, Sonora, Mexico. *Revista Mexicana de Ciencias Geológicas*, **16**, 35–62.

STITT, J. H. 1975. Adaptive radiation, trilobite paleoecology, and extinction, ptychaspid biomere, late Cambrian of Oklahoma. *Fossils and Strata*, **4**, 381–390.

STOUGE, S. and BAGNOLI, G. 1988. Early Ordovician conodonts from Cow Head Peninsula, western Newfoundland. *Palaeontographica Italica*, **75**, 89–179.

—— —— 1990. Lower Ordovician (Volkhovian–Kundan) conodonts from Hagudden, northern Öland, Sweden. *Palaeontographica Italica*, **77**, 1–54.

SWEET, W. C. 1995. Graphic assembly of a conodont-based composite standard for the Ordovician System of North America. 139–150. *In* MANN, K. O. and LANE, H. R. (eds). *Graphic correlation*. SEPM, Special Publication, **53**, 263 pp.

TAYLOR, E. 1977. Upper Cambrian of western North America: trilobite biofacies, environmental significance, and biostratigraphic implications. 397–425. *In* KAUFFMAN, E. G.

and HAZEL, J. E. (eds). *Concepts and methods of biostratigraphy.* Dowden, Hutchison and Ross, Stroudsburg, Pennsylvania, viii + 658 pp.

VANGUESTAINE, M., BREUER, P. and LEHNERT, O. *in press.* Discovery of an Early Ordovician conodont fauna in the Salm Group of the Stavelot Inlier, Belgium. *Bulletin de l'Institut Royal des Sciences Naturelles de Belgique, Sciences de la Terre.*

WEBBY, B. D., DROSER, M. L., PARIS, F. and PERCIVAL, I. G. (eds). 2004. *The great Ordovician biodiversification event.* Columbia University Press, New York, 408 pp.

WILGUS, C. K., POSAMENTIER, H. and ROSS, C. A. and KENDALL, C. G. S. C. 1988. Sea-level changes: an integrated approach. *Society of Economic Paleontologists and Mineralogists, Special Publication*, **42**, 407 pp.

WOLSKA, Z. 1961. Conodonts from Ordovician erratic boulders of Poland. *Acta Palaeontologica Polonica*, **6**, 339–365.

ZHANG, S. and BARNES, C. R. 2002. Late Ordovician–Early Silurian (Asgillian–Llandovery) sea level curve derived from conodont community analysis, Anticosti Island, Québec. *Palaeontology, Palaeoclimatology, Palaeoecology*, **180**, 5–32.

[Special Papers in Palaeontology, 73, 2005, pp. 135–141]

GRAPHICAL REFINEMENT OF THE CONODONT DATABASE: EXAMPLES AND A PLEA

by WALTER C. SWEET

Department of Geological Sciences, The Ohio State University, 155 South Oval Mall, Columbus, OH 43210, USA; e-mail: sweet4@mindspring.com

Abstract: An example from the North American Ordovician is used to contrast temporal and geographical scales developed biozonally and graphically. Other biostratigraphical studies that involve data assembled graphically are cited, and a plea is directed to students of conodonts to assemble and evaluate their data graphically, or at least to present the data in a form that can be used by others.

Key words: graphical correlation, Ordovician, biozones, conodont, composites, fossil record.

COLLATION of morphological and distributional data from millions of conodont elements in tens of thousands of collections derived from stratigraphical sections at hundreds of localities spread across all seven of the present-day continents requires a framework with temporal and spatial dimensions and wide and unambiguous geographical recognizability. The framework most commonly used today is a stacked sequence of biozones, which are variable in scope, and have been established at different times and places by investigators dealing for various reasons with collections of different composition and significance. Although this framework still has many gaps and continues to be refined, it has served us well as the basis for 'alpha-level' biostratigraphical integration of the world-wide Upper Cambrian–Triassic rock (and conodont) record.

Despite its current utility in assembling the stratigraphical history of conodonts, the biozonally based record is clearly biased in favour of those geographical areas and stratigraphical intervals that have been most intensively sampled and carefully studied. Furthermore, because there is now no globally comprehensive biozonal scale it is not possible to include in a single framework conodonts from many, even well-studied sections or areas. To this extent, then, the record that many regard as ultimately basic to an understanding of conodont faunas and their phylogenetic development and temporal distribution is far from complete.

Consideration of the ways the conodont database has been assembled suggests that its scope and content might be appreciably increased through use of a semi-quantitative methodology termed graphical correlation. This method of stratigraphical correlation was first described by Shaw (1964) and is discussed in detail by the several contributors to a volume edited by Mann and Lane (1995); however, it has been used to date by only a few students of conodonts. Even so, results of several published studies that utilize conodonts indicate that more widespread application might produce a temporally and geographically elegant framework within which to develop rigorously constrained taxonomic concepts and phylogenetic scenarios. In the following paragraphs, I describe steps that I have followed in assembling, by graphical means, a high-resolution biostratigraphical framework for the North American Ordovician; summarize some of the more significant results of conodont-based graphical correlation that involve rocks of post-Ordovician age; and conclude with an appeal to conodont workers to explore the graphical method of biostratigraphical assembly, or at least to present the results of their studies in a form that would enable others to use those results in effecting such collations.

GRAPHICAL ASSEMBLY OF A HIGH-RESOLUTION FRAMEWORK FOR THE NORTH AMERICAN ORDOVICIAN

A widely used conodont-biozonal framework resolves North American Ordovician rocks into a succession of 27 or 28 biozones of greatly different temporal and spatial significance. However, a conodont-based graphical collation of range-data, summarized in several recent reports (Sweet 1995*a*, *b*; Sweet and Tolbert 1997; Sweet *et al.* in press), uses essentially the same distributional information but arranges it into a framework of 1118 composite-standard units (or CSUs). The two frameworks are compared in Text-figure 1, which also shows the extent of the major components of the graphically assembled framework.

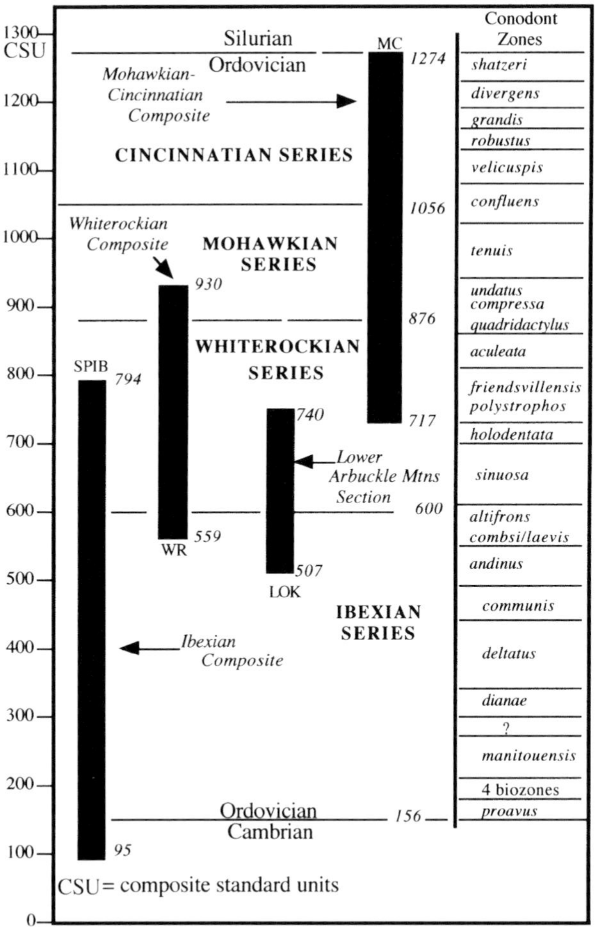

TEXT-FIG. 1. Major components of a North American Ordovician composite standard section compared with the 27 or 28 biozones into which the same stratigraphical interval is currently divided. SPIB is the Shingle Pass–Ibex CS of Sweet and Tolbert (1997); WR is the Whiterockian CS of Sweet *et al.* (in press); LOK is the lower 558 m of the Arbuckle sequence of Oklahoma; MC is the Mohawkian–Cincinnatian CS of Sweet (1984 *et seq.*).

The graphically assembled North American framework has three major components. The Mohawkian–Cincinnatian Composite (MC in Text-fig. 1) is a collation of information from stratigraphical sections at 124 localities in eastern, central and west-central United States. The standard reference section (or SRS) for this component is a continuous 368-m core drilled in Mason County, Kentucky, and preserved at the Department of Geological Sciences at The Ohio State University. Sweet (1979, 1984, 1995*a*, *b*) has shown how he and others have correlated additional sections with an evolving composite section based on the SRS. It is sufficient here to note that MC has for many years been a stable summary of relations between its 124 component sections. That is, subsequent correlations, like those of Upper Ordovician rocks from three central Nevada localities reported by Sweet (2000), have not altered any of the MC ranges of species used in determining correlation of component sections.

The Ibexian Composite (SPIB in Text-fig. 1) is a conodont-based graphical collation of information from stratigraphical sections at 12 localities in western Utah, east-central Nevada and the El Paso region of western Texas. SRS for this composite section is a long surface section along the crest of a ridge about a mile south of Shingle Pass in the southern Egan Range of Lincoln County, Nevada (Sweet and Tolbert 1997). With the Shingle Pass section as SRS, serial correlation (and recorrelation) of 13 sections in the Ibex area of western Millard County, Utah, was effected using scaled range data for conodont species abstracted from reports by Ethington and Clark (1982) and Hintze *et al.* (1988). Subsequently, SPIB was expanded eastward by graphical collation of conodont range-data from the Scenic Drive section in the southern Franklin Mountains near El Paso, Texas (Repetski 1982). Range-data and graphs that use these data to assemble SPIB are provided by Sweet and Tolbert (1997).

Although the Mohawkian–Cincinnatian Composite (MC in Text-fig. 1) has a substantial pre-Mohawkian (or 'Chazyan') interval at its base, and the upper third of the Ibexian Composite (SPIB in Text-fig. 1) is, by definition, Whiterockian, melding the two into a single, credible composite requires information from stratigraphically intermediate sections. Six sections in central Nevada that satisfy these requirements have been correlated graphically and results are in a report by Sweet *et al.* (in press). The network formed by these sections is the Whiterockian Composite (WR) of Text-figure 1.

Melding the three regional composite sections into a single North American Ordovician Composite Section has been effected in two steps. First, the Mohawkian–Cincinnatian and Whiterockian composites were compared graphically; the results are depicted in Text-figure 2. Although dispersion of FADs and LADs (first and last appearance datums) in Text-figure 2 is, at first glance, rather diffuse, a line of correlation (LOC) that expresses a stratigraphically realistic relationship between these two data sets is controlled in the midsection of the graph by the array of six FADs and LADs deemed to be those best controlled in the two sections. A cluster of five FADs anomalously to the right of the LOC in the upper third of the graph represents first occurrences of a group of species that are widely distributed in the central and eastern parts of North America but apparently appeared much earlier in western sections, which are now components of the Whiterockian composite. Ranges of these species must, of course, be adjusted downward in MC. This has not yet been done; however, ranges of none of these species have been used to effect correlations of component MC sections, so overall stability of MC remains unaltered.

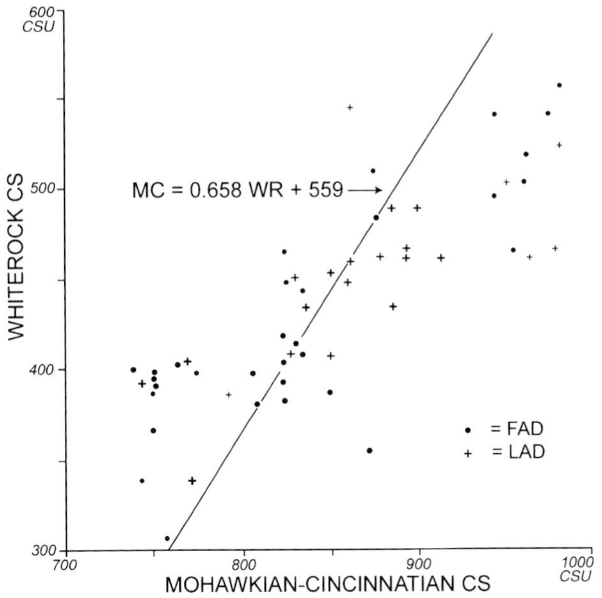

TEXT-FIG. 2. Graphical correlation of the Whiterockian composite section with the Mohawkian–Cincinnatian composite section.

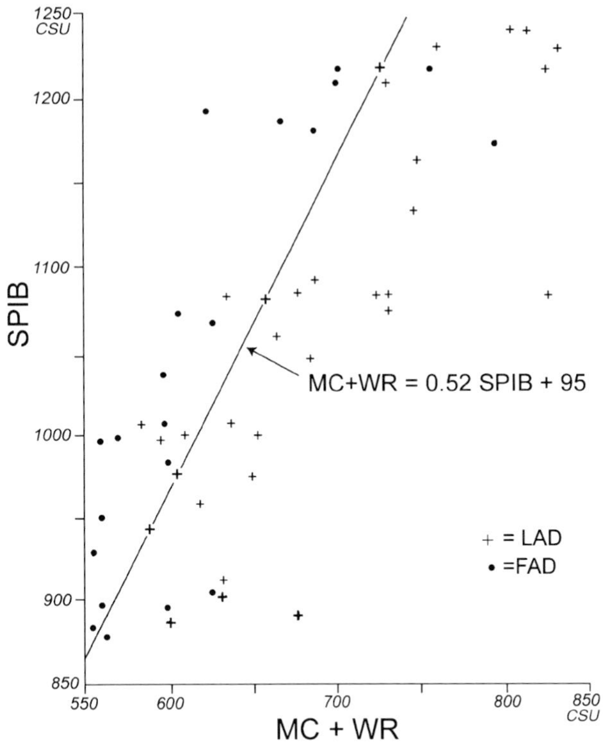

TEXT-FIG. 3. Graphical correlation of the Ibexian composite section (SPIB) with a Mohawkian–Cincinnatian composite section augmented by data from the Whiterockian composite section.

As indicated in Text-figure 1, there is considerable overlap between the Whiterockian composite section (WR) and the upper third of the Ibexian composite (SPIB). Distributional data plotted in the graph of Text-figure 3 suggests the relationship between SPIB, MR and MC shown in Text-figure 1. Because no part of either WR or MC projects below the 559-CSU level it should be obvious that the position and extent of the pre-559-CSU segment of the pan-Ordovician composite cannot be verified graphically. However, as Sweet and Tolbert (1997) have shown, there was apparently little regional variation in rock-accumulation rate in component sections of the Ibexian composite (SPIB); hence, I have assumed that downward continuation of the LOC sited in Text-figure 3 reasonably relates that part of SPIB to the North American Ordovician composite section.

Hundreds of metres of Lower and Middle Ordovician rocks are grandly exposed on the south flank of the Arbuckle Mountains anticline in south-central Oklahoma. In 1972 I sampled this important stratal succession at closely spaced intervals, assisted by Stig M. Bergström and Valdar Jaanusson. I subsequently processed and collected conodont elements from all those samples and, in 1984, regarded the section as a potential link between the better-known Middle and Upper Ordovician of eastern North America and less well-known Lower and lower Middle Ordovician rocks of western parts of the continent. Preliminary graphs (Sweet 1984, fig. 11) suggested, however, that the Arbuckle section included a number of unconformities, the extent and significance of which could not be determined from distributional data available in 1984.

In 1987, Bauer described conodonts of the McLish and younger formations of the Arbuckle sequence and showed how they enabled collation of that portion of the section into the Mohawkian–Cincinnatian composite section. The lower 558 m of the Arbuckle section, although abundantly fossiliferous, defied precise correlation. However, integration of the Ibexian, Whiterockian and Mohawkian–Cincinnatian composites enabled determination of the relative position of the lower Arbuckle succession. The graph of Text-figure 4, based on data that Bauer is also using in the report on Joins and Oil Creek conodonts he is preparing for publication, illustrates my current view as to the position of the lower part of the Arbuckle Mountains section (LOK) and Text-figure 1 shows the relationship of this section to the major components of the North American Ordovician composite section.

The graphically assembled biostratigraphical framework just described has been built up through use of the scaled ranges of more than 400 entities that have been treated as conodont species in reports on North American Ordovician stratigraphy and palaeontology. However, an equally large, or even larger, number of conodont species is typical of sections in other parts of the world and integration of their distribution into a

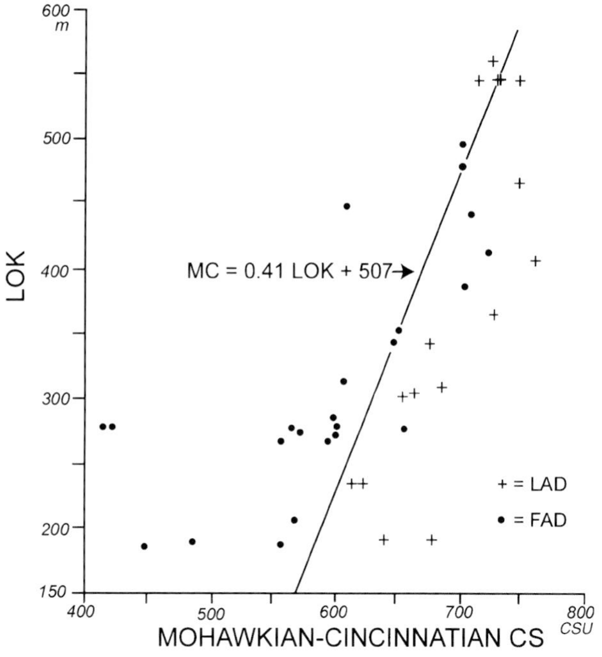

TEXT-FIG. 4. Graphical correlation of the lower 558 m of the Arbuckle Mountains section (LOK) with a Mohawkian–Cincinnatian composite section that also includes data from the Whiterockian composite section.

globally inclusive Ordovician composite section has yet to be effected. Surely such integration will be necessary before any framework, however assembled, will approach the 'completeness' needed for meaningful phylogenetic analysis.

It is not necessary that everyone using the data assembled in the North American Ordovician composite agree with the taxonomical concepts I and my colleagues have followed in assembling it. That is, the 'species' whose scaled ranges are involved in the numerous correlations that produced this framework are subjective interpretations of information published, or otherwise readily available, in reports on individual elements in more than 5000 samples taken at 142 North American localities. In the most comprehensive of those reports (e.g. Bergström and Sweet 1966; Ethington and Clark 1982) frequencies of elements in each sample are tabulated on an 'apparatus-component' (or 'form-species') basis. Thus anyone interested in exploring apparatus reconstructions may work with the raw data, organized in a high-resolution temporal and geographical framework, and not be restricted to considering distribution of species based on subjectively reconstructed apparatuses. Unfortunately, editorial policies have in some cases placed limits on the amount of tabulated information authors may include and the purely biostratigraphical interests of other authors have discouraged detailed sample-by-sample and element-type reporting of their materials.

OTHER CONODONT-BASED GRAPHICAL NETWORKS

Armstrong (1995) has used the ranges of conodonts and graptolites graphically to derive a notion of their distribution in North American rocks of Late Ordovician and Early Silurian age. Evaluation of Armstrong's study must be deferred, however, until all the range-data he used in effecting the several correlations have been reported and questions answered about certain of the procedures he seems to have used in making the graphical assembly.

Miller and Flokstra (1999) have also employed graphical correlation techniques in an attempt to show that there are serious problems with global recognition of the basal Ordovician stratotype now sited in the Green Point section of Newfoundland. Unfortunately, Miller and Flokstra do not present the basic data they used to derive the several graphs in their report; they do not attempt to relate any of their range-data sets to a common section (SRS) such as SPIB, which was reported in detail by Sweet and Tolbert (1997); and the different scales on the *x*- and *y*-axes of their graphs make visual interpretation of the plots very difficult. The Green Point section may, indeed, pose many of the problems identified in Miller and Flokstra's study; however, their curious use of graphical correlation does little to solve (or even illuminate) those problems.

The commonly used framework for Silurian rocks still consists of just 13 or 14 conodont-defined biozones even though Zhang and Barnes (2002) have refined biozonation of its Llandoverian portion. Kleffner (1989, 1995) has shown, however, that graphical collation of the same range-data permits recognition within the Silurian of some 92 composite-standard units, thus increasing resolution greatly. Kleffner and Barrick (2002) have recently incorporated range-data for chitinozoans and many of the graptolite species traditionally considered in Silurian biostratigraphical schemes and this has substantially increased precision of the graphically derived scale.

Conodont-based biozonal biostratigraphy is probably most advanced and is claimed to have the greatest temporal and spatial resolution in rocks of Devonian age. However, Klapper *et al.* (1995), Klapper (1997) and Klapper and Becker (1998) have shown that a Frasnian framework assembled and evaluated graphically not only resolves at a much higher level than do the 8–13 biozones in standard biostratigraphical schemes, but also that FADs of some of the biozone-defining species in those schemes are clearly diachronous.

The Middle Devonian has customarily been divided into 12 conodont-defined biozones. Gouwy and Bultynck (2002), however, report that they have used coral, brachiopod and conodont range-data from Eifelian and

Givetian rocks at nine Belgian localities to assemble a regional composite section within which they recognize 1114 composite-standard units. Details of their correlation schemes have yet to be published, but results summarized to date suggest that biostratigraphical resolution in this part of the Devonian will ultimately be greatly increased. Clearly use of graphical-correlation techniques will permit additional biostratigraphical resolution in the Devonian.

Finally, I concluded (Sweet 1988) from graphical collation of conodont-range information at 13 localities that Lower Triassic rocks could be divided into 135 composite-standard units and that successive 5-CSU 'packages', termed Standard Time Units by Shaw (1964), could be recognized confidently in sections at all 13 of the localities considered. The same stratigraphical interval had previously been divided into only nine conodont-defined biozones, not all of which could be recognized in sections at the 13 localities considered. Subsequently, I used (Sweet 1992) the scaled ranges of 32 conodont, seven ammonoid and a single bivalve species to effect a graphical correlation of sections through the Permian/Triassic boundary interval at seven localities in China, Kashmir, Pakistan and Iran. Results of this study clarified relations at the base of the Lower Triassic, and Sweet *et al.* (1992) used the framework to show close congruence of oxygen-isotope logs between three of the sections used in compiling it. This independent use of the conodont-based graphical framework indicates its reliability and high resolving power but raises serious questions about the resolution of traditional ammonoid-based biozones in this important stratigraphical interval. These questions have not yet been completely answered, although it should be noted that conodonts, not ammonoids, now have pride of place in biostratigraphical arguments in the Permian/Triassic boundary interval.

A NOTE ON RESOLUTION

The 143 sections of the graphically derived composite section depicted in Text-figure 1 vary considerably in thickness, sample spacing and stratigraphical extent, and they involve rocks that differed somewhat to greatly in their relative rates of accumulation. Thus it is difficult to determine a numerical value for resolution that would apply to all component sections. Shaw (1964) used regression analysis to derive equations for the LOCs that related sections of his study to the CS and, for the arrays considered in calculating those LOC equations, he computed the Standard Error of Estimate to determine the size of the smallest package of CSUs that could be recognized in all component sections at various levels of statistical confidence. These packages were termed Standard Time Units (or STUs). In 1984 I also fitted LOCs by least-squares regression but recognized that computations such as the Standard Error of Estimate were inappropriate. Hence, to give a notion of the value of STUs, I devised an index, *W*, which measured the width of LOC arrays as their boundaries intersected the *x*-axis. Subsequently, I and most other authors have abandoned least-squares solutions and fit LOCs by non-statistical means. No one has yet suggested how to quantify the concept of resolution in systems so devised, although it seems intuitively obvious that the graphically derived CS depicted in Text-figure 1, which is divisible into 1118 CSUs, represents greater potential resolution than does the zonally derived scale, which is divisible into only 27 or 28 biozones.

CONCLUSION, AND A PLEA

Shaw (1964) emphasized that the complete ranges in time of fossil species are essentially unknowable, although limits of those ranges might be approached by application of the graphical methodology he introduced. By extension, it might also be argued that we shall never know if the stratigraphical record of biological events is complete, at least until we have a compete record with which to compare it! Thus it could not be argued that replacing zonal-biostratigraphical procedures with graphical-correlation procedures would ultimately result in revealing a complete record, and no such claim is made here. Through use of graphical correlation methodology, however, we can bypass the biases introduced by *a priori* establishment of correlation units (i.e. biozones and their many variants) and use instead the non-biozonal scale that is established as samples are collected from measured sections in the field and processed in the laboratory.

Building a database for conodonts that resolves at a high level both temporally and globally can be accomplished graphically, as demonstrated in the examples cited in previous paragraphs of this report. This will require understanding and insight, as well as the investment of an immense amount of time on the part of conodont workers. However, widespread access to sophisticated computational equipment and programs, as well as the internet connections that provide the ability to share basic information widely, should make assembly of such a system far easier than would have been the case a few years ago. To begin, it would help if authors of studies on conodont faunas would cite in their reports the measured position of the samples from which they derived their specimens and the frequency and nature of those specimens in each sample. Tabulations of this sort might then be used graphically (or biozonally) by other workers and fitted graphically into an evolving high-resolution database. Ultimately, it might be an important project for the Pander Society to develop some general rules for assembling this database, determining how to make it accessible to members, and monitoring its use and growth.

Acknowledgements. Data used to build up the North American CS are from published reports by many authors, whose contributions are specifically acknowledged in reports using those data. I am particularly indebted to Raymond Ethington and Anita Harris for supplying the raw data on which the yet-to-be-published Whiterockian CS is based, and to Jeffrey Bauer of Shawnee State University for advice on taxonomic and range changes he has made in his study of our conodont collections from the Joins and Oil Creek formations of south-central Oklahoma. Perceptive comments by Gilbert Klapper and an anonymous referee identified, and have helped me correct, a number of problems in an earlier version of this report.

REFERENCES

ARMSTRONG, H. A. 1995. High-resolution biostratigraphy (conodonts and graptolites) of the Upper Ordovician and Lower Silurian. Evaluation of the Late Ordovician mass extinction. *Modern Geology*, **20**, 41–68.

BAUER, J. A. 1987. Conodonts and conodont biostratigraphy of the McLish and Tulip Creek Formations (Middle Ordovician) of south-central Oklahoma. *Oklahoma Geological Survey Bulletin*, **141**, 58 pp.

BERGSTRÖM, S. M. and SWEET, W. C. 1966. Conodonts from the Lexington Limestone (Middle Ordovician) of Kentucky and its lateral equivalents in Ohio and Indiana. *Bulletins of American Paleontology*, **50** (229), 271–441.

ETHINGTON, R. L. and CLARK, D. L. 1982. Lower and Middle Ordovician conodonts from the Ibex area, western Millard County, Utah. *Brigham Young University Geology Studies*, **28**, 1–160.

GOUWY, S. and BULTYNCK, P. 2002. Graphic correlation of Middle Devonian sections in the Ardenne, Belgium: implications on stratigraphy and establishment of a regional composite. *Strata, Series 1*, **12**, 86.

HINTZE, L. F., TAYLOR, M. E. and MILLER, J. F. 1988. Upper Cambrian–Lower Ordovician Notch Peak Formation in western Utah. *United States Geological Survey, Professional Paper*, **1393**, 30 pp.

KLAPPER, G. 1997. Graphic correlation of Frasnian (Upper Devonian) sequences in Montagne Noire, France, and western Canada. 113–129. *In* KLAPPER, G., MURPHY, M. A. and TALENT, J. A. (eds). *Paleozoic sequence stratigraphy, biostratigraphy and biogeography: studies in honor of J. Granville ('Jess') Johnson*. Special Papers of the Geological Society of America, **321**, 401 pp.

—— and BECKER, R. T. 1998. Comparison of Frasnian (Upper Devonian) conodont zonations. *Bollettino della Società Paleontologica Italiana*, **37**, 339–348.

—— KIRCHGASSER, W. T. and BAESEMANN, J. F. 1995. Graphic correlation of a Frasnian (Upper Devonian) composite standard. 177–184. *In* MANN, K. O. and LANE, H. R. (eds). *Graphic correlation*. SEPM (Society for Sedimentary Geology), Special Publication, **53**, 263 pp.

KLEFFNER, M. A. 1989. A conodont-based Silurian chronostratigraphy. *Bulletin of the Geological Society of America*, **101**, 904–912.

—— 1995. A conodont- and graptolite-based Silurian chronostratigraphy. 159–176. *In* MANN, K. O. and LANE, H. R. (eds). *Graphic correlation*. SEPM (Society for Sedimentary Geology), Special Publication, **53**, 263 pp.

—— and BARRICK, J. E. 2002. Newly revised and calibrated conodont-, graptolite-, and chitinozoa-based Silurian chronostratigraphy developed using the graphic correlation method. *Geological Society of America, Abstracts with Programs*, **34**, 27.

MANN, K. O. and LANE, H. R. (eds) 1995. *Graphic correlation*. SEPM (Society for Sedimentary Geology), Special Publication, **53**, 263 pp.

MILLER, J. F. and FLOKSTRA, B. R. 1999. Graphic correlation of important Cambrian-Ordovician boundary sections. *Acta Universitatia Carolinae – Geologica*, **43**, 81–84.

REPETSKI, J. E. 1982. Conodonts from the El Paso Group (Lower Ordovician) of westernmost Texas and southern New Mexico. *New Mexico Bureau of Mines and Mineral Resources, Memoir*, **40**, 121 pp.

SHAW, A. B. 1964. *Time in stratigraphy*. McGraw-Hill, New York, 365 pp.

SWEET, W. C. 1979. Late Ordovician conodonts and biostratigraphy of the western Midcontinent Province. *Brigham Young University Geology Studies*, **26**, 45–85.

—— 1984. Graphic correlation of upper Middle and Upper Ordovician rocks, North American Midcontinent Province, U.S.A. 23–25. *In* BRUTON, D. L. (ed.). *Aspects of the Ordovician System*. Palaeontological Contributions from the University of Oslo, **295**, 228 pp.

—— 1988. A quantitative conodont biostratigraphy for the Lower Triassic. *Senckenbergiana Lethaea*, **69**, 253–273.

—— 1992. A conodont-based high-resolution biostratigraphy for the Permo-Triassic boundary interval. 120–133. *In* SWEET, W. C., YANG ZUNYI, DICKENS, J. M. and YIN HONGFU (eds). *Permo-Triassic events in the eastern Tethys*. Cambridge University Press, Cambridge, 181 pp.

—— 1995*a*. A conodont-based composite standard for the North American Ordovician: Progress report. 15–20. *In* COOPER, J. D., DROSSER, M. L. and FINNEY, S. C. (eds). *Ordovician Odyssey: short papers for the Seventh International Symposium on the Ordovician System*. Pacific Section, Society for Sedimentary Geology (SEPM), 498 pp.

—— 1995*b*. Graphic assembly of a conodont-based composite standard for the Ordovician System of North America. 139–150. *In* MANN, K. O. and LANE, H. R. (eds). *Graphic correlation*. SEPM (Society for Sedimentary Geology), Special Publication, **53**, 263 pp.

—— 2000. Conodonts and biostratigraphy of Upper Ordovician strata along a shelf to basin transect in central Nevada. *Journal of Paleontolgy*, **74**, 1148–1160.

—— and TOLBERT, C. M. 1997. An Ibexian (Lower Ordovician) reference section in the southern Egan Range, Nevada, for a conodont-based chronostratigraphy. *United States Geological Survey, Professional Paper*, **1579-B**, 51–84.

—— ETHINGTON, R. L. and HARRIS, A. G. in press. A conodont-based standard reference section in central Nevada for the lower Middle Ordovician Whiterockian Series. *Paleontographica Americana.*

—— YANG, Z., DICKINS, J. M. and YIN, H. 1992. Permo-Triassic events in the eastern Tethys – an overview. 1–8. *In* SWEET, W. C., YANG, Z., DICKINS, J. M. and YIN, H. (eds). *Permo-Triassic events in the Eastern Tethys.* Cambridge University Press, Cambridge, 181 pp.

ZHANG SHUNXIN and BARNES, C. R. 2002. A new Llandovery (Early Silurian) conodont biozonation and conodonts from the Becsie, Merrimack, and Gun River formations, Anticosti Island, Quebec. *Paleontological Society, Memoir*, **57**, 46 pp.

[Special Papers in Palaeontology, 73, 2005, pp. 143–157]

THE LIKELIHOOD OF STRATOPHENETIC-BASED HYPOTHESES OF GENEALOGICAL SUCCESSION

by PETER D. ROOPNARINE

Department of Invertebrate Zoology and Geology, California Academy of Sciences, 875 Howard St., San Francisco, CA 94103, USA; e-mail: proopnarine@calacademy.org

Abstract: Understanding the microevolutionary processes underlying patterns of morphological and systematic relationships in the fossil record often requires information derived from stratigraphically ordered samples of fossils. This stratophenetic inference has formed the historical basis of the study of microevolutionary processes. The use of stratigraphic succession as a means to order sequences of events, and in the process identify samples comprising individuals ancestral to individuals in later samples, however, is an implict and generally untested assumption of stratophenetic methods. The failure to evaluate the relative and independent contributions of stratigraphy, morphology and palaeoenvironments/palaeoecology to hypotheses of microevolutionary genealogy, and in fact the hypothesizing of genealogies themselves, is often viewed as problematic to the traditional evolutionary and stratophenetic approaches to phylogeny reconstruction. This paper argues that the use of various types of information is in fact a strength and necessity of genealogical reconstruction. The independence of these data must be recognized, however, thereby allowing probabilistic evaluation of stratigraphic and morphological data, and the relative ranking of multiple genealogical hypotheses. A simple likelihood ratio-based approach to microevolutionary resolution is used to illustrate the complexity underlying stratophenetic assumptions. An example is presented using the Early Devonian ozarkodinidid conodont *Wurmiella wurmi*. Use of stratigraphic data only results in the unbranching and uninterrupted genealogical chain expected of stratophenetic studies of a single species. Morphometric data, by contrast, produce a more complicated, branching pattern of relationships among samples. The two hypotheses may be combined to produce a single stratogenealogy, which although conforming largely to stratigraphic order, does incorporate some of the complexities resulting from morphological comparisons. The conclusion is that stratophenetic patterns do not necessarily reflect the most supportable hypotheses of genealogical descent; those hypotheses must be formulated independently of stratigraphic order, and may be evaluated in the contexts of stratigraphic, morphological and palaeoenvironmental data.

Key words: stratophenetics, genealogy, ancestors, likelihood, microevolution.

DARWIN (1859) suggested that the evolutionary history of lineages would be preserved as the geological succession of fossil groups related by ancestry and descent. This view was adopted as the framework within which to document the results of descent with modification, whether between stratigraphically ordered members of the same (presumed) species, or between higher taxa (e.g. Trueman 1922). The historical palaeontological literature is replete with such examples of stratigraphically ordered genealogical representations (refered to here as 'stratogenealogies'). This historical database formed the basis of Eldredge and Gould's (1972) hypothesis of punctuated equilibrium, which uses the recurrent pattern of within-lineage morphological stability, or stasis, followed by abrupt morphological evolution, as one of its central arguments. Gingerich (1974, 1976, 1979) subsequently formalized the construction and purpose of stratogenealogies, termed stratophenetics, arguing that lineages could be established on the basis of stratigraphic occurrence and phenetic similarity, and evolutionary hypotheses subsequently constructed on the basis of morphometric variation within and among lineages. Gingerich intended stratophenetics to be a tool both to examine patterns of microevolution and to serve as an alternative to cladistics in the discovery of phylogenetic relationships.

There are consequently two avenues of macroevolutionary thought that depend on the implicit assumption of stratogenealogical continuity. First, stratophenetics asserts that stratogenealogies represent primary data for the construction and testing of phylogenetic hypotheses, and implicitly accepts the primacy of stratigraphic order in the derivation of stratogenealogies. Second, macroevolutionary theory depends on the examination of stratogenealogies for the establishment of patterns of stasis and punctuated speciation. Beneath these phylogenetic and macroevolutionary levels, however, lies the implicit assumption of stratogenealogical continuity between stratigraphically successive samples, where stratigraphic adjacency is equivalent to ancestor–descendant relationships. Ancestorization of taxa, specifically species, raises numerous issues in phylogenetic analysis, and similar questions should be asked of the justification for stratophenetic ancestorization at the microevolutionary scale. Ancestorization is certainly implied when the sample dots of a stratophenetic series are connected in orderly stratigraphic succession. *Ad hoc* arguments of stratigraphic succession and morphological similarity are commonly used to support arguments of evolutionary continuity, but evolutionary continuity itself is the *raison d'être* for presumed stratigraphic succession and morphological similarity.

The purpose of this paper is to suggest that the microevolutionary framework of stratophenetic analysis is based upon assumptions that are nevertheless qualitative assessments of the fossil record. These assumptions should therefore be amenable to evaluation in probabilistic frameworks, and a simple model is suggested. Evaluation relies on the independent assessment of morphological and stratigraphic data, in a sense formalizing decisions that are generally made by palaeontologists on the basis of experience and expectation. The paper does not question the validity of a stratophenetic approach or the existence of stratogenealogies, but does seek to highlight the assumptions made when 'connecting the dots'.

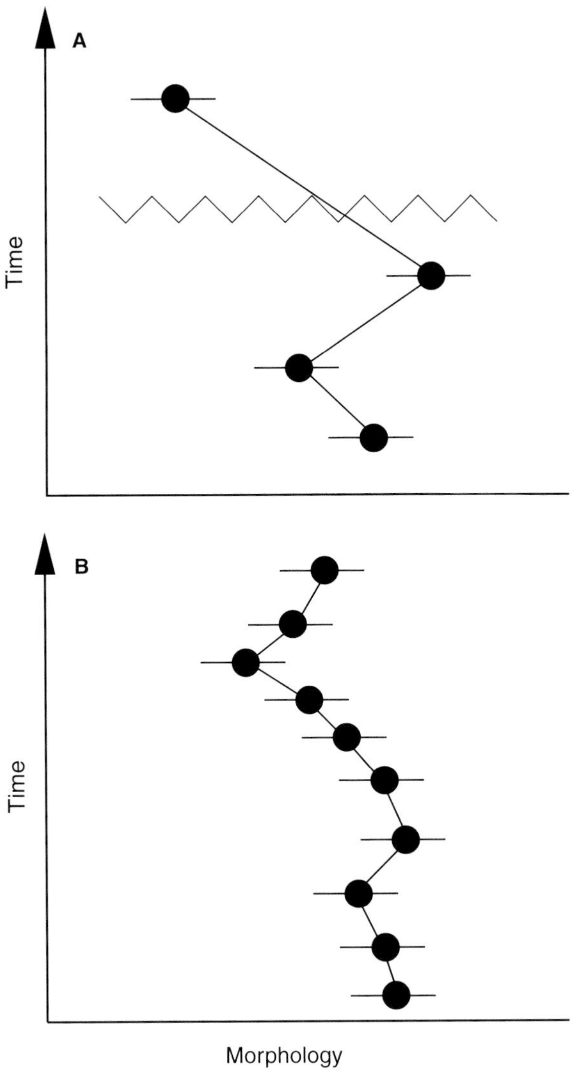

TEXT-FIG. 1. Two hypothetical single-section stratophenetic series. Time scale is the same for both series. Crooked line in A denotes depositional interruption.

The quality of stratophenetic data

The justification for stratophenetic ancestorization lies in implicit assumptions, which are in turn couched in a history of palaeontological experience. Whether a stratopheneticist accepts a stratogenealogy without explicit evaluation of support for ancestorization, or a cladist rejects it without evaluation of stratigraphic data on the basis of philosophical objections, workers in general tend to accept or neglect stratophenetic studies by judging the stratigraphic quality of various stratophenetic series. When presented with stratophenetic data, for example the two series shown in Text-figure 1, many workers will immediately agree that series B is of better quality and more supportable than series A, on the basis of morphological consistency, stratigraphic resolution, sampling density and sedimentological continuity. There are, however, the two critical factors inherent in such qualitative assessments. First, it is assumed that successive samples in the section belong to the same genealogy, that is, older samples are ancestral to younger samples. Stratigraphic data are therefore given primary consideration over morphological/character data. But when is the assumption of genealogy correct, and when is it in error? Second, judgments of the two series are commonly based on expertise, drawing upon three general domains of evidence: sedimentology, morphological description and biofacies analysis (see e.g. McCormick and Fortey 2002).

This is generally recognized when the information is used informally, being derived from considerations of taphonomic conditions, density and continuity of preservation, sedimentary conditions, statistical distributions of fossil morphologies (usually mean and variance), and the geographical distribution of the lineage of interest (e.g. Bell *et al.* 1985; Koch 1986; McCormick and Fortey 2002). The 'connect-the-dots' approach is a simplification and summarization of these data, giving the order of stratigraphic occurrence primacy over all the other data types. For example, if a stratogenealogy was constructed of the series in Text-figure 1A purely on the basis of phenetic similarity, the result differs from the hypothesis based solely on stratigraphic order. Connecting any two samples within a series may therefore be done on the basis of weighing support for the hypothesis of 'connection' as provided independently by the potentially independent domains of stratigraphic and morphometric data. Does this mean that the stratigraphic data are misleading in implying an unbranching evolutionary succession? Not necessarily, but it is important to note that in a strict stratophenetics framework, strato- is primary, and phenetic is decidedly secondary.

The true historical patterns of microevolutionary fossil data within a section must lie somewhere between stratigraphically guided ancestorization of successive samples and hypotheses of relationship that disallow ancestorization. Therefore, somewhere between a strict rejection versus implicit acceptance of stratogenealogies lies a distribution of hypotheses that are supported variably by different types of palaeontological and geological data. I hope to show in this paper that these types of data may support genealogical hypotheses differently and independently, but I will also suggest that their combined support allows the relative ranking of multiple competing hypotheses.

I will outline the basis for the argument in the following sections by considering the sedimentological/preservational and morphological areas of evidence and their relationship to the assumption of ancestry. Biofacies and palaeoecological information, though critical to this argument, are not yet incorporated into the argument. I will restrict the arguments to a consideration of microevolutionary series, thereby alleviating concern about any underlying models of speciation. The goal is to construct a framework within which to develop methods for the objective evaluation of stratogenealogical hypotheses; I suspect that any number of methods could be suited to such a framework. Here I will present a simple likelihood ratio-based model of relative support for alternative hypotheses of microevolutionary genealogical relationships.

Stratigraphically successive samples of the ozarkodinidid conodont *Wurmiella wurmi* (Early Devonian, Nevada) (Murphy *et al.* 2004) will serve as a test case. Conodont lineages are excellent examples of the implicit use of stratigraphic succession in genealogical reconstruction. Justification for the stratophenetic approach may be found easily in the superior quality of the conodont fossil record (Foote and Sepkoski 1999). Ignoring the incompleteness of even this record, however, can have significant consequences for phylogenetic and genealogical analyses, as demonstrated by Donoghue's apparatus-based cladistic reconstructions of relationships among genera within the conodont family Palmatolepidae (Donoghue 2001). The following section explains the logic underlying the construction of microevolutionary genealogical hypotheses, and the necessity of a likelihood approach.

A logical basis for likelihood support

There are extant individuals belonging to an evolutionary lineage that at any given time are descended from other, preceding individuals also belonging to that lineage. Therefore, all individuals comprising a population at any given time possess at least one ancestral population. Within the entire history of a species, such ancestors may be direct or indirect (ancestors of ancestors) (Foote 1996). If the lineage or genealogy G comprises all the individuals, s, of the ancestral population and its direct and indirect descendant populations,

$$G = \{S_t | t \geq 0\}$$

then G may be partitioned exhaustively as ancestor-descendant (A–D) subsets of populations n as

$$G = \{A(n_0, n_1), A(n_1, n_2), ..., A(n_{t-1}, n_t)\}$$

where n_t is a population and t is time elapsed since origin of the lineage (that is, origin of the ancestor). $A(n_i, n_{i+1})$ symbolizes the A–D relationship between n_i and n_{i+1} and consists of all individuals s_i assignable to population n_i.

The palaeontological record of G, however, is almost never a simple sampling of G because of geological and preservational factors. In other words, while the individuals comprising the fossil record of G were certainly members of at least one of the A–D subsets, the subsets or populations themselves can never be reconstructed from the fossil record. Furthermore, if our current understanding of microgeographical variation is any indication (e.g. Lieberman 2001), then we must accept the possibility that the arrangement of G's subsets at any given time were potentially spatially and reproductively complex. In order then to reconstruct a stratophenetic estimate of the lineage, denoted here as $\hat{G}$, we must rely on stratigraphic and morphological information, interpreting $\hat{G}$ independently within each context. The resulting stratogenealogical alignment of samples derived from G is intended to be indicative of the ancestor–descendant relationships of the populations from which those samples were derived. Therefore, the reconstruction of $\hat{G}$ requires a measure of support for all stratogenealogical A–D hypotheses, that is, the set of all possible A–D relationships among the fossil samples.

Biological models. Each stratophenetic series implies a broad range of genealogical models or microevolutionary 'trees' beyond the stratigraphically ordered, unbranching hypothesis. Text-figure 2 gives three possible examples for a three-taxon series. The first is the traditional model, with each sample comprising individuals ancestral to individuals in the stratigraphically successive sample (Text-fig. 2A). Alternatively, more complex situations often prevail in nature, the results of spatial and temporal heterogeneity

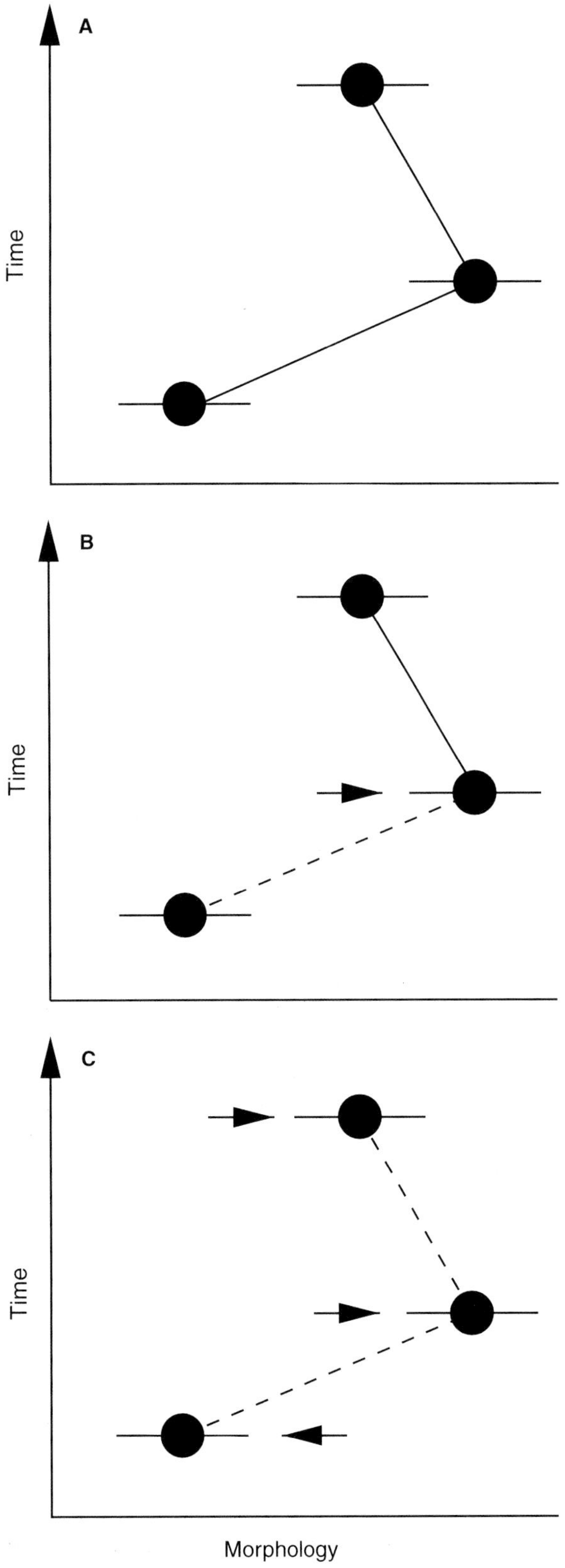

TEXT-FIG. 2. Hypothetical three-sample stratophenetic series. A, continuous genealogy of ancestral and descendant samples. B, genealogical discontinuity between the first and second samples, the arrow indicating an immigration event after local extinction of the taxon. C, the series is not genealogical and all samples are the result of episodic immigration.

caused by, for example, barriers to dispersal, habitat fragmentation or geographical distance (Lande *et al.* 2003). One could imagine a situation where local extinction led to the introduction of individuals from another conspecific deme or metapopulation (Levins 1969; Hanski 1999), or a morphologically similar sister-species (Text-fig. 2B), or a situation where the depositional locality was seeded intermittently but populations were ephemeral (Text-fig. 2C). Strict morphological similarity would probably favour one model over the others, and sedimentary information could support hypotheses of continuity or interruption. Biogeographical complexities could only be implied, not resolved, unless data from additional localities were available.

Various modes of evolution also affect genealogical hypotheses (Text-fig. 3). A mode of constrained stasis (Roopnarine 2001) should result in samples that are essentially morphologically indistinguishable and hence temporally uninformative, in which case stratigraphic position is of primary importance in generating the stratogenealogy. Significant anagenetic directionality, on the other hand, will generate morphological hypotheses that are largely congruent with stratigraphic ordering, because morphological dissimilarity increases with increasing stratigraphic distance (the degree of congruency increasing with increasing directionality). As many stratophenetic series probably conform statistically to a random walk, however (Raup 1977; Bookstein 1988; Roopnarine *et al.* 1999; Roopnarine 2001), corresponding stratogenealogical hypotheses will be the combined results of support derived independently from stratigraphic and morphological data. The topology of those hypotheses will be complex at small stratigraphic scales because of the expected morphological overlap, but stratigraphic ordering will dominate at longer time scales because of the natural time-dependent divergence of random walks. In addition, modes of evolution may vary both among characters of a single taxon (mosaic evolution), as well as during the geological lifetime of a species.

Stratogenealogical hypotheses

Stratigraphic continuity. The single criterion for lineage reconstruction from stratigraphic data is that samples containing ancestral individuals precede those containing descendants, stratigraphically. It is of course possible that the *in situ* arrangement could be reversed, but such sedimentary and taphonomic overprints should be dealt

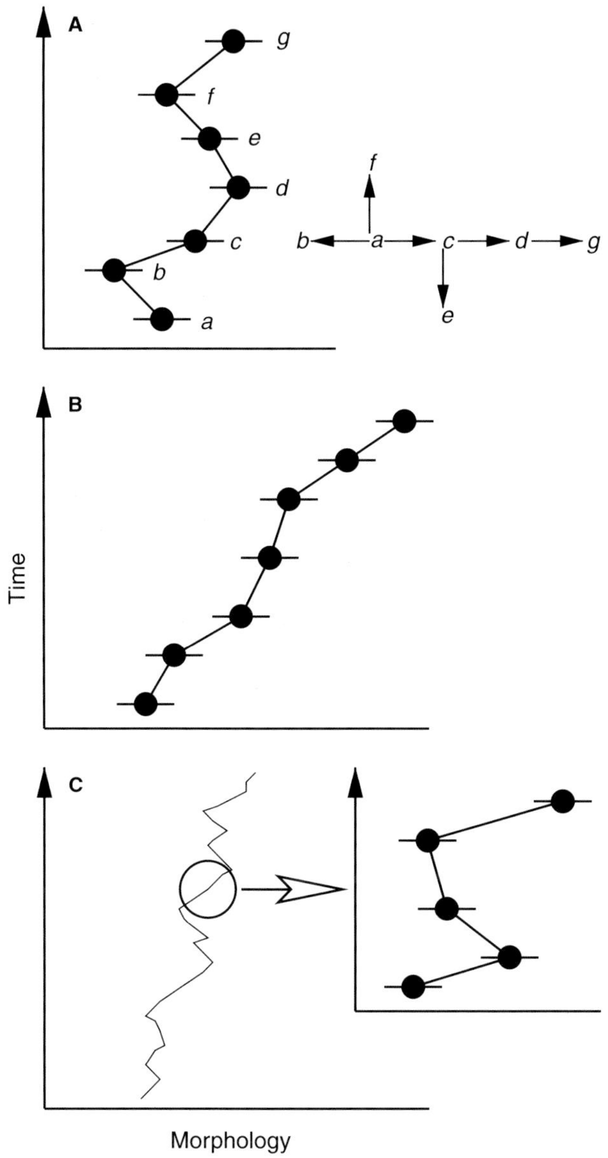

TEXT-FIG. 3. Stratophenetic series generated by different modes of evolution. A, a lineage in morphological stasis where phenetic similarity is not in accord with stratigraphic order. Samples are statistically indistinguishable. The branching diagram on the right is the genealogical tree based solely on morphometric information and statistical similarity. B, lineage exhibiting directional change, resulting in congruence between stratigraphic order and phenetic similarity. C, a lineage viewed over a significantly longer time scale than A or B, conforming to an unbiased random walk. Phenetic similarity at a coarse sampling resolution will result in general congruence between stratigraphic order and morphology because of the relationship between the variance of random walks and time elapsed, but sampling at higher resolution yields the figure on the right, similar to the static situation in A.

with independently of the lineage under consideration. Also, it is possible for ancestral and descendant individuals to be present within a single time-averaged sample, but the objective here is to arrange samples genealogically as a reconstruction of *G*. If this criterion is adhered to, the result is the simple, unbroken stratigraphic (chronological) alignment of samples common in stratophenetic studies. The consideration of each A–D link between samples is a hypothesis to be supported though, which means that information additional to chronological ordination becomes important. This information generally serves intuition well when assessing the sedimentary quality of a section. For example, given a stratogenealogy of three samples, $a \rightarrow b \rightarrow c$, and the knowledge that a depositional hiatus lies between samples *b* and *c*, a formal consideration of $a \rightarrow b$ and $b \rightarrow c$ as hypotheses would result in the former being supported more strongly by the sedimentary record.

Morphological consistency. Evolutionary variation and change can be expected to be largely continuous and quantitative as temporal intervals approach generational timescales. Gingerich (1979), in his formalization of the stratophenetic method, outlined a criterion of morphometric similarity required for the establishment of genealogical succession. The criterion has certainly been applied implicitly if not explicitly in stratophenetic studies, and has been the major vehicle for testing evolutionary mode since Eldredge and Gould (1972) (e.g. Gingerich 1974; Malmgren and Kennett 1981; Chiba 1996). A statistically significant difference, or lack of statistical similarity, between two stratophenetic samples should not be taken, however, as proof against an A–D hypothesis, because the probability of this hypothesis (that the samples could have been drawn from the same statistical population), no matter how vanishingly small, is never zero. This probability then may be an appropriate measure of the morphometric basis of an A–D hypothesis, and can be used to develop a comparison of competing hypotheses.

One caveat to this approach is the fact that the magnitude of morphometric dissimilarity or statistical distance is not necessarily a reliable guide to genetic, genealogical or phylogenetic relationships, in much the same way that genetic distance fails as a metric for taxonomic delimitation (Ferguson 2002). If we consider, however, that morphometric dissimilarity often reflects the magnitude of evolutionary change separating two populations, then it may serve as the best morphological criterion available for stratogenealogical reconstruction. It is also important to note that this metric does not reflect the amount of evolutionary time elapsed between two samples; time does not enter into this criterion.

Likelihood support. Given *n* stratigraphic samples ($n > 1$), there are exactly *n* ancestor–descendant (A–D) hypotheses that can be constructed for every sample beyond the first (stratigraphically oldest); sample j has i possible ancestral samples in the stratigraphic column, and there is the additional possibility that none of those samples comprises an actual ancestral individual of j (see Appendix). Because we can never reconstruct actual A–D relationships of the genealogy (or at least we can never know if our reconstruction is correct), we must adopt a probabilistic approach. One goal of this paper is to use likelihood ratios to attach a measure of relative support to each A–D hypothesis. The general likelihood of the n A–D hypotheses is represented as

$$L(H_1, H_2, ..., H_n|R)$$

which is read as the likelihood of a hypothesis given the observed data R. It is challenging to establish these likelihoods without knowledge of prior probabilities or the observation of repeated events. The former would require appropriate models of morphological evolution, which we simply do not currently have (though developmental genetics and functional genomics provide hope; e.g. Porter and Johnson 2002; Getz 2003), or prior knowledge of evolutionary mode; but mode itself is one of the properties which we wish to derive from genealogical hypotheses! Use of repeated events is impossible because the ancestry of a group is a single, unrepeated evolutionary event. Yet likelihood support remains a viable approach, as likelihood in this instance may be interpreted as the 'probability' that an event which has already occurred would give rise to the observed situation; in this case, the events would be both evolutionary and preservational, and the observed situation is of course the stratophenetic hypothesis. Calculating appropriate probabilities may therefore be a valid alternative to establishing the likelihoods.

Likelihood support for a given hypothesis is the ratio of its likelihood and the likelihood of a competing hypothesis. Because the likelihood of a hypothesis given observed data is proportional to the probability of the data given the hypothesis, the ratio of the likelihoods of two hypotheses is equivalent to a ratio of their corresponding probabilities (Edwards 1992). Therefore, the likelihood ratio of two hypotheses, H_1 and H_2, given observed data R is

$$L(H_1, H_2|R) \equiv \frac{P(R|H_1)}{P(R|H_2)}$$

where $P(R|H_i)$ is the probability of observing R given hypothesis H_i. This ratio is often refered to as a 'Bayes factor' (Good 1965; Kass and Raftery 1995), being considered by Turing to be the 'factor in favour of a hypothesis H, provided by evidence E…' (where E are the data equivalent to R in the above formula) (Good 1979, p. 393). The difference between the likelihoods and probabilities is subtle yet important: a probability as stated above is measured as the probability of observing the data given a fixed hypothesis of how events (in this case evolution or preservation) unfold. In other words, the probability refers to future events on the basis of observed events and is not entirely appropriate to stratogenealogical or other historical reconstruction. A likelihood, on the other hand, acknowledges that the only fixed quantities are the observed data, and evaluates any number of alternative hypotheses accordingly. The equivalence between the two statistics, however, is amenable to the stratogenealogical situation, because all alternative hypotheses can be stated for a single stratigraphic section and set of samples, and the probability of R calculated for each hypothesis.

Assume that there are three successive stratophenetic samples, a, b and c. We wish to compare the hypotheses $A(a, c)$ and $A(b, c)$ (see Appendix), that is, that $a \rightarrow c$ represents an A–D hypothesis exclusive of b, or $b \rightarrow c$ represents an A–D hypothesis more direct than $A(a, c)$. Quantitative likelihood support (S) for each hypothesis may be formulated as the sum of the log likelihood ratios of the hypotheses from the two domains of stratophenetic information considered here (morphology and stratigraphy/sedimentology) (Edwards 1992) as

$$S = \sum_{m,s}^{1\rightarrow n-1} L(A(i, n)) = \prod_{m,s}^{1\rightarrow n-1} I(A(i, n))$$

where $I(A(a, c))$ is a likelihood ratio, $L(A(i, n))$ its log transform, and m and s represent morphology and stratigraphy/sedimentology, respectively. This summation of likelihood ratios is termed variously 'likelihood support' (Edwards 1992) or 'weight of evidence' (Peirce 1878; Good 1979), and has significant information theoretic implications for sequences of data (Good 1979). The likelihood ratios themselves are 'odds', or the ratios of probabilities of the occurrence of the same event under different hypotheses. Therefore, for the example given above, one of the morphometric or stratigraphy/sedimentology likelihood ratios may be written as

$$I(A(b, c)) \equiv \frac{P(R_{bc}|H_{A(b,c)})}{P(R_{ac}|H_{A(a,c)})}$$

where $P(R_{bc}|H_{A(b,c)})$ is the probability of the observed data, R_{bc}, given an A–D hypothesis, $H_{A(b,c)}$. Derivation of the conditional probabilities are outlined separately below for morphometric and stratigraphic/sedimentological data ('Material and methods'). While the summing of likelihood ratios derived from different types of data may seem to be a mixing of 'apples and oranges', the summation is a measure of support for one hypothesis versus another, on the basis of observed data of different types. This is a precise formulation of the stratophenetic criteria. Care must be taken to avoid confusing hypothesis and data; the hypothesis is one of evolution, while the data are simply measures of morphology or depositional history.

A final alternative should be mentioned briefly, and that is the calculation of $P(H_i|R)$, the probability of a hypothesis given the observed data. This type of posterior probability may be approached using Bayes' formula, but is not pursued in the current paper (Appendix).

Wurmiella wurmi

The systematics of the genus *Ozarkodina* Branson and Mehl, 1933 was reviewed recently (Murphy *et al.* 2004) in which a group of forms that have an apparatus similar to that of *O. excavata* (Branson and Mehl, 1933) were collected in the genus *Wurmiella*. These authors envisaged four long-ranging lineages as constituting the main taxonomic units in the genus (Murphy *et al.* 2004), of which one is represented by the middle Lochkovian *W. wurmi*. The genus is characterized by a simple blade with rather even-sized denticles and generally small expansion of the basal platform of the P1 element and the similarity of its apparatus, especially the P2 elements with their very high cusps, the even curvature of the M elements and the incrementally larger posterior denticles of the Sc element. *W. wurmi* is further described in Roopnarine *et al.* (in press), where it is discriminated from several other contemporaneous species of *Wurmiella*.

Roopnarine *et al.* (in press) examined the morphometry of both ontogenetic and evolutionary variation of P1 elements of *W. wurmi* from a single section in central Nevada (see 'Material and methods'). The absence of reliably recognizable geometric landmarks along the P1 basal platform margin prevented the application of standard geometric morphometric techniques for the morphometric comparison of this character, so a new method for the quantitative description of margin shape was developed using standardized cubic splines. A subsequent examination of stratophenetic pattern in the *W. wurmi* samples, using a random walk-based technique (Roopnarine 2001), demonstrated that three independent aspects of margin shape evolved differently (Roopnarine *et al.* in press). The overall arch and relative position of the basal cavity were highly constrained and static, while the shape of the margin anterior to the basal cavity underwent an episode of significant directional evolution. Although Roopnarine *et al.* (in press) ascribed these changes to microevolutionary processes, that conclusion, as do all stratophenetic interpretations, presumed that the stratigraphically ordered samples represent a genealogical succession. One purpose of this paper is to evaluate the validity of that presumption by ranking all competing stratogenealogical hypotheses on the basis of likelihood support.

MATERIAL AND METHODS

Sample collection

Samples were collected from more than 70 superposed beds within the Lower Devonian Windmill Limestone (Lochkovian) (Johnson 1970) at the SP-VII section, Coal Canyon, northern Simpson Park Range, central Nevada (Berry and Murphy 1975; Murphy and Berry 1983). The SP-VII section exposes a regressive sequence that begins in the upper part of the deep-water, graptolitic Roberts Mountains Formation and the overlying allodapic carbonate and interbedded shale slope deposits of the Windmill Limestone (Johnson and Murphy 1984). SP-VII has been correlated previously to the Copenhagen Canyon section of the Monitor Range (COP IV–V), the latter section serving as a standard reference for the region (Murphy and Berry 1983). The section spans approximately 10 myr of the Early Devonian (*uniformis*, *praehercynicus* and *hercynicus* graptolite zones).

Specimens of *Wurmiella* were recovered from more than 50 beds, of which 15 with adequate sample sizes (> 20 individuals) were selected as the basis of the stratophenetic work (Roopnarine *et al.* in press). The lowest stratigraphic level sampled is denoted as sample 200, but thereafter stratigraphic levels (samples) are enumerated 8–15 in ascending stratigraphic order, with sublettering (e.g. 9H) indicating successively higher samples. Morphometric and statistical analyses of specimens and samples are described in detail in Roopnarine *et al.* (in press), but brief descriptions are given below. Samples 200 through 10Li were analysed for the present work.

Standard spline analysis

The smooth morphology of the element basal platform margins, and the absence of all but two geometrically homologous Type I landmarks (Bookstein 1991) (Text-fig. 4) means that much of the information of morphological variation is encompassed in the outline of the margin. A cubic spline was used to describe the platform margin outline and hence capture the variation (Roopnarine *et al.* in press). Individual elements were imaged in lateral view with a Pulnix electronic CCD camera mounted on a Wild dissecting microscope. The lateral margin of the platform was digitized using TnImage for Linux. The digitized points were then converted to Bookstein-style shape coordinates and the shape of each margin was reconstructed by fitting a series of cubic splines to the shape coordinates. The splines were standardized by computing new coordinates at fixed x-axis intervals (Text-fig. 5). Use of a cubic spline to reconstruct the margin outline ensures a

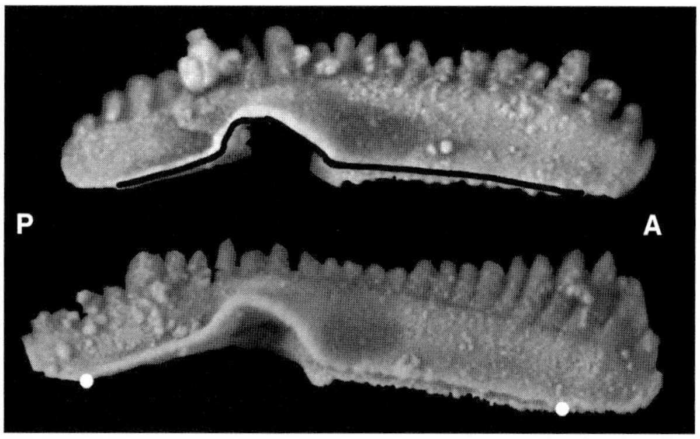

TEXT-FIG. 4. *Wurmiella wurmi* (Murphy *et al.* 2004). SP-VII section, Coal Canyon, Nevada; Windmill Limestone; P1 element, lateral and basal views; black line along base of lateral view outlines margin digitized for morphometric analysis, and white dots represent Type I landmarks used for standard spline reconstruction (Roopnarine *et al.* in press); P denotes posterior, and A anterior; ×90.

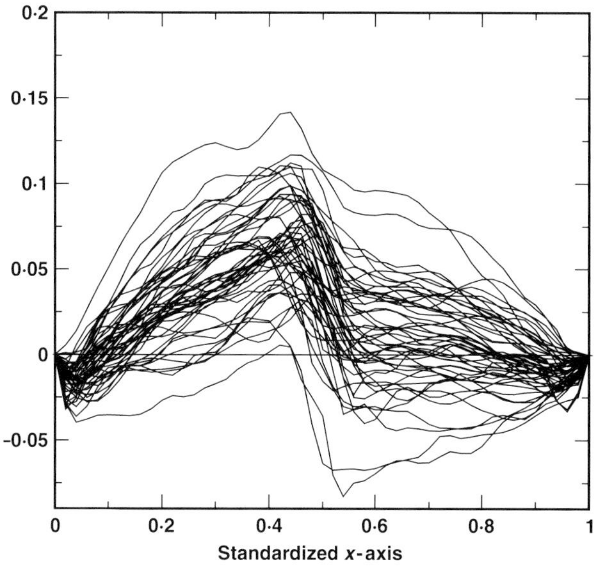

TEXT-FIG. 5. Standardized splines of platform margin profile, sample 10Li. The x-axis is a standard (0,0) (1,0) baseline, while variation along the y-axis has been rescaled according to baseline length (Roopnarine *et al.* in press) and represents variation in the lateral profile of individual *W. wurmi* P1 elements.

smooth reconstruction while minimizing the curvature between points; the result is that no extraneous information is added to the description of the outline. Sample homogeneity/heterogeneity was assessed using singular value decomposition (SVD) of mean centred standard spline matrices (Sampson *et al.* 1996). Each specimen was given a score or location on each SVD eigenvector (a function of the variable loadings on the eigenvector), the value of which reflects the shape of the spline.

Mode of evolution

Describing the mode of evolution of *W. wurmi* during the time represented by the SP-VII section was a three-step procedure. First, all the samples were aligned stratigraphically and placed on a composite stratigraphic scale. Second, the samples were ordinated within a common multivariate morphometric space to facilitate standard intersample comparisons. This was accomplished with a canonical variates analysis (CVA) of the sample SVD scores, which tests both intersample discrimination [with a prior multivariate analysis of variance (MANOVA)] and ordinates specimens according to their canonical or Mahalanobis distances. Finally, a random walk and randomization technique (Roopnarine 2001) was used to search for non-randomness in the SVD I–III stratophenetic series. Evolutionary mode of *W. wurmi* in the SP-VII section was determined by analysing the temporal passage of the samples through the conodont lower profile morphospace as described by SVD analysis. The analysis used here measures the deviation of the stratophenetic series from the expectation of a random walk through the morphospace. It proceeds by measuring the pattern, or the information content of the series with an estimation of the Hurst component, which is in fact a measure of the fractal dimension of the passage. The relationship of the calculated Hurst estimate (h) to that expected of a random walk is tested with comparison of the estimate to a distribution of estimates derived from repeated randomization and re-analysis of the stratophenetic series. Any stratophenetic series so analysed may then be classified as significantly directional, random or constrained (static). The entire procedure and results of this example are described in greater detail in Roopnarine (2001) and Roopnarine *et al.* (in press), respectively.

Stratigraphic continuity

Constructing a stratophenetic genealogy assumes that both ancestor and descendant have probabilities of preservation greater than zero. Given a sequence of fossiliferous samples, there are two bits of data relevant to the construction of a probability measure of an A–D hypothesis, namely the total number of fossiliferous samples in the section, and the number of samples that contain members of the lineage under consideration. From these observations we may derive two very simple measures; first, the frequency with which the lineage is preserved, and second, the number of samples between a putative ancestor and its descendant. If we assume that our sampling of the stratigraphic section has been thorough, then the number of fossiliferous samples represents the complete sample space of our problem. Proceeding from this assumption, a simplistic and conservative estimate of A–D probabilities can be constructed.

Let N be the total number of fossiliferous samples in the section, and k the number that contain the lineage of interest. A frequentist estimate of a probability of preservation would be simply $p = k/N$. Assume that $A(i,n)$ encompasses r samples (inclusive of the putative ancestral and descendant samples). Then if as above the descendant sample is denoted by n and the putative ancestor as i, there are exactly $r - 1$ fossiliferous samples subsequent to i, $r - 2$ of which are assumed not to contain individuals ancestral to those in n. The probability that i is the most direct ancestor of n is therefore measured as the probability of lineage (i,n) being preserved only once in $r - 1$ trials, at the last 'event', which is given as

$$\rho(A(i,n) = p(1-p)^{r-2} \Rightarrow P(R|H_{in}).$$

The resulting probabilities complement the implicit assumptions of stratophenetic series and fossil lineage succession, namely that: (1) the closer two samples are stratigraphically, the more reasonable the assumption that they represent a more direct A–D relationship, and (2) this probability approaches the overall probability of preservation as the number of intervening samples approaches zero, that is, $r \rightarrow n$. Therefore if we now proceed to calculate the likelihood ratios of the competing A–D hypotheses, support will increase with decreasing stratigraphic gap, but they will be weighted by the magnitude of the gap. Furthermore, the stratogenealogy produced by such a consideration of stratigraphic data only will always be the simple unbroken and unbranching genealogy generally expected of microevolutionary stratophenetic series, because it is expected that probabilities will always be greatest between stratigraphically proximal or adjacent samples. Those probabilities will by no means be equal, however, varying as the 'gappiness' of the section varies. Their contributions to the overall stratogenealogy (morphology included) will therefore vary accordingly. The likelihood support for hypothesis H_j, given several alternative hypotheses, is then

$$S_S = \ln \frac{P(R|H_{jn})}{P(R|H_{in})}$$

Support is calculated for all alternative hypotheses in this manner using the probability of a single hypothesis as a common denominator [$P(R|H_{in})$ in this case].

The effect of varying sampling intensity, or sample density, must be considered in the above formulation, but does not change the conclusions drawn from the likelihood ratios. For example, increasing the number of samples between i and n will alter the probabilities of preservation and continuity as calculated above. If none of the new samples contains the lineage of interest, then the probability of preservation p will decrease as r increases. The greatest probabilities will still lie between proximal samples, however, and support will therefore remain greatest for proximal samples. The stratigraphic contribution to the overall support function will change though, and this is discussed below. If, on the other hand, one of the new samples does contain the lineage of interest, then the previous most likely ancestral sample is displaced, as support is now greatest for the new sample and sample n. This of course again matches the expectation that stratigraphically proximal samples are more likely to represent A–D relationships.

Morphometric consistency

$P(R_{in}|H_{A(i,n)})$ is the probability, on the basis of morphology, that sample i comprises individuals ancestral to individuals in sample n. As species identifications of fossil taxa are almost exclusively morphology-based, and Gingerich's (1979) criterion of statistical similarity among stratophenetic members of a single lineage is a reasonable one, it is proposed here that a suitable measure of $P(R_{in}|H_{A(i,n)})$ is equivalent to the probability that both samples i and n were drawn from the same statistical population. An immediate objection to this proposal would be that it assumes strict morphometric stasis between the hypothesized ancestor and the descendant. Postulating any other sort of evolutionary relationship, however, such as oscillating evolution (unbiased random walk) or directional evolution requires the modelling of evolutionary change between the two samples. That in turn requires the involvement of elapsed time in the calculations, which violates the earlier proposal that morphology and time (stratigraphy) be treated separately in the likelihood support function. So 'no evolution' in this regard is the most appropriate model for this situation. The result is the assumption that the two samples under consideration were derived from the same population (statistically speaking). This is a conservative criterion for genealogical reconstruction, because the use of this measure of global phenetic similarity means that any difference between the two samples, whether based on actual evolution or mere measurement error, counts against a hypothesis of genealogical relationship. The criterion could be made more strict by allowing $P(R_{in}|H_{A(i,n)})$ to have only two possible values, 0 and 1. Zero would result from a statistically significant difference between the two samples, and 1 from the inability to distinguish them.

There are a number of alternative methods for the measurement of the probability that two samples were drawn from the same population. Given the vagaries of fossilization, which yield samples of varying sizes, variances and different time-averages, plus our lack of knowledge of the true parametric distributions underlying populations of fossil taxa, an empirical resampling approach will be used here. Efron and Tibshirani (1993) present a bootstrap approach for determining the probability that two univariate samples were drawn from the same population. The test uses a Studentized correction to account for unequal sample sizes, and its achieved significance level (ASL) is used as an approximate measure of $P(R_{in}|H_{A(i,n)})$. The three morphometric characters identified as the most significant aspects of shape variation among the samples (singular value decomposition eigenvectors I, II and III) (Roopnarine *et al.* in press) represent independent aspects of shape variation. The morphological likelihood ratio was calculated as ratios of the joint probabilities of the results of the bootstrapped analysis, described above, applied to each character individually. Likelihood support for A–D hypotheses derived from the morphometric data is therefore a ratio of products,

$$S_m = \ln \frac{\Pi P(R|H_{jn})_i}{\Pi P(R|H_{in})_i}$$

where $P(R|H_{jn})_i$ is the probability associated with samples j and n on the ith SVD eigenvector.

Likelihood support

Text-figure 6 illustrates an example of the calculation of the stratogenealogy for the simple series of Text-figure 2. Samples *a* and *b* contain the potential ancestors of individuals in the final, youngest sample *c*. The morphometric and stratigraphic probabilities are shown, and are both greater for sample *b*. The likelihood supports for the alternative branches may be calculated using either set of probabilities as the ratio denominators, for example

$$S(A(c,a)) = \ln \frac{0{\cdot}3}{0{\cdot}5} + \ln \frac{0{\cdot}1}{0{\cdot}6} = -2{\cdot}3$$

which is the total support for sample *a* being the ancestor, while support for sample *b* is greater (equalling zero) (Text-fig. 6B). The situation becomes slightly more complicated when increased sampling adds a new lineage-bearing sample, *b*.1 between samples *b* and *c*. The morphometric probabilities are the same for *b* and *b*.1 as ancestors of *c* because they are morphometrically identical, but the stratigraphic probability is greater for *b*.1 because it is now the sample most proximal to *c*. Recalculated support values are, based on the probabilities shown in Text-figure 6C,

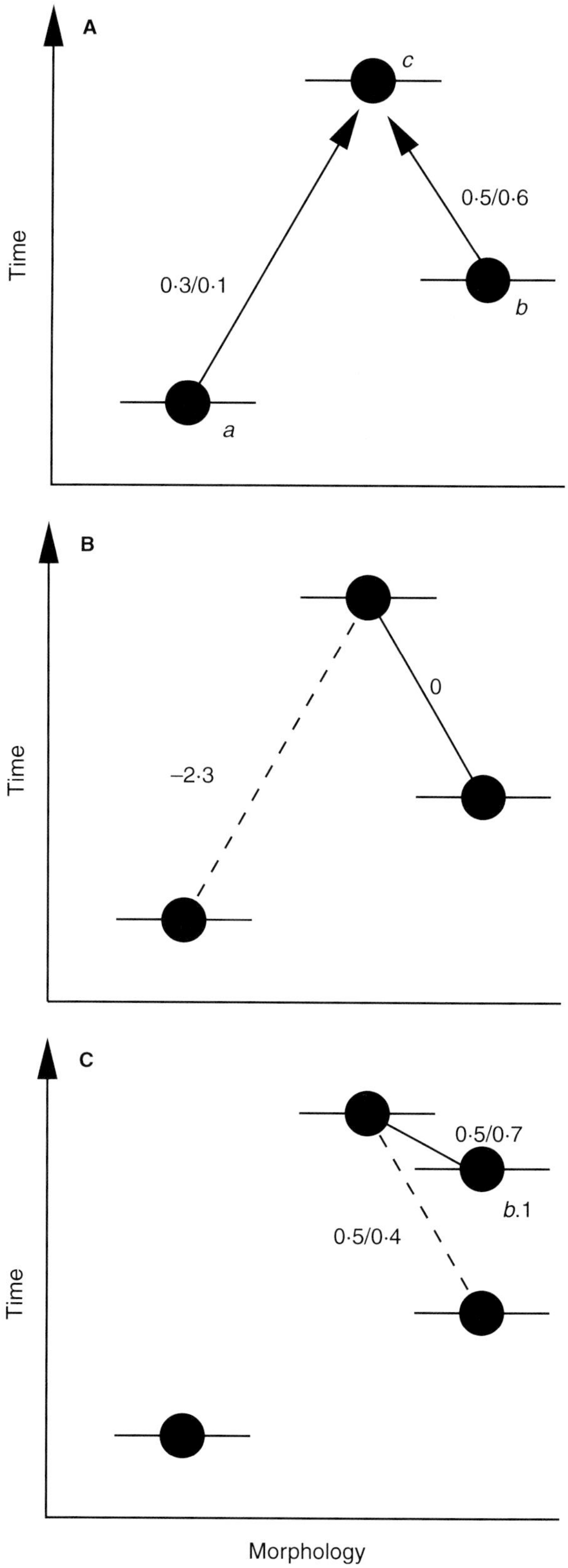

TEXT-FIG. 6. Calculation of likelihood support for series illustrated in Text-figure 3. A, calculation of support for hypotheses *A*(*c*,*a*) and *A*(*c*,*b*) using both morphometric and stratigraphic probabilities; those probabilities are listed next to each branch, as morphometric/stratigraphic. B, final support for each hypothesis; support is greater for *A*(*c*,*b*). C, the addition of a new sample between *b* and *c* alters the probabilities of stratigraphic continuity.

$$S(A(c,b)) = \ln\frac{0\cdot5}{0\cdot5} + \ln\frac{0\cdot4}{0\cdot7} = -0\cdot56$$

and

$$S(A(c,b.1)) = \ln\frac{0\cdot5}{0\cdot5} + \ln\frac{0\cdot7}{0\cdot7} = 0$$

Sample *b*.1 is now supported most strongly as containing individuals most directly ancestral to individuals in sample *c*.

RESULTS

Morphometric analyses

The results of the morphometric analyses (Roopnarine *et al.* in press) are reiterated here briefly. *Wurmiella wurmi* can be distinguished easily from other species of *Wurmiella* in beds of the SP-VII section on the basis of its lateral margin profile. Standard cubic spline analysis summarized P1 element shape in lateral profile with three major aspects of shape variation, the most significant (statistically) being the overall arch or concavity of the element itself. The other aspects, in order of decreasing significance, were a contrast between concavity of the element anterior and posterior to the basal cavity, and finally the relative position of the basal cavity itself.

Analysis of the modes of evolution exhibited by these three characters was performed with the Hurst estimation technique described by Roopnarine (2001). Only one character, the contrast of concavity between anterior and posterior sections of the element, exhibited any significant evolutionary directionality, between beds 8A and 9H. The other characters fluctuated between random oscillatory evolution and constrained stasis.

Stratigraphy-based genealogy

There are 18 fossiliferous samples in the SP-VII section between the first and last *Wurmiella wurmi* samples (samples 200 and 10Li, respectively) analysed here. Nine of those samples actually contain individuals of *W. wurmi*, yielding an estimate of preservation probability of 0·5. The sequence of fossiliferous and non-fossiliferous samples is illustrated in Text-figure 7. The corresponding unbranched preservation-based genealogy is presented in Text-figure 8A. As expected, this genealogical hypothesis

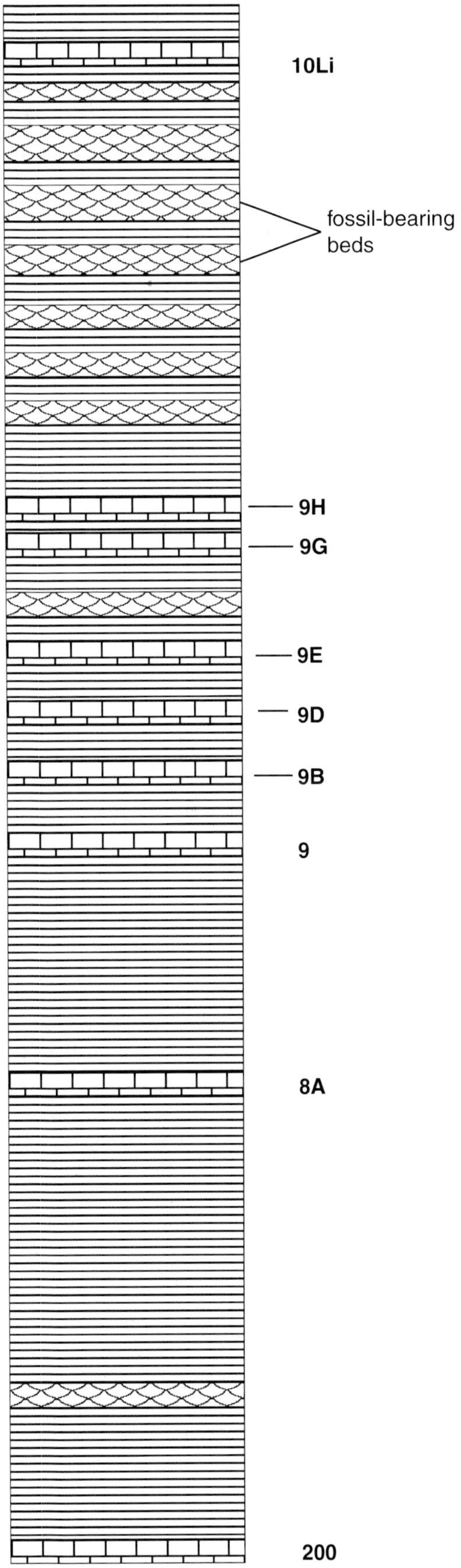

TEXT-FIG. 7. Schematic of the lower SP-VII section illustrating samples used in this study. Numbers refer to *Wurmiella*-bearing samples, while cross-hatched beds mark the location of fossiliferous samples where *Wurmiella* is absent.

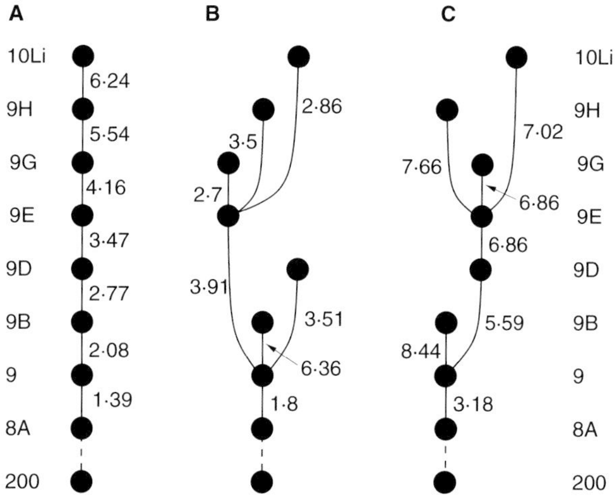

TEXT-FIG. 8. Genealogical diagrams of *Wurmiella* samples, SP-VII section. Numbers along each branch are the likelihood support values (ratios of probabilities) for those branches. The first (lowest) branch of each diagram is dashed and without a support value because it is the only hypothesis possible for the second sample of the series and therefore cannot be compared to any alternatives. A, stratigraphy-only tree, based on comparisons of preservation probabilities; this tree conforms to the traditional stratophenetic topology, but support values of individual branches vary according to the gappiness of the intervening section. B, morphometrics-only tree based on statistical similarity between samples; note that the topology is not ordinated stratigraphically, resulting in branching. C, summary tree with branches (hypotheses) of greatest total support, combining stratigraphic and morphometric likelihood ratios; this tree is not a consensus of A and B, but rather a representation of those branches with the greatest summed support.

is the familiar stratophenetic one, but note that the support (log transforms of likelihood ratios) varies among the branches of the genealogy. This variability stems from the variation of preservation between putative ancestral and descendant sample beds. Furthermore, the variability means that the contribution of preservation to overall likelihood support for various A–D hypotheses will vary.

Morphometric-based genealogy

The morphology-based genealogy is, as expected, not a simple unbranching chain congruent with the preservation-based genealogy (Text-fig. 8B). There are two points of disagreement between the results. First, sample 9 is supported most strongly as ancestral to three subsequent samples, 9B, 9D and 9E. Furthermore, sample 9E is supported most strongly as ancestral to the remaining three samples from beds 9G, 9H and 10Li. These results are expected based on the lack of any strong directional stratophenetic change among the samples, and reflects the statistical and morphometric similarity among the various samples. For example, it is more probable that samples 9E and 10Li were drawn from the same statistical population, than samples 9 and 10Li.

The final stratogenealogical hypothesis best supported on the basis of both preservation and morphology is formulated by summing the likelihood supports for every possible A–D hypothesis among the samples; that is, comparing the likelihood support of branches to a sample from all possible ancestral samples. The branch with the greatest support is selected as the particular A–D hypothesis depicted for that sample. The result (Text-fig. 8C) is a compromise between the preservation and morphology-based trees, but it is not a consensus hypothesis. This final hypothesis has the greatest summary support among all possible hypotheses. There is some resolution of the branching implied by the morphology-only tree, but it is also evident that morphological variation, at least for this example, is not ordered stratigraphically.

DISCUSSION

Given the philosophical and methodological limitations on the identification of ancestors (Hull 1979) in the fossil record, what is the value of constructing a stratogenealogical hypothesis? Philosophical restrictions notwithstanding, patterns of phylogenetic sister-relationships inevitably restrict us to consideration of processes operating at or above the level of the phylogenetic analysis (e.g. selectivity of mass extinctions based on terminal-taxon patterns), which is a comfortable fit with many exercises in macroevolutionary theory. Alternatively, phylogenetic patterns alone may be informative to the derivation of secondary pattern-based hypotheses of rates of diversification, biased diversification, extinction, and so on. Processes operating along the branches and at the nodes of the phylogenetic hypothesis remain largely invisible, however, because we cannot derive any continuity among the constituent data points, that is, stratigraphically ordered samples of single species or lineages. The aim of the stratophenetics program has been to access information at precisely those levels by studying the evolutionary dynamics implied by the stratigraphically ordered samples of single species and postulated sister-species (Gingerich 1979). There are two necessary and implicit assumptions to carrying out this program, first that stratigraphic ordering is genealogically meaningful, and second that samples ordered in such fashion represent ancestor–descendant relationships. No amount of *ad hoc* arguments based on sedimentary continuity, phenetic similarity or biofacies stability, however, can be substituted for objective evaluation of those assumptions.

This paper is not so much an attempt to present a user's method for objective evaluation of stratophenetic assumptions, as it is an examination of the formulation of stratogenealogical hypotheses. Data of sedimentary continuity, phenetic variation and biofacies continuity all contribute to the assessment of stratophenetic series. But they may bear very different implications for resulting genealogical hypotheses. These data and their implications can be used to formulate independent hypotheses, and if formulation takes a probabilistic approach, then it is possible to evaluate all potential A–D genealogical hypotheses exhaustively using the methodology of likelihood support or Bayes factors.

The paper has done precisely that by presenting two analyses, albeit simple ones, of preservational and morphometric data of the conodont *Wurmiella wurmi*. The resulting genealogy is a complex, branched tree, and it is the result predicted on the basis of earlier analyses of morphological variation within this section. Stratophenetic congruence between stratigraphic and morphological data should be expected only under conditions of significant directional evolution, or on a long time-scale under conditions of unbiased random walk. Random variation or stasis at relatively short time-scales will result in incongruencies, and there is absolutely no justification nor necessity for giving stratigraphic order precedence over morphological variation. In the case of *W. wurmi*, overall morphological evolution was interpreted stratophenetically to be one of relative stasis (Roopnarine *et al.* in press). The result is a stratogenealogical hypothesis that is branched, and suggests discontinuity of the lineage, at the population or subpopulation level, at the SP-VII locality. Although this may be an uncomfortable conclusion for many, the hypothesis may be regarded as incomplete. The lineage discontinuity suggests the 'coming and going' of certains types of morphology, or portions of the morphological range of the taxon, but there are several alternative explanations.

First, temporal branching at the microevolutionary scale could be an artefact of preservation and/or sampling. It is possible that higher resolution sampling, if fossiliferous beds were available, would resolve the morphometric branches into an unbroken, stratigraphically congruent chain. Second, terminal branches could indeed represent real events of emigration, local extinction and subsequent replacement, or immigration. Population biology and the ecology of Recent populations and species support these ideas of spatial and temporal heterogeneity (Levins 1968; Lande *et al.* 2003). Such data relevant to stratogenealogical hypotheses are available from biofacies analysis, palaeocommunity composition of all samples within a section, and the analysis of multiple sections within the geographical range of the taxon. Finally, true morphological stasis will produce incongruities between stratigraphic and morphological data. A reasonable assumption of a species in stasis is that its overall temporal-morphological distribution is a normal one, and samples drawn randomly from that distribution may be representative of the species at any time during its geological life. There is therefore no mathematical relationship between age and morphology.

The value of recognizing the proposed independence of various types of data in a stratophenetic analysis is that stratogenealogical hypotheses become testable and falsifiable on the basis of those data. Every proposed ancestor–descendant relationship at the stratophenetic level may be evaluated and ranked relative to other possible A–D relationships. It remains to be seen if this approach will yield useful results when applied to longer stratophenetic series, and across a broad range of taxa and preservational conditions. Addressing the previously implicit assumptions of stratophenetic analysis, however, will I hope generate renewed interest in studying evolutionary processes at fine palaeontological scales. The crucial test of the stratogenealogical approach will be its application to putative cladogenetic events.

Acknowledgements. I am greatly indebted to Mike Murphy for luring me into the world of conodonts, and for a great deal of helpful work, discussion and comments on this manuscript. A great deal of thanks also to Nancy Buening for her work on this project. Many people have contributed to my thoughts on this topic, including Laurie Anderson, Ken Angielczyk, Benoit Dayrat, Paul Fitzgerald, Lindsey Leighton, Carol Tang, Ignacio Valenzuela-Ríos and Gary Vermeij. I also wish to acknowledge comments from Norm McLeod and an anonymous reviewer. Thanks to Phil Donoghue and Mark Purnell for their patience, and for organizing this volume, and thanks to the organizers of ECOS VIII. This project was supported by NSF-EAR 9814354.

REFERENCES

BELL, M. A., BAUMGARTNER, J. V. and OLSON, E. C. 1985. Patterns of temporal change in single morphological characters of a Miocene stickleback fish. *Paleobiology*, **11**, 258–271.

BERRY, W. B. N. and MURPHY, M. A. 1975. Silurian and Devonian graptolites of central Nevada. *University of California, Publications in Geological Sciences*, **110**, 109 pp.

BOOKSTEIN, F. L. 1988. Random walk and the biometrics of morphological characters. *Evolutionary Biology*, **9**, 369–398.

—— 1991. *Morphometric tools for landmark data*. Cambridge University Press, New York, 435 pp.

BRANSON, E. B. and MEHL, M. G. 1933. Conodont studies, 3: Conodonts from the Harding Sandstone of Colorado;

Bainbridge (Silurian) of Missouri; Jefferson City (Lower Ordovician) of Missouri. *University of Missouri Studies*, **8**, 5–72.

CHIBA, S. 1996. A 40,000-year record of discontinuous evolution in island snails. *Paleobiology*, **22**, 177–187.

DARWIN, C. 1859. *The origin of species*. Penguin Books, London, 477 pp.

DONOGHUE, P. C. J. 2001. Conodonts meet cladistics: recovering relationships and assessing the completeness of the conodont fossil record. *Palaeontology*, **44**, 65–93.

EDWARDS, A. W. F. 1992. *Likelihood*. The Johns Hopkins University Press, Baltimore, 275 pp.

EFRON, B. and TIBSHIRANI, R. J. 1993. *An introduction to the bootstrap*. Monographs on Statistics and Applied Probability, **57**. Chapman & Hall, New York, 456 pp.

ELDREDGE, N. and GOULD, S. J. 1972. Punctuated equilibria: an alternative to phyletic gradualism. 82–115. *In* SCHOPF, T. J. M. (ed.). *Models in paleobiology*. Freeman, Cooper, San Francisco.

FERGUSON, J. W. H. 2002. On the use of genetic divergence for identifying species. *Biological Journal of the Linnean Society*, **75**, 509–516.

FOOTE, M. 1996. On the probability of ancestors in the fossil record. *Paleobiology*, **22**, 141–151.

—— and SEPKOSKI, J. J. 1999. Absolute measures of the completeness of the fossil record. *Nature*, **398**, 415–417.

GETZ, W. 2003. "Evo-devo" and the conundrum of sympatric speciation. *Bioscience*, **53**, 313–314.

GINGERICH, P. D. 1974. Stratigraphic record of Early Eocene *Hyopsodus* and the geometry of mammalian phylogeny. *Nature*, **248**, 107–109.

—— 1976. Paleontology and phylogeny: patterns of evolution at the species level in early Tertiary mammals. *American Journal of Science*, **276**, 1–28.

—— 1979. Paleontology, phylogeny, and classification: an example from the mammalian fossil record. *Systematic Zoology*, **28**, 451–464.

GOOD, I. J. 1965. *The estimation of probabilities: an essay on modern Bayesian methods*. The M. I. T. Press, Cambridge, Massachusetts, 109 pp.

—— 1979. Studies in the history of probability and statistics. XXXVII. A. M. Turing's statistical work in World War II. *Biometrika*, **66**, 393–396.

HANSKI, I. A. 1999. *Metapopulation ecology*. Oxford University Press, Oxford, 512 pp.

HULL, D. L. 1979. The limits of cladism. *Systematic Zoology*, **28**, 416–440.

JOHNSON, J. G. 1970. Great Basin Lower Devonian Brachiopoda. *Geological Society of America, Memoir*, **121**, 421 pp.

—— and MURPHY, M. A. 1984. Time-rock model for Siluro-Devonian continental shelf, western United States. *Geological Society of America, Bulletin*, **95**, 1349–1359.

KASS, R. E. and RAFTERY, A. E. 1995. Bayes factors. *Journal of the American Statistical Association*, **90**, 773–795.

KOCH, P. L. 1986. Clinal geographic variation in mammals: implications for the study of chronoclines. *Paleobiology*, **12**, 269–281.

LANDE, R., ENGEN, S. and SAETHER, B. 2003. *Stochastic population dynamics in ecology and conservation*. Oxford Series in Ecology and Evolution. Oxford University Press, New York, 224 pp.

LEVINS, R. 1968. *Evolution in changing environments*. Monographs in Population Biology, **2**. Princeton University Press, Princeton, New Jersey, 120 pp.

—— 1969. Some demographic and genetic consequences of environmental heterogeneity for biological control. *Bulletin of the Entomological Society of America*, **15**, 237–240.

LIEBERMAN, B. S. 2001. Applying molecular phylogeography to test paleoecological hypotheses: a case study involving *Amblema plicata* (Mollusca: Unionidae). 83–103. *In* ALLMON, W. D. and BOTTJER, D. J. (eds). *Evolutionary paleoecology*. Columbia University Press, New York, 357 pp.

MALMGREN, B. A. and KENNETT, J. P. 1981. Phyletic gradualism in a Late Cenozoic planktonic foraminiferal lineage; DSDP Site 284, southwest Pacific. *Paleobiology*, **7**, 230–240.

McCORMICK, T. and FORTEY, R. A. 2002. The Ordovician trilobite *Carolinites*, a test case for microevolution in a macrofossil lineage. *Palaeontology*, **45**, 229–257.

MURPHY, M. A. and BERRY, W. B. N. 1983. Early Devonian conodont-graptolite collation and correlations with brachiopod and coral zones, central Nevada. *Bulletin of the American Association of Petroleum Geologists*, **67**, 371–379.

—— VALENZUELA-RÍOS, J. I. and CARLS, P. 2004. On classification of Pridoli (Silurian)–Lochkovian (Devonian) Spathognathodontidae (conodonts). *University of California, Riverside, Campus Museum Contribution*, **6**, 1–25.

PEIRCE, C. S. 1878. The probability of induction. *Popular Science Monthly*, **12**, 705–718.

PORTER, A. H. and JOHNSON, N. A. 2002. Speciation despite gene flow when developmental pathways evolve. *Evolution*, **56**, 2103–2111.

RAUP, D. M. 1977. Stochastic models in evolutionary paleontology. 59–78. *In* HALLAM, A. (ed.). *Patterns of evolution as illustrated by the fossil record*. Elsevier, Amsterdam, 591 pp.

ROOPNARINE, P. D. 2001. The description and classification of evolutionary mode: a computational approach. *Paleobiology*, **27**, 446–465.

—— BYARS, G. and FITZGERALD, P. 1999. Anagenetic evolution, stratophenetic patterns, and random walk models. *Paleobiology*, **25**, 41–57.

—— MURPHY, M. A. and BUENING, N. in press. Microevolutionary dynamics of the Lower Devonian conodont *Wurmiella* from the Great Basin of Nevada. *Paleontologia Electronica*.

SAMPSON, P. D., BOOKSTEIN, F. L., SHEEHAN, H. F. and BOLSON, E. L. 1996. Eigenshape analysis of left ventricular outlines from contrast ventriculograms. 211–233. *In* MARCUS, L. F., CORTI, M., LOY, A., NAYLOR, G. J. P. and SLICE, D. E. (eds). *Advances in morphometrics*. Plenum Press, New York and London, 620 pp.

TRUEMAN, A. E. 1922. The use of *Gryphaea* in the correlation of the Lower Lias. *Geological Magazine*, **49**, 256–268.

APPENDIX

Symbol glossary

1. $(a|b)$ – read as 'a given b', a conditional dependence. This means that b has been observed, or is assumed to be an event which has occurred, and a is one hypothesis proposed as an explanation of b. Alternatively, b may represent an underlying hypothesis responsible for producing the observed outcome a.
2. $G = \{A(n_0,n_1), A(n_1,n_2), \ldots, A(n_{t-1},n_t)\}$ – the complete or exhaustive set of genealogical relationships within lineage G. As such, the relationships are not independent, as $A(n_0,n_1)$ and $A(n_1,n_2)$ imply that n_0 is the indirect ancestor of n_2.

Bayesian approach

Bayes' formula would permit the calculation of $P(H_i|R)$, the probability of hypothesis given the observed data R. The formula states

$$P(H_i|R) = \frac{P(R|H_i)P(H_i)}{\sum_{j=0}^{j=n} P(R|H_j)P(H_j)}$$

where $P(H_i)$ is the probability of hypothesis H_i, which in this case is simply $1/n$. The difficulty in applying Bayes' formula here lies in the calculation of $P(H_i)$ when H_i is the hypothesis stating that none of the $n - 1$ observed putative ancestral samples is actually the ancestor. The approach taken in this paper is to calculate $P(H_i)$ empirically by comparison of putative ancestor and descendant sample distributions. No such distribution is available for the 'missing ancestor', and in actuality the missing ancestor concept represents a realistically unknowable number of alternative hypotheses.

Use of likelihood ratios and the support or weight of evidence (S) approach allows a partial avoidance of this problem. The weight of evidence in favour of any particular hypothesis may be compared to that in favour of another, without enumeration of all n hypotheses, if the denominator probability is common to all the ratios. It remains unknown if S (missing hypothesis) would exceed S for any of the other $n - 1$ hypotheses, but those latter hypotheses may still be evaluated relative to each other.

[Special Papers in Palaeontology, 73, 2005, pp. 159–183]

THE CHRONOPHYLETIC APPROACH: STRATOPHENETICS FACING AN INCOMPLETE FOSSIL RECORD

by JERZY DZIK

Instytut Paleobiologii PAN, Twarda 51/55, 00-818 Warszawa, Poland; e-mail: dzik@twarda.pan.pl

Abstract: Palaeontological evidence on the course of evolution is represented by fossil samples of ancient populations arranged according to their objective time-and-space coordinates. In the method of stratophenetics, morphological differences between successive samples that accumulate along a geological section are accepted as evolutionary in nature. Evolution is then reconstructed as a series of hypotheses of the ancestor–descendant relationship. Assuming a strict enough correspondence between morphological and molecular evolution, the lack of any statistically significant difference between samples neighbouring in time and taken from the same geographical location (a geological section) suggests a genetic continuity between the populations represented by them. With increasing time and space separating samples, the strength of such inference decreases, but the reasoning (referred to as chronophyletics) remains, in principle, the same. Different hypotheses of ancestry are in an unavoidable logical conflict because any lineage remains rooted in only one ancestral lineage although it may split into several descendant lineages. Testing phylogenetic trees with fossil evidence thus requires that a cladogram or phenogram is transformed into a set of hypotheses on the ancestor–descendant relationship (evolutionary scenario) and the inference has to proceed back in time (by retrodiction). The proposed methodology is illustrated with data on the Ordovician balognathid and Devonian palmatolepidid conodonts.

Key words: evolution, phylogeny, testing, methodology, conodont apparatuses.

THERE can be no doubt that the fossil record is awfully incomplete. Most organisms lack a mineralized skeleton and have little chance to be preserved as fossils; those that have such skeletons suffer from the incompleteness of time recorded in sedimentary strata. Time is missing from rocks because of either non-deposition or their subsequent removal by tectonics and erosion. What makes the situation even more painful to palaeontologists is that continental or shallow-water marine environments, which are taxonomically the most diverse, abound in sedimentary gaps. Although the record in deep oceanic sediments is more complete (McKinney 1985), it is continuously destroyed by subduction and so no deep ocean sediments older than Mesozoic are readily preserved. Nevertheless, despite all these shortcomings, in some regions and taxonomic groups the existing knowledge of ancient populations is good enough to be comparable in its completeness with data on Recent organisms. Finding a new species in a well-sampled rock succession may not be much easier than is enjoyed in invertebrate zoology or phycology.

In such extraordinary cases, the fossil evidence may be rich enough to dispense with speculation in deciphering the course of evolution. One has only to order the data stratigraphically to see the change. If at least simple biometric analyses are undertaken, such procedure is called 'stratophenetics', the term introduced by Gingerich (1979). Reliability of this inductive inference on the course of evolution suffers strongly from incompleteness of the record. It is no wonder that many neontologists go so far as to question the absolute significance of stratigraphic data and palaeontologically based reconstructions of evolution (e.g. Schaeffer *et al.* 1972; Patterson 1981). Only trees based solely on morphological data are considered by them to be of scientific value. Being so dependent on rarely achievable completeness of the fossil record, stratophenetics does not attract much interest even among palaeontologists. It is becoming completely superseded by cladistics. This is why it is rather difficult to compare efficiency of stratophenetics in approaching the real course of evolution with that of other attitudes to the fossil evidence.

The published cladistic analyses of fossils usually refer to high-rank taxa with a poor fossil record. Among rare cases of relatively complete fossil evidence used to such purpose is the work by Hengsbach (1990) on the evolution of ectocochliate cephalopods. He correctly noticed that conchs of the nautiloid *Aturia* and the ammonoid *Cymaclymania* are virtually identical in most aspects generally accepted to be of diagnostic value. The obvious conclusion, in terms of cladistics, is that the clymeniids (known only from the Famennian) and aturiids (known only from the Eocene–Miocene) are 'sister taxa' and thus had a common ancestor similar to Recent *Nautilus* (or, alternatively, to the early Ordovician *Lituites*). The only problem is that stratigraphically ordered findings document relatively well the evolutionary origin of both lineages and this hardly corresponds to the results of purely morphological analysis. The lineage of *Aturia* is rooted in generalized Cretaceous nautilids (e.g. Dzik and Gaździcki 2001) and *Cymaclymenia* belongs to a lineage initiated in the early Famennian from the tornoceratid goniatites (Korn 1992). Their common ancestor lived as early as in the early Ordovician and was morphologically distinct from either *Aturia* or *Cymaclymenia* (Dzik and Korn 1992). The methodologically interesting classical cladistic analysis by Hengsbach (1990) is thus a good case of *reductio ad absurdum*, showing the danger of neglecting stratigraphical order of fossils as the basic evidence. In the present review I use another opportunity to confront the efficiency of cladistic and stratophenetic approaches offered by the more sophisticated analysis of Donoghue (2001).

In fact, the problem of how not to lose stratigraphical information has been extensively discussed from the morphological (cladistic) point of view. Some solutions have been proposed to incorporate geological time into morphology-based phylogenetic trees as evidence additional to morphology (e.g. Harper 1976; Fisher 1994; Wagner 1998). Whenever an inconsistency between the resulting distribution of morphologies in the tree and their stratigraphical distribution emerges, this is considered to be a stratigraphical debt to be compensated for with morphological evidence of enough strength (e.g. Fox *et al.* 1999; Fisher *et al.* 2002). This attitude to palaeontological evidence, referred to as 'stratocladistics' (Fisher 1994), uses taxa defined exclusively on a morphological basis as the units of evolution. The stratigraphical extent of taxa is included subsequently (mostly because of technical reasons), at the stage when the cladogram is transformed into a phylogenetic tree with branching determined by assuming a bifurcating pattern of evolution (e.g. Benton *et al.* 1999; Benton 2001). In fact, the sister taxa relationship, i.e. the concentration of evolutionary change in speciation events, producing not necessarily an existing 'ghost range' in the introduced sister lineage, is plainly contradicted by palaeontological data (Dzik 1999). Thus, like any other variety of cladistic methodology, stratocladistics does not refer to evolution as a real world process with lineages composed of a continuity of specimens or populations sampled by palaeontologists as fossils or fossil assemblages. Instead, this is a rather abstract presentation of the pattern of distances in kinship (blood relationship) as a series of bifurcations creating 'sister taxa' (although there are attempts to make conclusions derived from cladistic analysis more realistic, e.g. Smith 1994).

All this makes the cladistic ways of reasoning involved in a rather complex interplay with the raw palaeontological data. If the method is to be used to infer the real course of past evolution from the morphology of organisms, the assumption that there is a correspondence between time and morphological difference is unavoidable. If so, fossil organisms, being geologically older and thus closer in time to the common ancestor than their Recent relatives, have a greater chance to be closer to the ancestor also morphologically, however imprecise the nature of correspondence between time and morphology. Circular reasoning, thus, emerges whenever data on organisms of different geological age are included in considerations (note that this has nothing to do with the circularity that is allegedly introduced by any use of stratigraphical evidence as claimed by Schaeffer *et al.* 1972, p. 39).

I am not ready to resolve this inherent difficulty with the method of stratocladistics. Instead of entering such methodological complexities, I propose rather to improve the opposite approach of Simpson (1976): to refer directly to the time, space and morphological dimensions of the process of evolution, i.e. to formulate hypotheses on the real course of evolutionary change ('vertical' ancestor–descendant hypotheses, instead of estimating distances in kinship-'horizontal' blood relationships) and confront them directly with the fossil evidence. This methodologically rather traditional attitude, with time and space considered the objective and definitive coordinates of palaeontological data, is discussed below. I attempt in particular to determine how much evolutionary palaeontology suffers from the incompleteness of the fossil record. Its influence on reliability of the methodology used is illustrated with examples. The fossil record of evolution of the palmatolepidid conodonts, the celebrated late Devonian guide fossils in the marginal area of the East European Platform, has been chosen for this purpose.

THE METHOD OF STRATOPHENETICS

The idea that fossils collected bed-by-bed from successive strata should allow restoration of the evolution of the lineage they represent was simple enough to grasp the attention of palaeontologists from almost the establishment of evolutionary theory (Reif 1983). Perhaps the oldest published case is the phylogeny of oppeliid Jurassic ammonites proposed by Waagen (1869). This approach immediately gained much popularity; one of the most stratigraphically strict evolutionary studies of those days is one on the early Palaeozoic hyoliths by Holm (1893). Although the use of biometrics soon followed, most of these early works have not survived close scrutiny; there are, however, a few exceptions, the famous Peterborough succession of the Jurassic ammonite *Kosmoceras* by Brinkmann (1929) being at the top of the list. Notably, such empirical studies actually pre-date the introduction of the genetically meaningful concept of biological populations. Measuring great numbers of fossil specimens, required by studies of this kind, is both time consuming and tedious and therefore in the majority of cases only rather limited numbers of characters are employed. Preferred characters are those that do not change during ontogeny, such as the size of mammalian teeth (e.g. Gingerich and Gunnell 1995), even if the information content in such traits is not especially impressive. In some cases, however, it has been possible to demonstrate profound changes in the dentition of mammals sufficient to distinguish genera (Rose and Bown 1984) or document the expansion of evolutionary novelties across the moulting stages in arthropods (Olempska 1989).

To be successfully applied, stratophenetics requires a rock section (1) that exhibits continuous sedimentation and, thus, offers a complete record of time (2) in which taphonomic conditions did not change significantly during its deposition, and (3) in which the environment was sufficiently stable that inhabiting populations were not forced to emigrate. Thus, the record has to be reliable from geological, ecological and taphonomic perspectives. This combination is rarely met in sedimentary strata, although not so rarely as is commonly assumed. Such studies are limited by patient collecting and fossil measurement and have to be undertaken in the context within which the data are to be analysed.

Unstable sedimentation

A complete record of geological time does not necessarily imply that the rate of sedimentation was uniform. Sedimentation rate variation may be dramatic in parallel with changes in primary productivity, when producers of rock-making calcareous skeletal remains (e.g. coccoliths) are replaced by those with organic skeletons (e.g. dinoflagellates). Decay of their remnants in the sediment increases its acidity, which dissolves calcareous grains (Ernst 1982; Ekdale and Bromley 1984). The effect is a misleading exaggeration of the rate of ecological or evolutionary change recorded in the rock. There are several cases of such distortion of the time record connected with black clay episodes within limestone successions (e.g. Dzik 1997). To some degree an increased density of sampling may help in coping with the unstable sedimentation rate if there is just stratigraphic condensation and not a complete lack of fossil sediment.

Taphonomic bias

The main difficulty with stratophenetics is that there is a significant difference between a complete record of geological time and a complete record of evolution. Taphonomy is the second obstacle. Only a small part of skeletal remains that were originally present on the sea bottom are fossilized. Of mineralogically different skeletons, those with the greatest potential to be fossilized are phosphatic teeth and bones that preserve well in carbonates but also in siliceous and clay-dominated sediments as long as the sedimentary environment was not too acidic. This makes vertebrates of much potential value in evolutionary studies. Unfortunately, with the exception of conodont and mammalian teeth, their dispersed skeletal elements are not sufficiently distinctive and numerous to allow the application of stratophenetics. Calcitic tests of foraminifers have been widely used in such studies (e.g. Grabert 1959; Berggren and Norris 1997) but aragonitic ammonite conchs, although restricted in their occurrence to specific facies, are somewhat more informative morphologically (e.g. Murphy and Springer 1989; Dzik 1990*a*). Well-preserved siliceous fossils are relatively rare; among them radiolarians are the most convenient subjects of stratophenetic studies (e.g. Kellogg 1975). Some uncertainty remains, however, regarding whether the biological species concept is applicable to them if interbreeding has not been documented in their Recent relatives. The same uncertainty exists for collagenous skeletons of pelagic graptolites (e.g. Lenz 1974; Springer and Murphy 1994), which are known to have lost sexual dimorphs (bithecae), probably representing males, early in their evolution.

Migrations

There are also sudden environmental changes that punctuate the distribution of organisms even in those parts of the stratigraphic column where neither apparent stratigraphical discontinuity nor taphonomic change is visible. This is because gaps in the fossil record of evolution may also result from ecologically controlled migrations (Text-fig. 1; Dzik 1990*b*). This aspect of incompleteness of the fossil record can be overcome to some degree by increasing sample sizes: unless the faunal change is drastic and truly instaneous, immigrants appear first as rare specimens contributing little to the fossil assemblage. Frequency distributions of lineages within the stratigraphic column tend to have a fusiform shape, with numbers of specimens gradually increasing with immigration and similarly

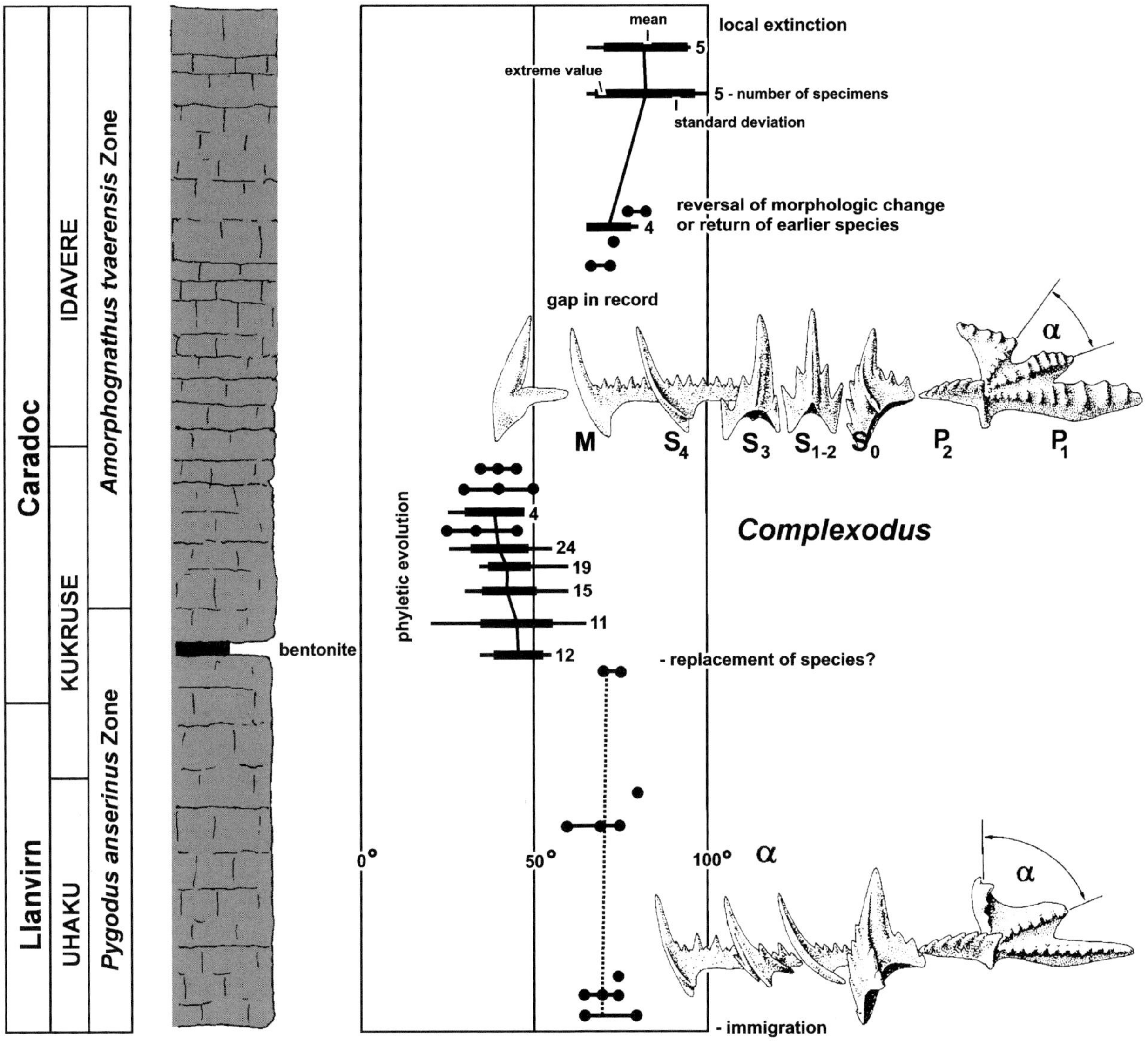

TEXT-FIG. 1. Record of evolution in a stratigraphically complete section discontinuous in effect of ecologically controlled migrations, as exemplified by the *Complexodus* lineage from Mójcza in the Holy Cross Mountains, Poland (from Dzik 1994, modified).

decreasing with migration of the habitat to another geographical location (Dzik 1984). In fact, distinguishing ecological change from evolution is not easy even in stratophenetic studies on fossil groups with an extremely good fossil record (e.g. Dzik and Trammer 1980).

Basic assumptions of stratophenetics

Whenever a section more or less complete in all aspects is available, a series of samples taken bed-by-bed of the rock offers the raw material for a stratophenetic study. Each of the ancient populations represented by fossils from neighbouring beds is in the same geographical place but separated by some distance of geological time. Although the method looks so obvious and simple, there is some implicit philosophy behind it. It has to refer to a series of assumptions, especially when sexual organisms are considered. Thinking in terms of population variability and its presentation in any possible way is then necessary. Stratophenetics requires not only stratigraphically dense sampling but also samples large enough to show a range of morphological variability in ancient populations arranged in lineages.

While interpreting the raw evidence it has to be assumed that (1) the unimodal distribution of all taxonomically significant characters proves a free interbreeding (panmixy) within the population represented by a sample; (2) a morphological similarity of samples close in time and space (neighbouring samples) results from a gene flow from the older one to the younger; and (3) a significant difference between the first and last samples of a stratigraphically ordered series is an expression of the evolution. This is based on an understanding of the population biology of Recent organisms and cannot be substantiated by palaeontological evidence alone. In fact, similar (if not the same) assumptions are necessary to undertake any taxonomic work based on morphology. In neontology, not unlike palaeontology, virtually all our knowledge of living populations is derived from studies of samples, not uncommonly taken at times different enough to introduce the problem of time averaging, or stored long ago in a museum. There is thus hardly any fundamental difference in methods of study of fossil and Recent organisms, although the fossil evidence has obvious limitations. As long as one accepts this as reasonable, if a series of insignificant differences between neighbouring samples accumulate along a geological section to result in a substantial difference between the basalmost and topmost populations, one is dealing with the process of evolution. This is how evolution can be observed from fossils.

Extraction of the evolutionarily meaningful information from a continuous fossil record of evolution at the population level is relatively easy in principle. Despite all the preoccupation of evolutionary biology with taxa ('the taxic approach'; Levinton 2001) the course of evolution can be palaeontologically documented without reference to any discrete units. Taxonomic nomenclature is irrelevant to stratophenetics. Only samples are of importance: more precisely, the information they offer on unimodal units of variability that correspond either to ancient populations or to discrete polymorphs, for instance sexes (palaeophena; Dzik 1990*a*). Their identification and presentation technically can be made in quite an intuitive way, but also by counting frequencies in morphological classes (Text-fig. 2) or by applying more elaborate morphometrics (see literature data recently reviewed in, e.g. Dzik 1990*a*, Sheldon 1996 and Levinton 2001). Obviously the process remains the same irrespective of the approach to the raw data used in its reconstruction. The way of measuring and presenting results may only help in understanding what actually happened in the evolution of a lineage and to make the case more convincing.

To overcome limitations of the method of stratophenetics while choosing an object of study one has to look for fossils that occur in great numbers, in rocks possibly complete stratigraphically, being also possibly immune to local ecological changes and sedimentary regime controlling taphonomy. This is why pelagic marine organisms with well-mineralized skeletons generated interest from the beginning of evolutionary studies in palaeontology. Initially ammonites occupied the centre of this research (e.g. Waagen 1869; Brinkmann 1929; Dommergues 1990), but were subsequently replaced by microscopic foraminifers (e.g. Grabert 1959; Pearson 1996; Berggren and Norris 1997), radiolarians (Kellogg 1975) and finally conodonts (e.g. Murphy and Springer 1989), the last of which have appeared unbeatable as a source of evolutionarily meaningful information. These early chordates owe their special value to easy chemical extraction from the rock matrix, more than 300 myr duration in the fossil record (Sweet 1988), the almost cosmopolitan distribution of many species and the great morphological information content of their statistically reconstructed apparatuses. Stratophenetically studied temperate and cold-water conodont lineages from the Ordovician (Dzik 1990*b*, 1994) and tropical lineages from the Carboniferous (Dzik 1997) have provided valuable information on the pattern of evolution at the population level. No correspondence between changes in environment recorded by fossil associations and evolutionary change in particular lineages has been identified (Dzik 1990*b*). It does not appear that environment or climate had much influence on the pattern of evolution, although there are claims to the contrary (Sheldon 1990). Problems with completeness of the fossil record are especially apparent in the rock sections representative of the tropical Late Devonian (Dzik 2002). This is why they have been chosen here to illustrate various aspects of the fossil record.

APPLICATION OF STRATOPHENETICS TO FAMENNIAN PALMATOLEPIDID CONODONTS

The palmatolepidids show the most structurally complex apparatuses among all the post-Ordovician conodonts. They are convenient objects for evolutionary studies also because of their taxonomic diversity, being represented in the Famennian by several sympatric species. In traditional biostratigraphical studies only the posteriormost platform P_1 elements of the apparatus are used to determine species, all of them being classified in the single genus *Palmatolepis*. Ironically, it has been convincingly shown that the fastest evolving and taxonomically most sensitive are not the platform elements (Klapper and Foster 1993; Metzger 1994; Dzik 2002) but those technically most difficult to collect in reasonable numbers, the anteriormost M elements and the medial S_0 element. Nevertheless, complete apparatuses undoubtedly offer much more biologically significant information than single robust elements, even if the latter are easier to collect. In terms of the standard apparatus taxonomy, as used for the Ordovician or Triassic conodonts, the palmatolepidids deserve separation into at least a few genera (Dzik 1991*b*). Some of the non-platform elements in their apparatuses show a profound morphological difference in number and orientation of processes, as well as their denticulation

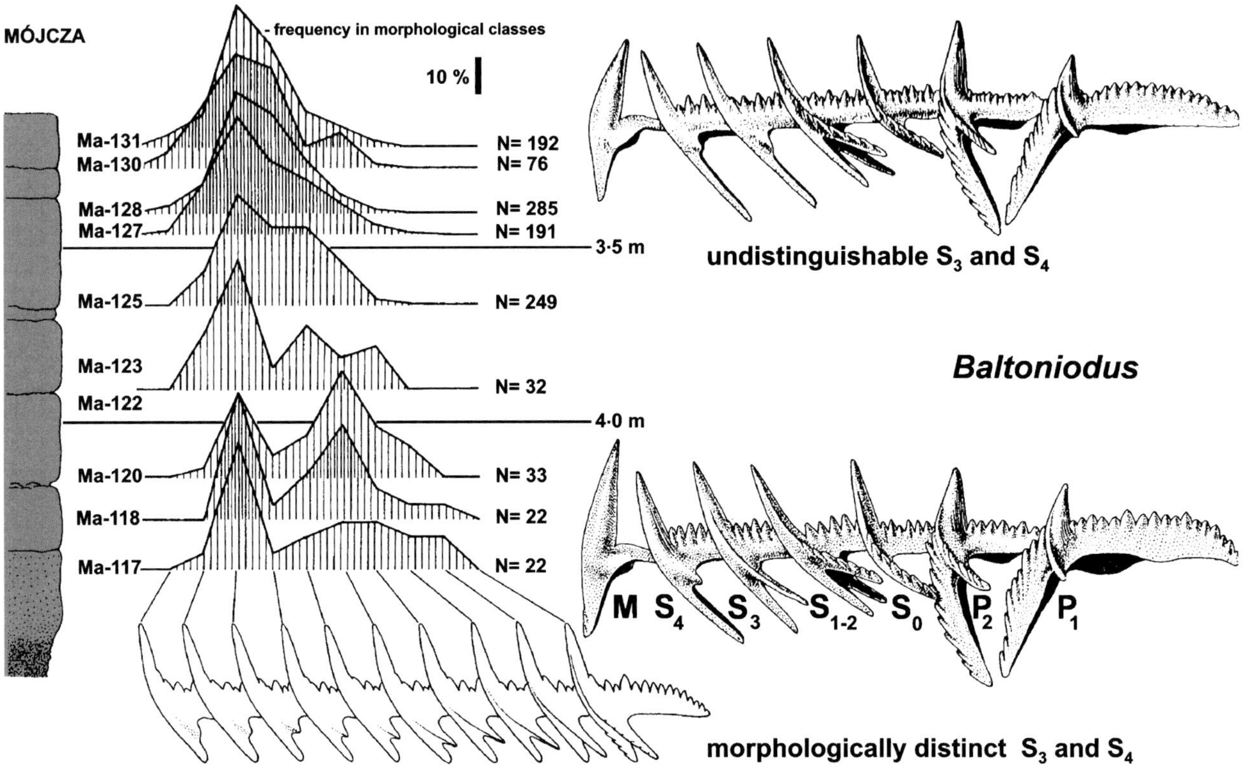

TEXT-FIG. 2. Gradual evolution of the conodont apparatus structure shown at the population level, without any metrics but by counting the frequency of particular classes within a morphological series, as exemplified by the early Ordovician (late Arenig) *Baltoniodus* lineage from Mójcza in the Holy Cross Mountains (from Dzik 1994, modified). Note that initially the frequency distribution is bimodal, with S_3 and S_4 locations morphologically distinct; in the course of evolution they became more and more alike, and in strata above this section are virtually undistinguishable.

(Text-fig. 3), far exceeding that observed in other Late Devonian genera, even those most widely defined.

The small area of the Holy Cross Mountains in central Poland, about 20 km wide, is one of many places where the complex evolution of the Palmatolepididae is well recorded (Text-fig. 4). This marginal part of the East European Platform was tectonically quiet in the late Devonian and its limestone strata are rich in conodonts, relatively little altered thermally and well preserved. Some of the sections there are thick enough to rely exclusively on the principle of superposition in stratigraphy. They represent various sedimentary environments and therefore may differ strongly from each other in composition of conodont assemblages because of the ecological sensitivity of many species. This makes homotaxy unreliable even over short distances, although this method of correlation may otherwise allow a high time resolution. The age correlation has to be based on probable phyletic transitions in lineages of index fossils, a type of reasoning which is reliable but of low resolution (Dzik 1995).

The stratigraphical condensation and punctuation of the record by numerous gaps in sedimentation limit evolutionary studies in the Famennian of the Holy Cross Mountains. The record is complete in the deeper parts of the local basin but fossils of conodonts are not common enough there to allow apparatus studies, probably both a result of a higher sedimentation rate and lower biological productivity of the environment. Immigration of new lineages and terminations of others, possibly replaced as a result of ecological competition or simply by random lateral environmental shifts, is a feature of the succession of assemblages (Dzik 2002).

Despite all the shortcomings of the empirical evidence on evolution, some examples of a successful application of stratophenetics to the Famennian conodonts of the Holy Cross Mountains can be offered. Among the lineages most persistent and well represented by numerous specimens is that of the early *Tripodellus* (Text-fig. 5), at the stage of evolution prior to the development of its diagnostic triramous P_2 elements (the generic affiliation of these populations thus remains arbitrary). Some measuring has been done on the platform P_1 elements from a section at Jabłonna. The strata there are poorly exposed, being deeply weathered and overgrown with forest vegetation. As a result, only sets of samples separated by gaps are available. Noteworthy within each of the sets is that

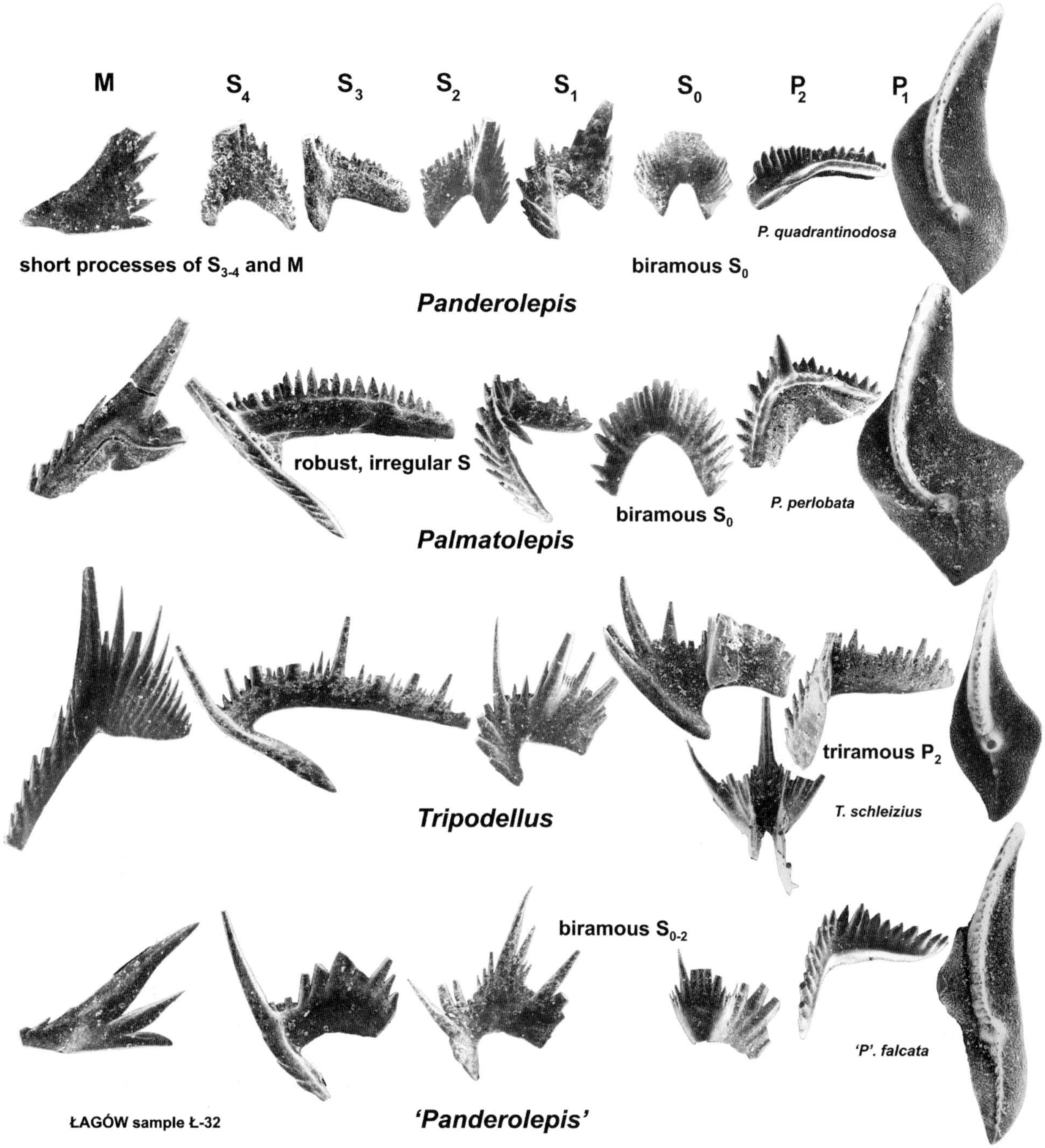

TEXT-FIG. 3. Apparatuses of the Famennian genera of the Palmatolepididae; statistical reconstructions based on material from the Holy Cross Mountains.

the change in the distribution of platform width is continuous. All sets together show the apparent general trend: an increase in elongation, which continues well above the segment of the lineage represented at Jabłonna.

Even if the sparsely distributed single examples of *Tripodellus* or their sets are considered, the differences between neighbouring samples do not appear to be especially significant, but the extreme samples are quite dissimilar. In many cases it is enough to arrange them in a stratigraphical order to see this (Text-fig. 6). Sometimes, despite the stratigraphical distance between samples, the variability of some characters overlaps. For instance, rare triramous P_2 elements occur significantly below the level of their exclusive occurrence.

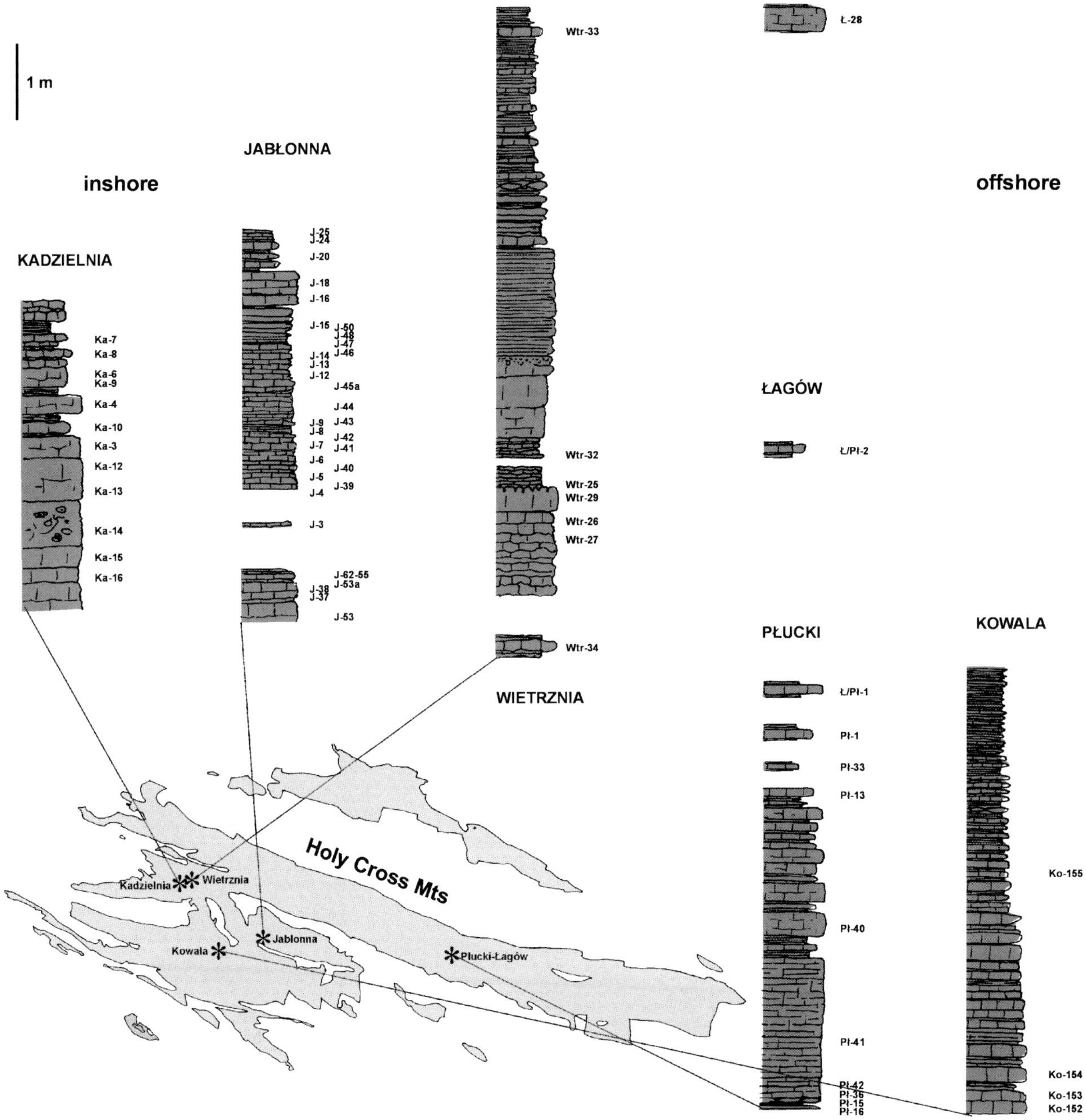

TEXT-FIG. 4. The sampled sections of early Famennian deposits in the Holy Cross Mountains, with their relative position indicated on a map showing the extent of Devonian exposures. Lithological columns are arranged according to their time relationship; no formal zonation is attempted. Note a profound facies differentiation over short distances; condensed sections in the centre of the area show the geological time record punctuated with gaps; in more complete sections to the south fossils are generally rare, probably because of low biological productivity and/or a high sedimenation rate.

The evolution within particular lineages of *Tripodellus* or any other Famennian conodont genus can thus be represented as a series of ancestor–descendant hypotheses concerning pairs of samples possibly close in time and space. Any such hypothesis can be tested by increasing the density and size of samples if the ancient populations are represented in rocks with fossils (frequently they are). It is possible to increase resolution to the level at which stratophenetics can be applied. Of course, this requires a lot of work, and so does not seem practical, but potential testability of an ancestor–descendant hypothesis is obvious in this particular case.

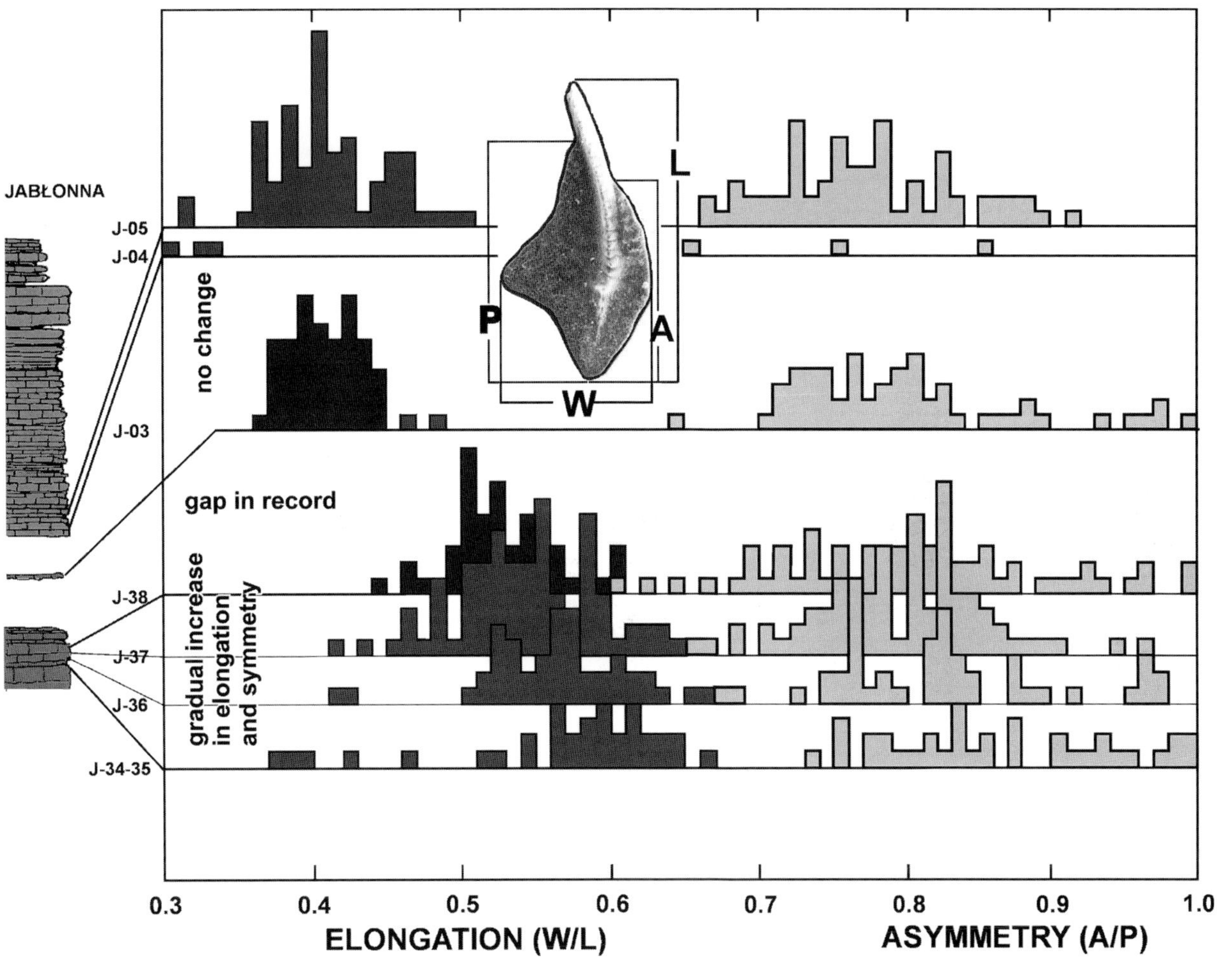

TEXT-FIG. 5. Stratophenetics of a segment of the early *Tripodellus* lineage in the Jabłonna section, Holy Cross Mountains. Note that the sampling is generally incomplete and only isolated sets of samples separated by gaps (mostly a result of poor exposure) are available; continuous changes are documented within particular sets and they seem to be consistent with the apparent general trend.

The conclusion most important to the subject of this review from cases such as Jabłonna or Mójcza (Text-figs 1, 5) is that there is no methodological difference between methods of reconstructing evolution based on complete and incomplete fossil evidence. It is just a matter of limitations in the availability of data. Stratophenetics appears to have a more general application to the extreme case.

Wide temporal gaps in the record may hide reversals and changes in the direction of evolution. However, there is no reason to restrict evolutionary studies to single geological sections and the missing evidence can be recovered potentially by additional sampling in other locations. At the very least, data from exposures in proximity have to be assembled. However, this introduces a spatial dimension to considerations, which forces the limits of stratophenetics to be crossed.

GEOGRAPHICAL DIMENSION OF EVOLUTION

The most obvious aspect of the geographical dimension of evolution is the phenomenon of migration of lineages. In any single section this produces a record that looks as if a cloud of organisms passed overhead, dropping to the sediment a rain of skeletal remnants that give the quantitatively presented stratigraphical range of a lineage its fusiform aspect (Dzik 1990*b*). Migrations thus influence the record in any place, but their documentation requires data from several localities. This aspect of the spatial distribution of organisms is of particular importance in the evolution of sexual organisms. Expansions and contractions of the geographical range of an originally panmictic population may result in its spatial split into daughter lineages that separately evolve within their own habitats. This may

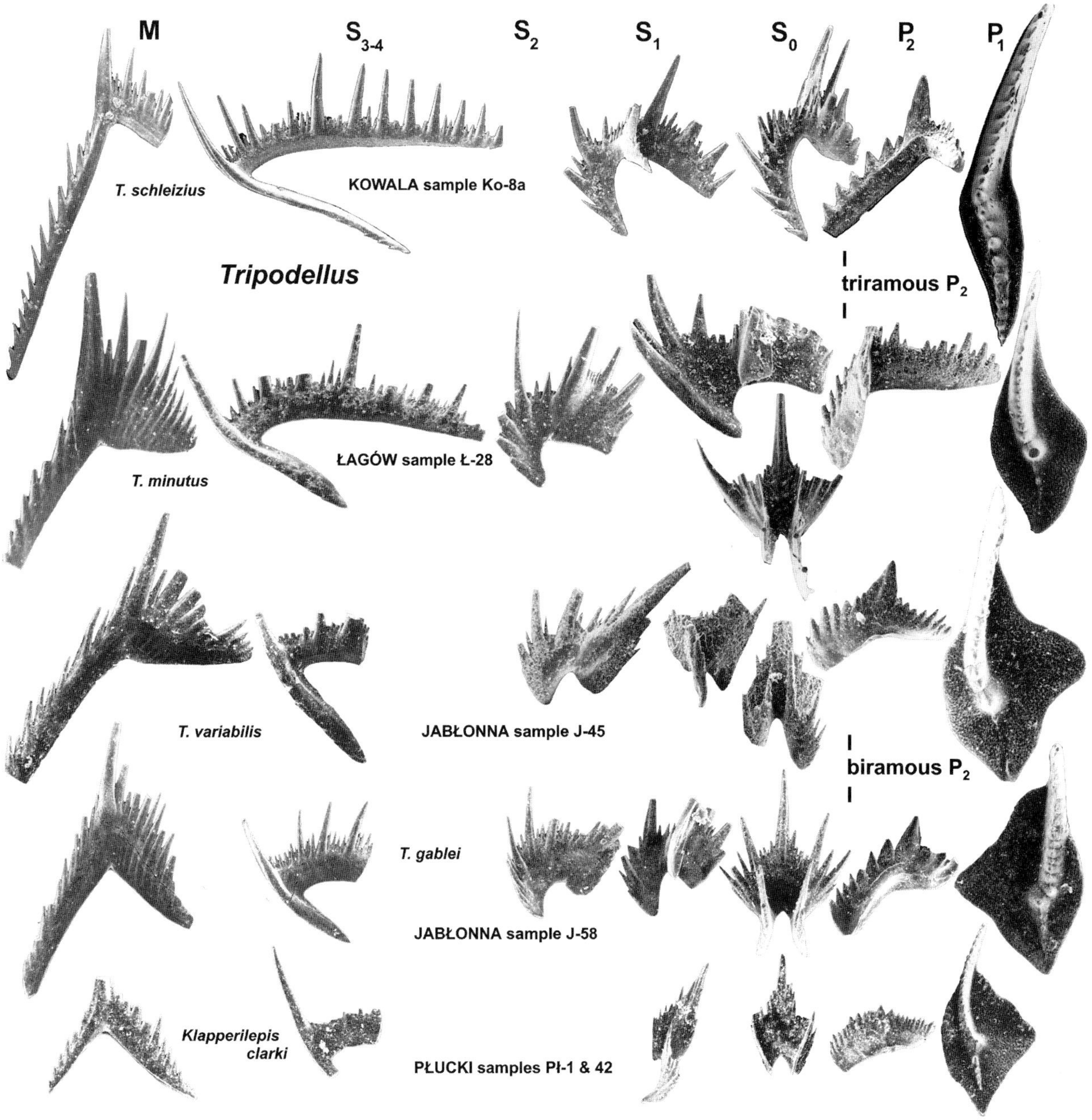

TEXT-FIG. 6. Succession of the *Tripodellus* lineage apparatuses arranged according to their stratigraphical order in the Holy Cross Mountains. Although in places the phylogeny of this clade is rather complex, with up to three sympatric species represented in the area, a general pattern of the main chronomorphocline is apparent. Usually in the populations their variability overlaps between neighbouring pairs of samples, suggesting genetic continuity.

be temporal, followed by subsequent homogenization of the populations through hybridization, but may also continue for long enough to allow a genetic barrier to develop, making reunification impossible. This is actually a description of the classic model of allopatric speciation (e.g. White 1968). Other ways to develop genetic isolation are a possibility but the available evidence (e.g. Bush 1994) remains controversial; however, there can be little doubt that in the real world of highly differentiated environments, the process of evolution of sexual organisms is very complicated in its geographical dimension (e.g. Avise *et al.* 1998).

Stratophenetics is a method of studying the phyletic evolution of lineages; their splitting is not accessible as long as only geological time and morphology are considered, not geographical space. An allopatric speciation cannot be observed in any single section. The final effect of a speciation can be noticed only when the newly established lineage immigrates. It has to be borne in mind, however, that the replacement of one population by another may as well occur before as after the allopatric speciation event they were subjected to. This distinction does not need to be expressed in morphology. If the populations can interbreed despite already developed morphological differences (but not genetic isolation), this smooths the change, but a sudden change in morphology may take place within the lineage, a change having nothing to do with speciation. The opposite occurs when one of the sibling species is replaced by another; no morphological change exists to be detected, although speciation has already taken place. In fact, the process of speciation is thus out of reach of not only stratophenetics but also palaeontology as a whole (Dzik 1991*a*).

Are species (and speciation) really of so great a significance in attempts to understand the phylogeny? This depends on what one actually wants to know. It could be argued that it is most important to see the process of anatomical and physiological transformations, mostly expressed in the morphology of organisms, not just count units of interbreeding.

The choice of one of these attitudes may depend on what material is dealt with. The biological species concept offers in principle the objective unit of diversity for students of Recent organisms. In palaeontology species have a dual, objective/arbitrary nature being 'objective evolutionary units on a time plane and at the same time arbitrary units crossing time planes' (Gingerich 1985, p. 29), which means that as long as one is studying fossil organisms from the same sample, same locality or from different sites not significantly different in geological age (thus, essentially on the same time plane) the concept of species may be used in its objective sense. Procedures of applying population and species rank taxonomy for palaeontological purposes are widely used (e.g. Dzik 1990*a*). However, this can be done only when obvious morphological differences developed between species, which tends to be the case when sympatric species assemblages are considered (e.g. Brown and Wilson 1956). If one wants to classify allopatric populations at the species level, numerous difficulties have to be faced. To overcome this, the precise age correlation of sections is necessary and a series of transitional localities available to show that morphologically different and geographically distant populations represent either end-members of a morphocline or spatially uniform separate species. This can be proven if their ranges overlap and both species occur sympatrically in marginal localities (Dzik 1979).

The important limitation on the usefulness of the biological species concept is that it refers to reproductive isolation, not morphology, as the defining aspect. In the case of allopatric units, speciation events and species distinction do not need to have anything to do with ecological adaptations or any morphological difference (de Vargas and Pawlowski 1998). The very existence of allopatric sibling species is thus of little importance to the evolution of ecosystems and the impact of 'new taxonomy' based on genetic instead of morphological distinctions may not be so great as is frequently claimed (e.g. Knowlton and Jackson 1994). Only after the species meet does niche partitioning becomes a must if more than one species is to survive. Although this may sound too radical, identification of speciation events in palaeontology is not only impossible technically but also of limited importance to the evolution of communities and their ecosystems.

In fact, speciation events are not necessary for evolution to occur. Fossil evidence convincingly shows that there is no connection between speciation events and evolution, although obviously no speciation is possible without evolutionary change. This is self-evident also on the basis of neontological observations: there is no correspondence between the number of species in a taxonomic group and the rate of its evolution. Our own monospecific lineage of *Homo* demonstrates this well.

There is no reason to assume that migrations occur immediately after speciation is completed. To determine the appearances and disappearances of taxa in rock sections is thus a waste of time from the evolutionary point of view. Both events are ecologically controlled and have nothing to do with evolution (Dzik 1994). At best they allow the correlation of ecological events in different sections. They definitely do not indicate the time and location of the speciation events. To see what happened in the evolution of a lineage before its immigration one has to look for a geological section in the geographical area of its origin. Obviously, to do this one has to determine their phyletic evolution stratophenetically.

Typically, any long-enough segment of the evolution of a fossil community appears to be a mixture of evolution in place and immigrations of allopatrically originating lineages (Text-fig. 1). Such is the Famennian evolution of the palmatolepidid conodonts in the Holy Cross Mountains. An interesting aspect of this interplay of local change and immigration is its influence on the population variability of morphological characters. This has been documented biometrically in the earliest Famennian, when the initial stage in the diversification of these conodonts took place.

SPECIATION AND MIGRATION OF FAMENNIAN PALMATOLEPIDIDS

Unlike their chronologically preceding relatives, the earliest Famennian palmatolepidid faunas, as shown by their apparatuses reconstructed by Schülke (1999), were of a rather low morphological diversity. Their diversification probably started from the three lineages documented in the upper part of the most complete section transitional from the Frasnian to Famennian at Płucki (Text-fig. 7, sample Pł-32). The extreme variability of platform elements in the early palmatolepidids makes biometric discrimination of species difficult. Some important characters, like the raising upward of element tips and more or less horizontal disposition of the platform margin are difficult to measure. Anyway, even an imperfect presentation allows an estimate of the extent of population variability in particular samples (Text-fig. 7). The morphology of the platform in P_1 elements ranges there from relatively narrow and planar (typical of later members of the lineage of *Tripodellus*), through sinuous and extended up to the dorsal end of the element (typical of the *Palmatolepis* lineage), to wide but short (*Klapperilepis delicatula*).

Below in the Płucki section, only two lineages are represented, documented unequivocally by the associated M elements of two kinds (Dzik 2002). One of them is of a generalized morphology possibly inherited from the Frasnian *Klapperilepis praetriangularis*, the other shows a fan-like arrangement of denticles on the external process, similar to those attributed to '*P.*' *arcuata* by Schülke (1999; the type population of the species is of significantly younger age). Although there is no doubt that two species are represented, there is a completely smooth transition in the morphology of platform P_1 elements. Specimens with a relatively narrow platform and transverse orientation of the angular platform lobe seem to form a separate cluster. Close to the base of the Famennian, where only one type of M element occurs, the frequency distribution of their shapes is clearly unimodal. The modal morphology is the same as in the related populations from higher samples. Nevertheless, the range of variability is much wider, encompassing not only most of the range occupied by the younger species with narrower platforms but also morphs with a very wide platform, which also occur in the latest Frasnian. In fact, the earliest Famennian and latest Frasnian populations of the *Klapperilepis* lineage do not differ from each other in their apparatus morphology (Dzik 2002).

Two aspects of this succession are of interest from an evolutionary point of view: the continuity across the Frasnian/Famennian boundary and the decrease in population variability within the same lineage after additional species appeared in the assemblage. The *Klapperilepis* population from the earliest Famennian, where it occurs alone without any other palmatolepidids, is morphologically identical to that of the latest Frasnian Upper Kellwasserkalk. The rather profound environmentally controlled faunal changes marked by the disappearance of the typically Frasnian lineages of *Lagovilepis* and *Manticolepis* had, thus, no influence on the phyletic evolution of the *Klapperilepis* lineage except for a somewhat delayed increase in its population variability. The latter may possibly be an effect of relief from competitive influence of other palmatolepidid species. They were rather distantly related and this is probably why their extinction from the assemblage had rather minor consequences. When the earliest Famennian assemblage was enriched in closely related species by immigration from the areas of their allopatric origin, the decrease in population variability of *K. praetriangularis* became more apparent. The platform elements of the species new in the area covered the range of shapes not much different from that represented originally by just the single ancestral species. The local population of *K. praetriangularis* was probably forced to adapt to the new conditions of partially overlapping ecological niches. The ecological phenomenon of character displacement (Brown and Wilson 1956) has already been invoked to interpret similarly profound changes in the population variability of Carboniferous conodonts (Dzik 1997, p. 70). Perhaps also in this case the competition between sympatric species reduced their variability.

Thus, in the evolution of Famennian palmatolepidids in the Holy Cross Mountains only one local lineage was represented at the beginning of the Famennian and new lineages emerged sequentially by immigration from their places of origination. Almost certainly they evolved there in a similar way as the lineage that shows a complete record in the Polish sections, that is by gradual morphological change and under competitive pressure from immigrants. This offers support for the traditional view of the evolution of the palmatolepidids: earliest Famennian recovery after extinction.

The evolution of the palmatolepidids was a process of ramification of their phylogenetic tree but the points of bifurcation invariably appear to be out of reach of the palaeontological method (Text-fig. 8). The method allows much confidence while tracing particular lineages but the origination of lineages remains obscure. To identify their origin one has to propose a hypothesis on their origin and look for ancestry in geographically distant places. If such a record is found, stratophenetics can be used. The inference on identity of the ancestor has to be based not only on its morphology but also on geological age (older than the base of the lineage documented elsewhere) and geographical location (close enough to make physical continuity between lineages likely). The inference proceeds back in time. This is one of the basic aspects of the method. The second is that tested hypotheses do not refer to taxa but to ancient populations. This general approach is specific for palaeontology in that it refers directly to ancestor–descendant hypotheses and uses geological time as the basic evidence. It is here referred to as 'chronophyletics'.

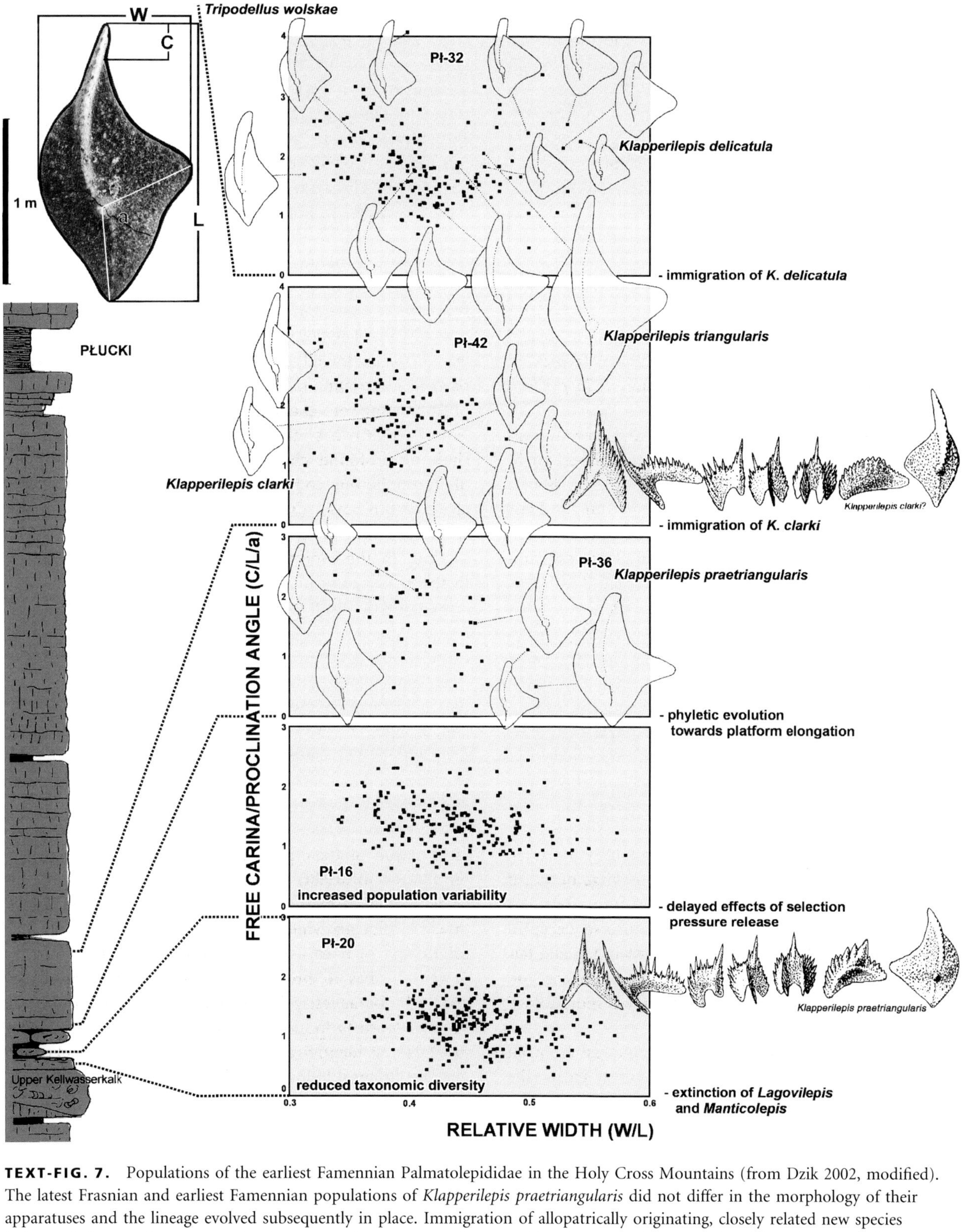

TEXT-FIG. 7. Populations of the earliest Famennian Palmatolepididae in the Holy Cross Mountains (from Dzik 2002, modified). The latest Frasnian and earliest Famennian populations of *Klapperilepis praetriangularis* did not differ in the morphology of their apparatuses and the lineage evolved subsequently in place. Immigration of allopatrically originating, closely related new species influenced the population variability (character displacement). The assemblage became richer in species although the complete range of morphologies initially did not increase very much. Eventually, as a result of subsequent divergent evolution generic rank differences emerged.

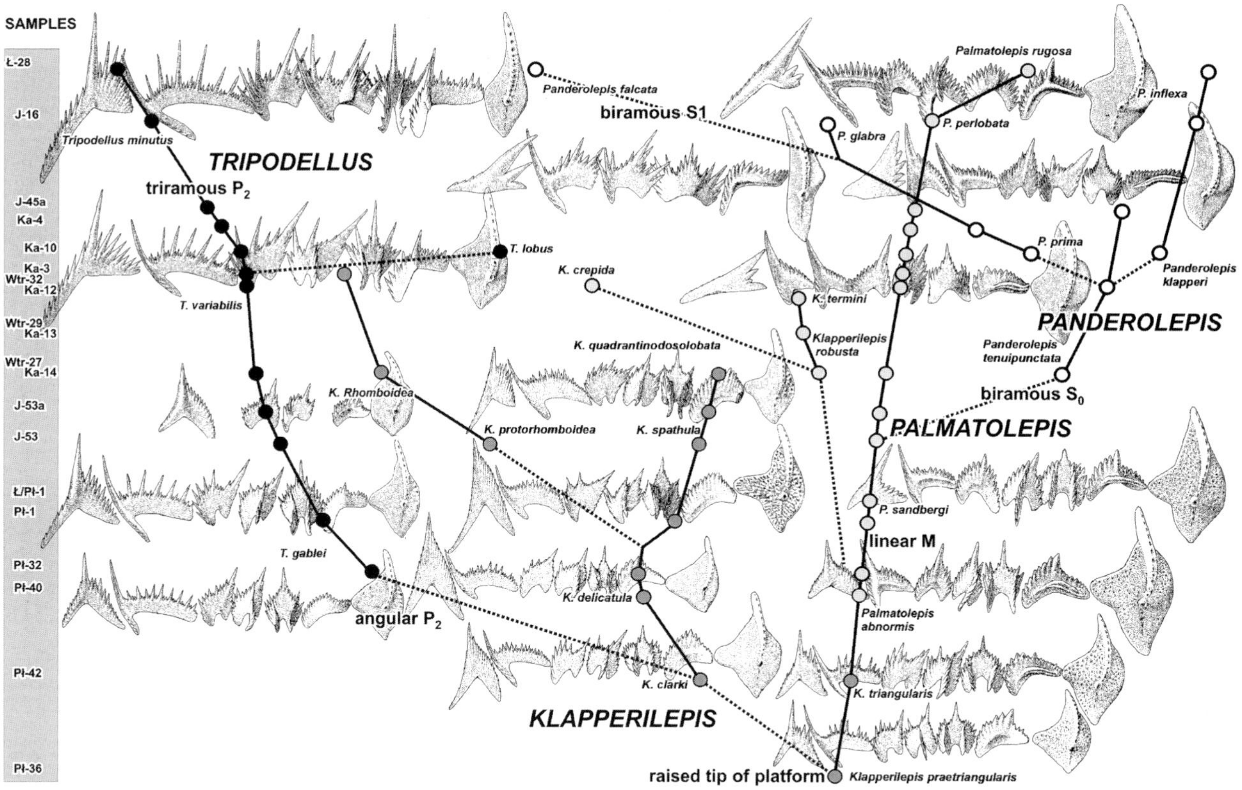

TEXT-FIG. 8. Famennian phylogeny of the Palmatolepididae based on fossil evidence from the Holy Cross Mountains. Reference populations (circles) with statistically well-documented apparatus structures are connected by hypothetical ancestor–descendant relationships (lines); hypothesized allopatric speciation and subsequent immigration events are indicated by broken lines. Note that only phyletic evolution can be potentially proven by increasing the density of sampling; speciation events are speculative irrespective of the quality of the record as they almost certainly occurred allopatrically in all cases. Their first appearances in geological sections are invariably the result of ecologically controlled immigration and have nothing to do with their evolutionary origin.

CHRONOPHYLETIC APPROACH TO THE FOSSIL RECORD

Stratophenetics may potentially offer definite evidence on the course of evolution at the population level. However, to be sure that one is dealing with evolution, a complete fossil record is required, from geological, ecological and taphonomic points of view. The normal situation in palaeontology is far from that. The evidence is frequently limited to single specimens, sparsely distributed in time and space and does not offer characters that are truly diagnostic. How to proceed then with such data to keep presentation of hypotheses on the course of evolution testable, despite the incompleteness of the record?

The answer offered here derives from the observation that there is no fundamental difference between complete sampling, sets of samples separated by gaps and quite isolated pieces of the record. In principle the hiatuses in time and space can be filled in future. Obviously, it would be unrealistic to expect that the assembled evidence will ever be complete enough to allow definite tests of hypotheses even on the phylogeny of the most privileged pelagic organisms equipped with mineralized skeletons. Only a small fraction of all those billions of individuals that lived in the geological past were fossilized and we do not have enough technical facilities even to document evolution of those that have a relatively complete fossil record. Evolutionary inference unavoidably has to be based on less abundant material, and commonly just on single crucial findings. Most of the description of the course of evolution will remain hypothetical or even conjectural. This should not result in any discomfort as long as the potential remains to test hypotheses on the ancestor–descendant relationship with the fossil evidence. Moreover, such hypotheses are not only testable but can even be refuted by evidence. They are falsifiable and this makes evolutionary studies in palaeontology truly scientific.

Falsifiability of ancestor–descendant hypotheses

To test an ancestor–descendant hypothesis the reasoning has to proceed back in time. This is because any organism may have uncountable successors and there is no way of deciding with which of those numerous lineages we deal while studying a particular fossil. However, only one hypothesis of ancestry is true, as any asexual organism may have only one ancestor and any sexual species only one ancestral species. The true course of evolution is being approached by increasing time, space and morphological proximity of data sets. Any new piece of evidence extending the lineage backward supports the hypothesis or contradicts it with power proportional to the dimensions of its departure from the expected. A hypothesis can be finally refuted if the restored succession of populations reaches the time horizon of the earlier proposed ancestor (Text-fig. 9). Such definite falsification is rarely reached but its possibility makes the method scientific (Dzik 1991*a*; see Engelman and Wiley 1977 for discussion from a cladistic point of view).

The straight line connecting populations of different age in a hypothesis on the ancestor–descendant relationship does not imply a linear course of evolution. This is just an application of Occam's Razor to the time and morphological dimensions of evolution. There is no need to violate the principle of parsimony by presenting the evidence in terms of hypothetical sister taxa. This would introduce an unnecessary ghost range, a succession of nonexisting populations of the 'sister lineage'. Only empirical evidence obtained later may force us to make the theory more complex.

The inferences by retrodiction, that is by proposing hypotheses on ancestry and potential testability of ancestor–descendant hypotheses, are thus crucial aspects of the chronophyletic approach to the fossil record of evolution. Evolution is therefore understood as an objective physical process with samples (populations) of different age connected to a series of hypothetical descent. To show how this can be done in practice a few examples from the Frasnian history of the palmatolepidids are discussed below.

CHRONOPHYLETICS OF THE PALMATOLEPIDID CONODONTS

Chronophyletics overcomes the methodological limitations of stratophenetics by introducing the geographical space and retrodiction to evolutionary considerations. This provides the basis for representing the course of evolution as a phylogenetic tree, here exemplified by the Famennian phylogeny of the Palmatolepididae (Text-fig. 8). As long

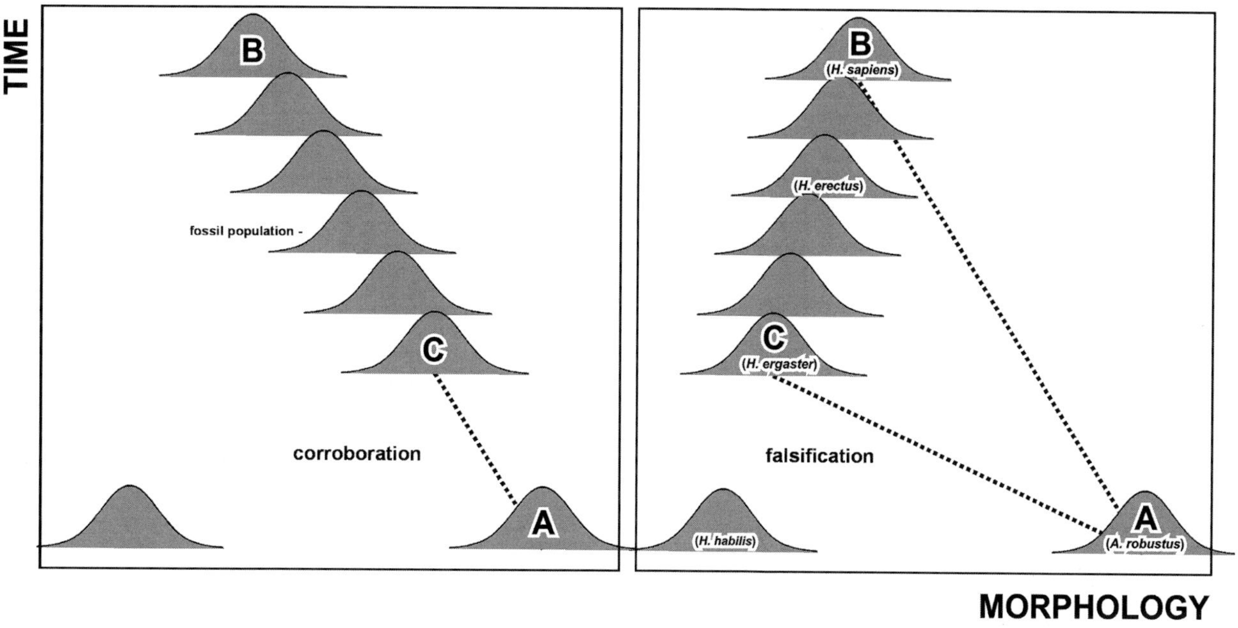

TEXT-FIG. 9. Testability of chronophyletic hypotheses based on the assumption that a species may have many successors but only one ancestor: hypotheses on ancestry are thus contradictory. The conclusive dismissal of a hypothesis can be achieved when retrodiction reaches an ancestor coeval to that originally proposed. Note that a departure of stratigraphically transitional new evidence from expectations may offer an estimate of the power of falsification. The evolution of hominids may serve as a simple explanation of the proposed way of reasoning: the claims that the Asian population of *Homo erectus* was derived from the African population of *Homo habilis* or, alternatively, *Australopithecus robustus*, are contradictory. The finding of a population ancestral to *H. erectus* (i.e. the population classified as *H. ergaster*), which is closer in morphology to *H. habilis* than to *A. robustus*, makes the second possibility weaker and its power is proportional to the time, space and morphological distance between the populations. After a continuous succession between *H. erectus* and *H. habilis* has been assembled, the *A. robustus/H. erectus* hypothesis is definitely refuted.

as it is based on materials from the restricted area of the Holy Cross Mountains, only the phyletic evolution of Polish populations can be potentially proven stratophenetically. The spatially disjunct speciation events have to remain speculative, as by definition they took place elsewhere. Even if the place of origin of those species in any single section can be identified, the process of speciation would be represented just by a phyletic change.

Going back in geological time, one can see an apparent decrease in the taxonomic diversity of the palmatolepidids and reduction of morphological differences separating sets of sympatric species. This aspect of the phylogeny is shown by the pattern of recovery after their terminal Frasnian decrease in diversity. Close to the base of the Famennian, the sympatric species differ almost exclusively in details of denticulation of the anteriormost element in the apparatus. Their lineages appear so similar to each other that proximity to the first split is apparent there, even if this cannot be documented in the area (Text-fig. 8).

The Frasnian in the Holy Cross Mountains (Text-fig. 10) is even less suitable for stratophenetic studies than the Famennian. The facies distribution is rather complex and

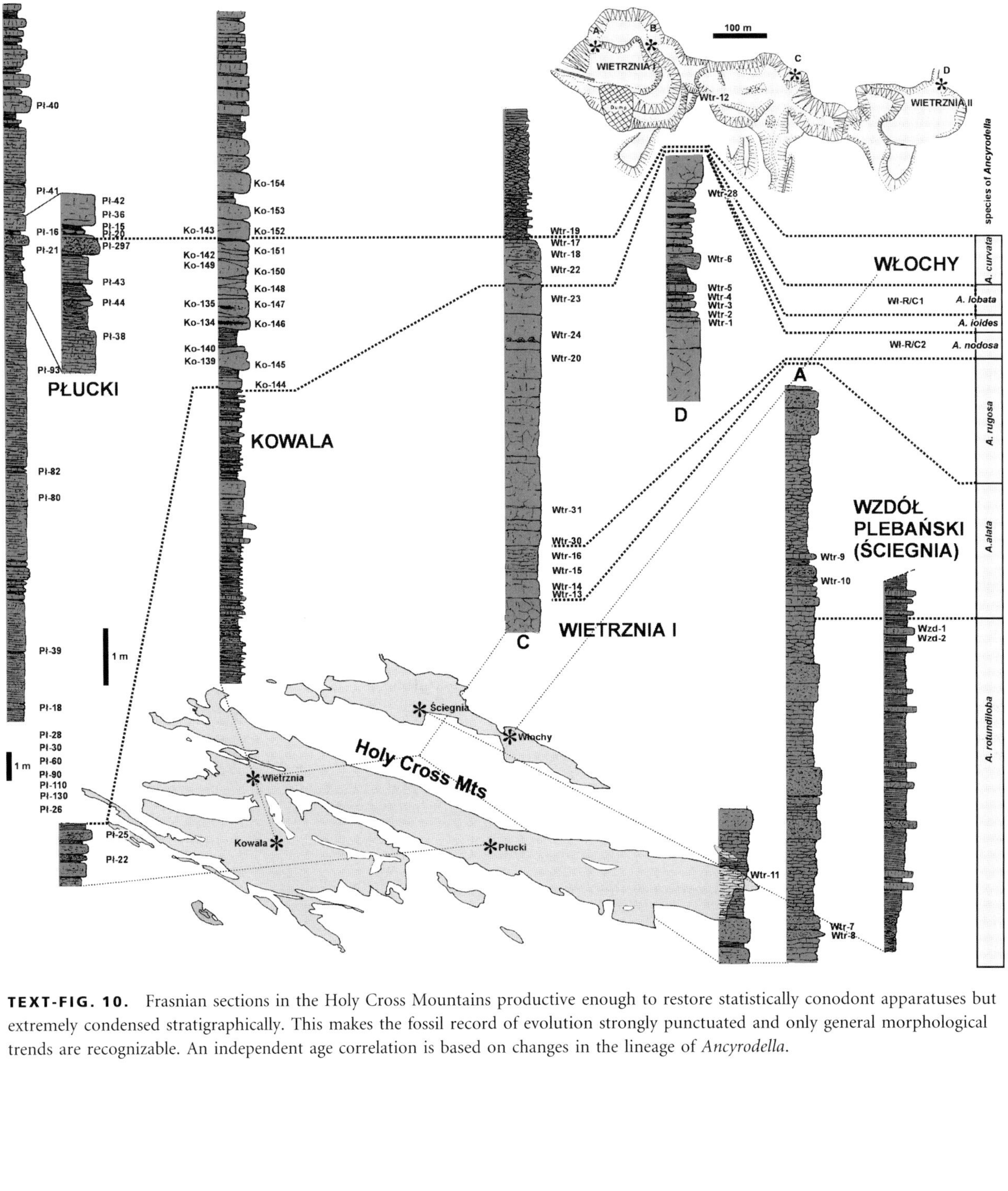

TEXT-FIG. 10. Frasnian sections in the Holy Cross Mountains productive enough to restore statistically conodont apparatuses but extremely condensed stratigraphically. This makes the fossil record of evolution strongly punctuated and only general morphological trends are recognizable. An independent age correlation is based on changes in the lineage of *Ancyrodella*.

no doubt the region was ecologically highly diverse. Homotaxy in this case is also of little use in the precise age correlation. Different palmatolepidid lineages may be restricted in their occurrences to sections less than 20 km apart. Those that are productive enough to offer material to evolutionary apparatus studies are extremely condensed stratigraphically. To base correlation on evidence independent of the evolution of palmatolepidids, the succesion of the polygnathid conodont *Ancyrodella* is chosen. This appears evolutionary in nature and as such seems a rather reliable basis for age correlation (Klapper 1990).

The fossil record of evolution of Frasnian palmatolepidids thus remains strongly punctuated. In such a situation only general morphological trends can be recognized. Series of samples arranged according to their geological age show a sequential introduction of evolutionary novelties, which subsequently marked major clades. It is most apparent in the symmetrical element of their apparatuses, originally having a median process that gradually disappeared to be replaced by the bifurcation of lateral processes.

Most interestingly, immediately below the Frasnian/Famennian boundary palmatolepidid diversity was closely similar to that in the middle Famennian (Text-fig. 11). The differences were expressed mostly in the morphology of the anteriormost M and symmetrical S_0 elements in the apparatus. The only systematic distinction of the Famennian lineages is bending of the tip of P_1 element platform: in the Frasnian it was bent upward except for a single species of *Klapperilepis*. In addition, in morphology of non-platform elements of the apparatus each of the geologically oldest members of the early Famennian palmatolepidid lineages resemble *Klapperilepis*, but not other late Frasnian conodonts. It appears thus to be the only lineage that survived to the Famennian and gave rise to all later palmatolepidids. Its roots are probably in an early *Manticolepis*, as suggested by its rather primitive M and generalized S_0 element morphology with bifurcation of lateral processes developing relatively late in ontogeny. Transitional populations were polymorphic. This kind of ramification of processes was initiated in the neighbouring, laterally located S_1 element before it expanded to the medial S_0. Apparently, the biramous S_0 of another latest Frasnian lineage, *Lagovilepis bogartensis*, represents a reversal to the ancestral status as its S_1 element is normally bifurcated. This species is thus unlikely to be ancestral to *K. praetriangularis*. Also, in the more anterior S_2 and S_{3-4} locations some temporally ordered changes can be identified, which were followed by a phylogenetic split into lineages that differ from each other mostly in the morphology of the M element (Text-fig. 11). The typical palmatolepidids possess highly arched M elements with straight processes. This trait had already developed in the early Frasnian *Mesotaxis* (Dzik 1991*b*) well before the medial process in the S_0 element disappeared (Text-fig. 12).

The geographical distribution of Frasnian palmatolepidids of the Holy Cross Mountains shows some regularity (Text-Fig. 13). As expressed by per cent contributions to samples, their frequencies are different in deeper- and shallower-water environments; some seem to be restricted to specific facies. For instance, the phylogenetically important *K. praetriangularis* lineage occurred in the extreme black shale environment in the marginal parts of the area (Text-fig. 13). It is not surprising that its origins remain cryptic.

The present picture of the phylogeny of palmatolepidids has been developed by arranging stratigraphically the data on populations and connecting them by a network of ancestor–descendant hypotheses. Particular hypotheses can be tested by increasing density of sampling and sample sizes within the area to meet the requirements of stratophenetics. Whenever the fossil record of a lineage is not good enough or terminates in local sections, sampling has to be extended to other areas. The lineages of allopatric origin have been attached basally to the morphologically closest of known lineages. The resulting hypotheses of ancestry have to be tested with evidence from elsewhere. If the basal extension of the lineage of the descendant meets another lineage, unrelated to that of the suggested ancestor, the hypothesis would eventually be refuted. Obviously, the whole phylogenetic tree has to be logically consistent. In cases of conflict between different interpretations, parsimony or common sense are the best guides. This means that any character distribution analysis is of much help, unless one enters circular reasoning.

The traditional palaeontological approach to evolution is fundamentally different from those preferred by neontologists. Although the fossil evidence is definitely less informative than data on recent organisms, only palaeontology may offer direct access to ancient evolutionary events. This makes the question of how to construct and test hypotheses on the course of evolution based on the fossil evidence (whatever methodology has been used to produce them) a matter of life-or-death for evolutionary palaeontology.

CLADISTICS VERSUS CHRONOPHYLETICS

There can be no doubt that some aspects of evolutionary history of organisms can be extracted from the pattern of their present diversity. In fact, much of our knowledge of evolution is based exclusively on Recent organisms. Zoological or botanical data on the distribution of characters among organisms is widely used to infer their phylogeny. Several methods have been developed to perform this task. Any such method has to assume some correspondence between the morphological similarity and time

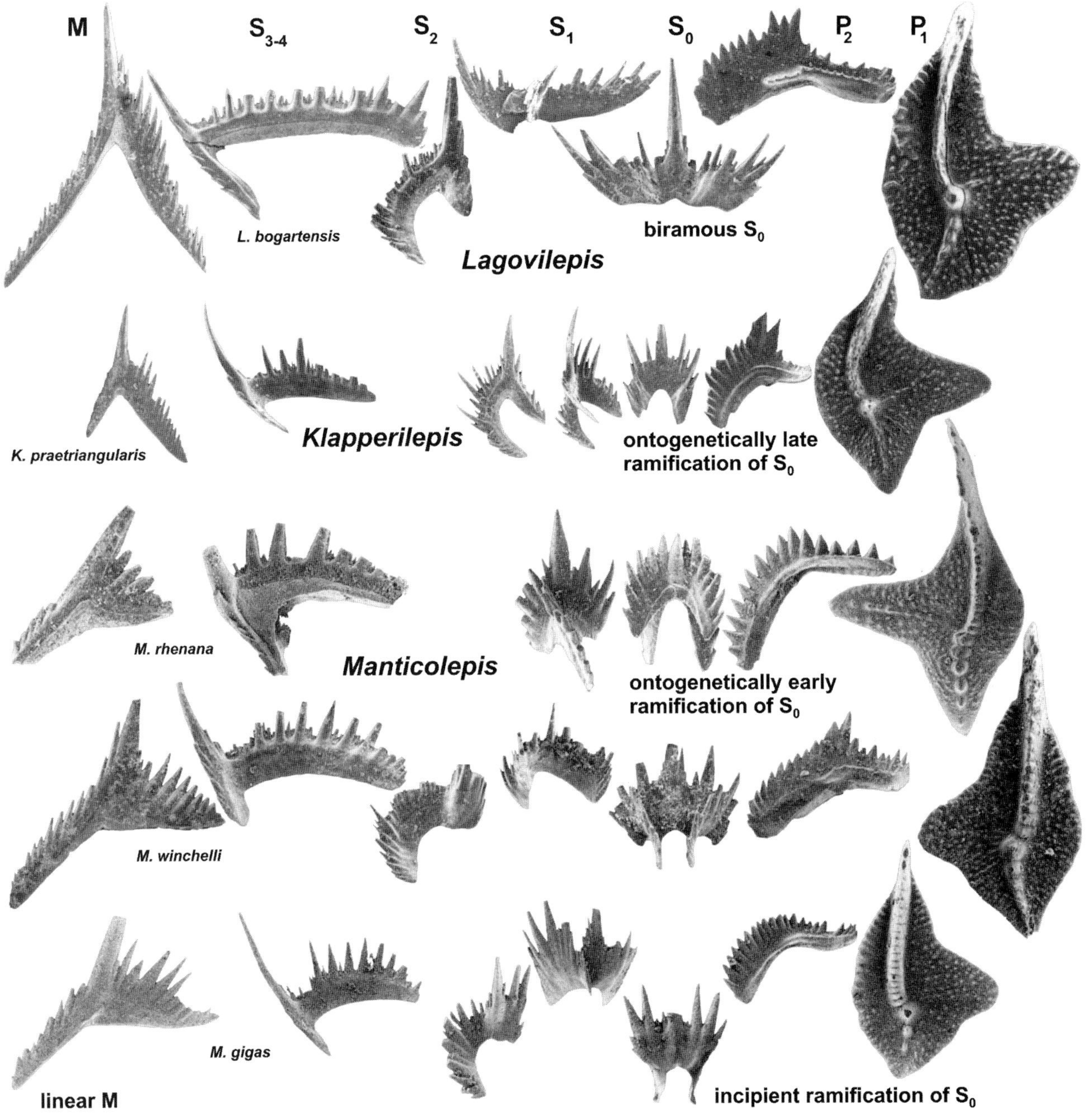

TEXT-FIG. 11. Apparatuses of the late Frasnian genera of the Palmatolepididae in the Holy Cross Mountains arranged partially according to their stratigraphic order of occurrence. Note that a gradual change in S_0 element marks their early evolution, followed by a split in morphology of the M element; the *Klapperilepis* lineage was confined to deeper-water facies.

that has passed since the separation of lineages of morphologically distinct members. The correspondence is hardly strict. Obviously, the rate of evolution may differ greatly between lineages. It may be quite irregular, resulting in misleading similarities from reversals, convergences and parallel evolution. As long as the purpose of the analysis is clearly stated (i.e. a restoration of the actual course of evolution) there is no disagreement between various methodological schools in dealing with these shortcomings of the evidence derived from Recent organisms. The tremendous recent progress in molecular phylogenetics, using both phenetic and cladistic ways of reasoning, exemplifies this very well.

Nevertheless, the situation in morphology-based and molecular phylogenetics is fundamentally different in several aspects. The most troublesome distinctions of the analyses of phenotypic differences and molecular sequences are: (1) morphological characters are not

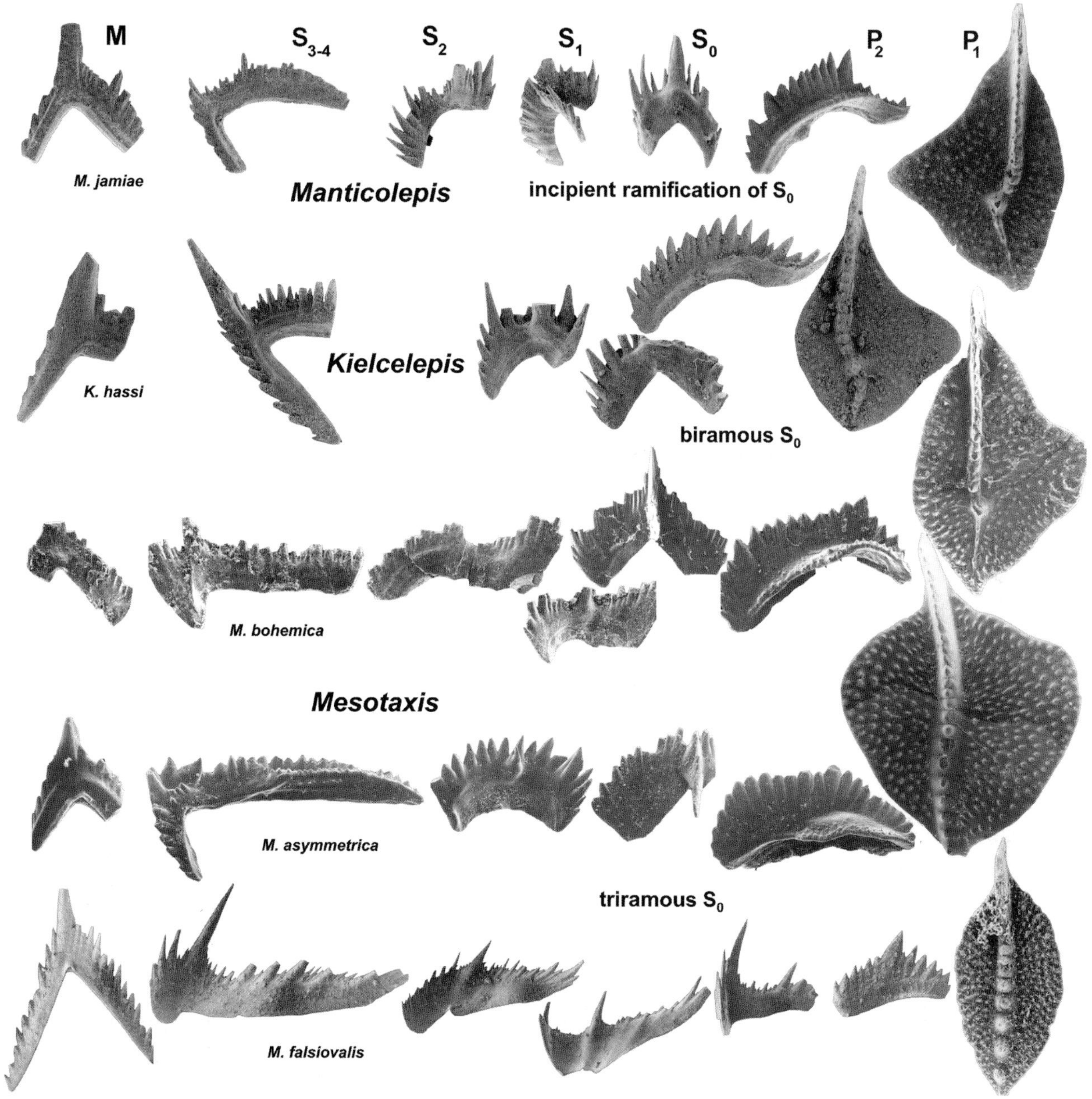

TEXT-FIG. 12. Apparatuses of the genera of the early Frasnian Palmatolepididae. Note that ramification of elements in the middle of the apparatus starts from S_1 and expands to S_0; in transitional populations S_0 is polymorphic. Temporally ordered changes also took place in the S_2 and S_{3-4} elements.

objectively discrete, unlike nucleotides or amino acids; and (2) unlike the molecular data, morphological evidence can also be obtained for organisms from the geological past.

The consequence of a failure to define morphological characters objectively has been considered damaging to numerical methods of analysing raw data by opponents of 'computer cladistics' (e.g. Wägele 1994). Unavoidably, one has to assume the equal value of morphological characters or arbitrarily give them a weight. Some *a priori* weighting techniques have been developed in molecular phylogenetics. In simple cases such as the inequality between nucleotide and amino acid sequences, this can be easily overcome. Even differences in rates of mutations can be statistically estimated and the necessary correction introduced (Felsenstein 1978, 1981). However, there is no way to do this in practice with morphological characters and character states. Their delimitation is rarely objective.

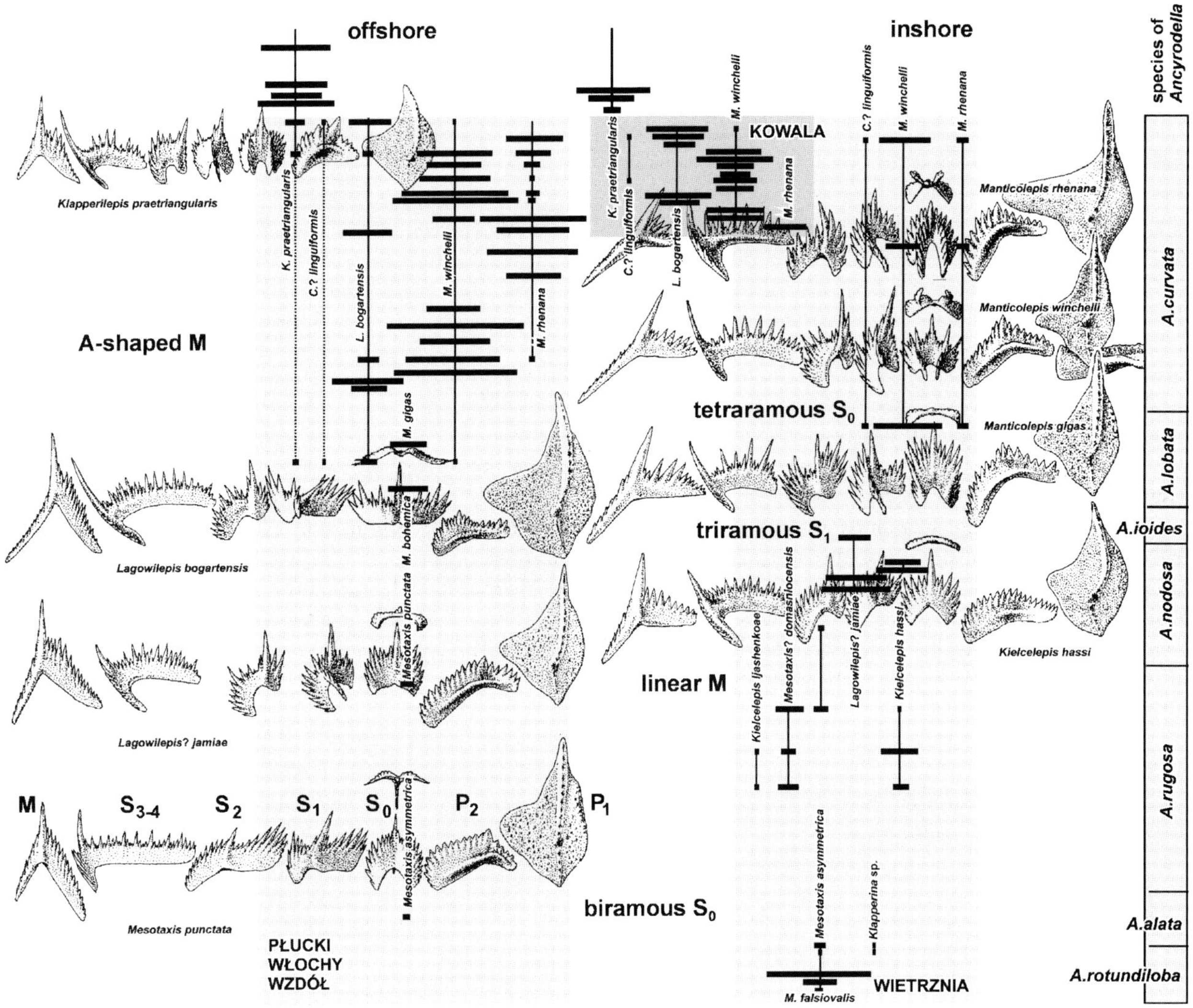

TEXT-FIG. 13. Succession of the palmatolepidid conodonts in various facies zones of the Frasnian in the Holy Cross Mountains. Note the punctuated distribution of most lineages (especially *Klapperilepis*), appearing and disappearing together with their specific environments.

Fossils cause difficulties of a different kind, and these are even more difficult to overcome. They involve the fundamental idea of correspondence between geological time and morphology. This assumption implies some regularity in distribution of morphologies in extinct organisms of different geological ages. In statistical terms fossil members of a lineage belonging to a monophyletic taxon (in cladistic terms, that is holophyletic: including all successors of the common ancestor) should be closer morphologically to each other than extant members of the same lineages. The similarity of their coeval sets should increase with the geological age of the horizon from which they come. This means that inclusion of fossil taxa in the same matrix of data as extant taxa must result in circular reasoning. This aspect of cladistics was addressed in a different way by Vermeij (1999), who pointed out that phylogenetic analyses derived from data matrices are not polarized, allowing data sets to be considered repeatedly even if they characterize long-separated branches of the evolutionary tree that could not possibly have interfered with each other's course of evolution. To be truly rigorous and logically consistent, one should restrict the analysis to organisms from the same time plane (it does not matter whether it is Recent or a segment of the past; Fortunato 1998).

The above objections refer equally to all methods of inference on the course of evolution based on morphology. The method of cladistics (or at least approaches of those cladists who are interested in restoring phylogeny, not just the order behind the diversity of organisms) also assumes certain patterns in the process of evolution that raise controversy. Some followers of the method believe that evolution is always divergent and that all the change is concentrated in a sudden speciation event. Both of these assumptions are contradicted by palaeontological evidence and considered either false or unnecessary by followers of methods of inference exposing stratigraphical order in the fossil record (e.g. Dzik 1991*a*). Perhaps all this would be of secondary importance if the testing of phylogenetic trees based on morphological evidence was strict enough. This does not seem to be the case, as most cladists consider parsimony and congruency between different sets of data to be sufficient for testing the trees.

Parsimony (the Occam's Razor Rule) and testability (or falsifiability) are truly the fundamental qualities of scientific theories. The most parsimonious formulation of ideas based on available facts makes it easier to test them with new data and helps in clearing science of unnecessary assumptions and redundant explanations. However, parsimony by itself does not guarantee access to truth and a more parsimonious phylogenetic theory does not necessarily describe reality better than a more complex one (e.g. Sheldon 1996). This has already been treated in depth in discussion on the maximum likelihood method in molecular phylogenetics (e.g. Felsenstein 1978; Stewart 1993). Panchen (1982) showed that what cladists claim is a test of their hypotheses is actually a repeated application of the principle of parsimony. It is not enough to choose a more parsimonious solution to approach the truth. To use parsimony alone is definitely a good strategy in theology but science requires more. Those who follow the Popperian attitude to science (most cladists claim to be among them) and consequently Alfred Tarski's concept of truth insist rather on confronting theories directly with the empirical evidence of a physical process. The language of science has to be checked for correspondence with the real (although inadequately known) world in every possible point. Evolution is a process of the physical world as long as it is understood to be a result of Darwinian changes in populations. The most direct evidence on the history of evolution is offered by fossils. They are not just a source of information on phylogeny but physically parts of evolving lineages. All that is necessary is their arrangement into a phylogenetic tree by filling unknown parts with hypothetical junctions. Whether or not any character analysis is performed is irrelevant. Such hypotheses on ancestor–descendant relationship among ancient populations can be tested by checking for correspondence with any kind of evidence that refers to populations of the same lineage in any logically consistent way (including cladistic analysis).

The cladistic test by congruency superficially looks similar to that derived from Tarski's concept of truth. It is assumed that any pattern of relationship based on particular sets of homologous characters must be congruent with patterns based on other homologues. In fact, even the basic conviction that homology can be identified without any reference to evolution is an illusion. Regardless, leaving aside the question of what degree such congruency must be followed by any evolving lineage, tracing new congruencies hardly has anything to do with hypothesis testing (O'Keefe and Sander 1999, p. 589). By no means is this a case of a deduction confronted by a basic statement about empirical evidence. Different hypotheses are simply compared. To truly test a hypothesis of the course of evolution some predictions (or retrodictions) derived from it on populations precisely located in time and space have to be matched with the fossil evidence.

In the original Hennigian form of cladistics, retrodiction on the course of evolution was to some degree possible, by arranging derived characters (synapomorphies) into a time series. This is what cladists call 'evolutionary scenarios' and consider this type of presentation inferior with respect to presentation of 'horizontal' (blood) relationships. Advanced ('computer') cladistics offers no way to confront directly diagrams of relationships with empirical (fossil) evidence. This taxic approach to palaeontological data requires that to make its trees comparable with the fossil evidence the branches of the cladogram have to be complemented with the observed ranges of taxa.

This was done for the palmatolepidid conodonts by Donoghue (2001), who calibrated a computer-generated cladogram with stratigraphical data on the first appearances of species rank taxa (thus unavoidably understood as chronospecies, unless their sudden appearance, stasis and exact fossil record of extinction are assumed). Unavoidably, the cladistic dogma of the dichotomous nature of evolution introduced nonexisting ghost ranges in each of the proposed sister lineages. Such a tree cannot thus be easily transformed into a series of ancestor–descendant relationship hypotheses. Only the succession in branching of the tree and dating of bifurcations can be compared with those represented in the chronophyletic diagram of phylogeny. As it appears, there is virtually no correspondence between the trees (Text-fig. 14). The main reason for this is not only that there are so many reversals and parallelisms in the chronophyletically documented evolution of the palmatolepidids. The major problem is the unequal value of characters. As already commented above, the quite trivial upward bending of the tip of the platform that originated in the late Frasnian at the beginning of the *Klapperilepis* lineage is the only aspect that differentiates its Famennian successors from virtually all Frasnian palmatolepidids. Probably by adding more and more data the cladogram could be made more congruent with the real course of evolution. This would not, however, remove distortions resulting from the fundamental flaws in the basic assumptions of the method (especially the concentration of all evolutionary change in speciation events). The alternative is to consider ancient populations

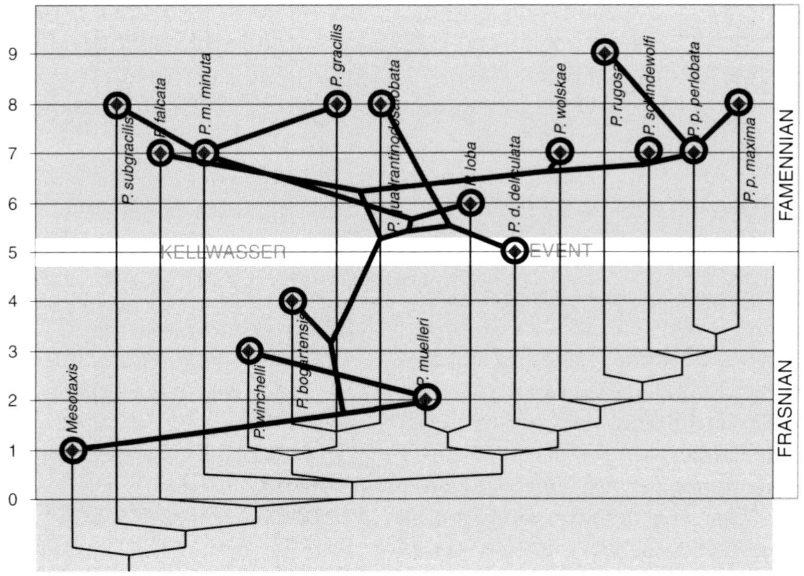

TEXT-FIG. 14. Phylogenetic tree by Donoghue (2001) based on a computer cladogram with stratigraphical ranges of taxa added, and superimposed lines of ancestor–descendant relationships derived from the chronophyletic tree (Text-figs 5, 11–12). Note that there is virtually no correspondence between the cladogram and documented events except for a few species of *Palmatolepis* of questionable distinction (they do not differ in their apparatus structure at all). The problems with the results of the cladistic analysis result mostly from the unweighting of characters and several misleading reversals and convergent changes within the clade. Unnecessary 'ghost ranges', which are in contradiction to the generally accepted pattern of palmatolepidid evolution at the Frasnian/Famennian boundary, also represent a violation of the principle of parsimony.

restored on the basis of fossil samples the basic units of phylogeny reconstructions (Dzik 1985, 1991*a*). The chronophyletic approach is simpler and more efficient in achieving the goal.

It appears thus that only the use of fossil evidence with its time and space coordinates allows identification of the true course of evolution and testing of phylogenetic trees. This requires that the morphological evidence is processed in a proper way and that the phylogenetic tree should be designed to correspond directly to the fossil evidence. Neither the methods of analysing the morphological evidence nor the way of using fossils to test the results need to be especially rigorous. Only the presentation of the tree has to be truly strict.

PROPOSED SOLUTION

The basic question is, of course, what does one want to attain: a precise natural system of classification expressing the design of the living world or an approximation of the pattern of events that resulted in the observed complexity of life? Whatever we do, this has to be consistent with the basic aspect of science; the presentation of the story should be testable. It is claimed here that the method of chronophyletics results in the presentation of descendant-ancestor hypotheses that fulfil requirements of scientific methodology.

Some problems of evolutionary biology can be resolved only with fossil evidence. These include such questions as: (1) how old geologically are major groups of organisms, (2) what was the anatomical organization of their ancestors and (3) what moved phylogeny in specific directions? They refer to the course of evolution which makes ancestor–descendant relationships the only objectively accessible aspect of evolution, the pattern of 'blood relationship' remaining very difficult to specify in objective terms and virtually impossible to test without reference to the actual course of evolution (Panchen 1982; Dzik 1991*a*; O'Keefe and Sander 1999).

The question emerges of how to proceed with methods of inferring evolution from the distribution of characters in Recent organisms to reach results that can be tested with direct palaeontological evidence on the course of evolution. This requires thinking in terms of samples, populations and lineages with their time, geographical and morphological dimensions. The basic unit of empirical evidence in palaeontology is a sample that represents an extinct population living in a specified geological time and having a specified position in geographical space. These time and space coordinates are objective and unchangeable. The time dimension of fossil evidence is not just an addition to morphological characters. The populations located in time and space are elements of the network of lineages that remains more or less hypothetical but is objective and potentially can be documented with the fossil evidence wherever it exists. To propose a phylogenetic (chronophyletic) hypothesis the populations are thus connected by ancestor–descendant relationships. To do this it is not necessary to define or describe populations. It is enough to point to particular samples, their time and space coordinates being inherently connected with them. The question of evolutionary species, which emerges in this context, has been satisfactorily solved on the grounds of neontology (Mayr 1969) and can be easily applied to the fossil material (e.g. Gingerich 1985).

In cases where the real pattern of evolutionary relationship is difficult to decipher, analysis of characters based on the assumed correspondence between time and morphological difference may help in restoring the network of transitions. The analysis has to be restricted to a single time slice. The pattern of nesting in cladograms and the sequence of branching in phenograms can then be expected to approximate the actual time sequence of events (appearances of characters). The set of characters attached to each of the branching points in the diagram characterizes a real organism from a time horizon older than that on which the analysis has been based. Potentially, at least some of the nodes of the tree characterized in this way can be matched with actual fossil ancestors or their relatives, close in time, geographical space and morphology. In cases where identification of the real extinct population fails, the hypothetical description of it that has been obtained can be incorporated into the basic evidence coming from the time horizon from which another set of morphological evidence comes. The next steps of the analysis can then be performed.

Morphological characters as discrete units are mostly products of the human mind. Any phylogenetic approach referring to such understood characters 'treats organisms as clusters of characters each one of which can be interpreted individually rather than as part of a functional complex' (Campbell and Barwick 1988, p. 207). This is an inherent bias of cladistics and phenetics, if used to infer phylogeny from fossils, but references to discrete characters can be avoided within the chronophyletic approach. One possibility is to represent morphology and variability of organisms in a diagrammatic way (as 'pictograms'; Dzik 1984, p. 11) to compare them as a whole. No formal algorithm to do this automatically is yet available, but there is probably no urgent need for it. It seems appropriate to quote in this place the maxim cited by Van Valen (1989 after Tukey 1962): 'Far better an approximate answer to the right question, which is often vague, than an exact answer to the wrong question, which can always be made precise.'

Acknowledgements. I am very grateful to Philip C. J. Donoghue, an anonymous reviewer and D. J. Batten for their numerous helpful comments and suggestions as to how to improve the style of this paper.

REFERENCES

AVISE, J. C., WALKER, D. and JOHNS, G. C. 1998. Speciation durations and Pleistocene effects on vertebrate phylogeography. *Proceedings of the Royal Society of London, B*, **265**, 1707–1712.

BENTON, M. J. 2001. Finding the tree of life: matching phylogenetic trees to the fossil record through the 20th century. *Proceedings of the Royal Society of London, B*, **268**, 2123–2130.

—— HITCHIN, R. and WILLS, M. A. 1999. Assessing congruence between cladistic and stratigraphic data. *Systematic Biology*, **48**, 581–596.

BERGGREN, W. A. and NORRIS, R. D. 1997. Biostratigraphy, phylogeny and systematics of Paleocene trochospiral planktic foraminifera. *Micropaleontology*, **43**, 1–116.

BRINKMANN, R. 1929. Statistisch-biostratigraphische Untersuchungen an mitteljurassischen Ammoniten bei Artbegriff und Stammentwicklung. *Abhandlungen der Gesellschaft der Wissenschaften zu Göttingen, Mathematisch-Naturwissenschaftliche Klasse, Neue Folge*, **1**, 1–249.

BROWN, W. L. and WILSON, E. O. 1956. Character displacement. *Systematic Zoology*, **5**, 51–64.

BUSH, G. L. 1994. Sympatric speciation in animals: new wine in old bottles. *Trends in Ecology and Evolution*, **9**, 285–288.

CAMPBELL, K. S. W. and BARWICK, R. E. 1988. Geological and palaeontological information and phylogenetic hypotheses. *Geological Magazine*, **125**, 207–227.

DE VARGAS, C. and PAWLOWSKI, J. 1998. Molecular versus taxonomic rates of evolution in planktonic foraminifera. *Molecular Phylogeny and Evolution*, **9**, 463–469.

DOMMERGUES, J. L. 1990. Ammonoids. 59–74. *In* McNAMARA, K. J. (ed.). *Evolutionary trends.* Belhaven, London, 368 pp.

DONOGHUE, P. C. J. 2001. Conodonts meet cladistics: recovering relationships and assessing the completeness of the conodont fossil record. *Palaeontology*, **44**, 65–93.

DZIK, J. 1979. Some terebratulid populations from the Lower Kimmeridgian of Poland and their relations to the biotic environment. *Acta Palaeontologica Polonica*, **24**, 473–492.

—— 1984. Phylogeny of the Nautiloidea. *Palaeontologia Polonica*, **45**, 1–255.

—— 1985. Typologic versus population concepts of chronospecies: implications for ammonite biostratigraphy. *Acta Palaeontologica Polonica*, **30**, 71–92.

—— 1990*a*. The concept of chronospecies in ammonites. 273–289. *In* PALLINI, G., CECCA, F., CRESTA, S. and SANTANTONIO, M. (eds). *Atti del secondo convegno internazionale Fossili Evoluzione Ambiente, Pergola 25–30 ottobre 1987.* Comitato Centenario Raffaele Piccinini, Pergola PS, 508 pp.

—— 1990*b*. Conodont evolution in high latitudes of the Ordovician. *Courier Forschungsinstitut Senckenberg*, **117**, 1–28.

—— 1991*a*. Features of the fossil record of evolution. *Acta Palaeontologica Polonica*, **36**, 91–113.
—— 1991*b*. Evolution of oral apparatuses in conodont chordates. *Acta Palaeontologica Polonica*, **36**, 265–323.
—— 1994. Conodonts of the Mójcza Limestone. *In* DZIK, J., OLEMPSKA, E. and PISERA, A. Ordovician carbonate platform of the Holy Cross Mountains. *Palaeontologia Polonica*, **53**, 43–128.
—— 1995. Range-based biostratigraphy and evolutionary geochronology. *Paleopelagos, Special Publication*, **1**, 121–128.
—— 1997. Emergence and succession of Carboniferous conodont and ammonoid communities in the Polish part of the Variscan sea. *Acta Palaeontologica Polonica*, **42**, 57–170.
—— 1999. Relationship between rates of speciation and phyletic evolution: stratophenetic data on pelagic conodont chordates and benthic ostracods. *Geobios*, **32**, 205–221.
—— 2002. Emergence and collapse of the Frasnian conodont and ammonoid communities in the Holy Cross Mountains, Poland. *Acta Palaeontologica Polonica*, **47**, 565–650.
—— and GAŹDZICKI, A. 2001. The Eocene expansion of nautilids to high latitudes. *Palaeogeography, Palaeoclimatology, Palaeoecology*, **172**, 297–312.
—— and KORN, D. 1992. Devonian ancestors of *Nautilus*. *Paläontologische Zeitschrift*, **66**, 81–98.
—— and TRAMMER, J. 1980. Gradual evolution of conodontophorids in the Polish Triassic. *Acta Palaeontologica Polonica*, **25**, 55–89.
EKDALE, A. A. and BROMLEY, R. G. 1984. Sedimentology and ichnology of the Cretaceous-Tertiary boundary in Denmark: implications for the causes of the terminal Cretaceous extinction. *Journal of Sedimentary Petrology*, **54**, 681–703.
ENGELMANN, G. F. and WILEY, E. O. 1977. The place of ancestor–descendant relationships in phylogeny reconstruction. *Systematic Zoology*, **26**, 1–11.
ERNST, H. 1982. The marl layer M 100 in the Maastrichtian of Hemmoor – an example of selective $CaCO_3$ dissolution. *Geologisches Jahrbuch A*, **61**, 109–127.
FELSENSTEIN, J. 1978. Cases in which parsimony and compatibility methods will be positively misleading. *Systematic Zoology*, **27**, 401–410.
—— 1981. A likelihood approach to character weighting and what it tells us about parsimony and compatibility. *Biological Journal of the Linnean Society*, **16**, 183–196.
FISHER, D. C. 1994. Stratocladistics: morphological and temporal patterns and their relation to phylogenetic process. 133–171. *In* GRANDE, L. and RIEPPEL, O. (eds). *Interpreting the hierarchy of nature*. Academic Press, San Diego, 298 pp.
—— FOOTE, M., FOX, D. L. and LEIGHTON, L. R. 2002. Stratigraphy in phylogeny reconstruction – comment on Smith (2000). *Journal of Paleontology*, **76**, 585–586.
FORTUNATO, H. 1998. Reconciling observed patterns of temporal occurrence with cladistic hypotheses of phylogenetic relationship. *American Malacological Bulletin*, **14**, 191–200.
FOX, D. L., FISHER, D. C. and LEIGHTON, L. R. 1999. Reconstructing phylogeny with and without temporal data. *Science*, **284**, 1816–1819.
GINGERICH, P. D. 1979. The stratophenetic approach to phylogeny reconstruction in vertebrate paleontology. 41–76. *In* CRACRAFT, J. and ELDREDGE, N. (eds). *Phylogenetic analysis and paleontology*. Columbia University Press, Irvington, 233 pp.
—— 1985. Species in the fossil record: concepts, trends, and transitions. *Paleobiology*, **11**, 27–41.
—— and GUNNELL, G. F. 1995. Rates of evolution in Paleocene–Eocene mammals of the Clarks Fork Basin, Wyoming, and a comparison with Neogene Siwalik lineages of Pakistan. *Palaeogeography, Palaeoclimatology, Palaeoecology*, **115**, 227–247.
GRABERT, B. 1959. Phylogenetische Untersuchungen an *Gaudryina* and *Spiroplectamina* (Foram.) besonders aus dem norddeutschen Apt und Alb. *Abhandlungen der Senckenbergische Naturforschende Gesellschaft*, **498**, 1–71.
HARPER, C. W. JR 1976. Phylogenetic inference in paleontology. *Journal of Paleontology*, **50**, 180–193.
HENGSBACH, R. 1990. Zur systematischen Stellung der Clymenien (Cephalopoda; Ober-Devon). *Senckenbergiana Lethaea*, **70**, 69–88.
HOLM, G. 1893. Sveriges kambrisk–siluriska Hyolithidae och Conulariidae. *Sveriges Geologiska Undersökning, Afhandlingar*, **C112**, 1–170.
KELLOGG, D. E. 1975. The role of phyletic change in the evolution of *Pseudocubes vema* (Radiolaria). *Paleobiology*, **1**, 359–370.
KLAPPER, G. 1990. Frasnian species of the Late Devonian conodont genus *Ancyrognathus*. *Journal of Paleontology*, **64**, 998–1025.
—— and FOSTER, C. T. JR 1993. Shape analysis of Frasnian species of the Devonian conodont genus *Palmatolepis*. *Paleontological Society, Memoir*, **32**, 1–35.
KNOWLTON, N. and JACKSON, J. B. C. 1994. New taxonomy and niche partitioning on coral reefs: jack of all trades or master of some? *Trends in Ecology and Evolution*, **9**, 7–14.
KORN, D. 1992. Relationship between shell form, septal construction and suture line in clymeniid cephalopods (Ammonoidea; Upper Devonian). *Neues Jahrbuch für Geologie und Paläontologie, Abhandlungen*, **185**, 115–130.
LENZ, A. 1974. Evolution in *Monograptus priodon*. *Lethaia*, **7**, 265–272.
LEVINTON, J. S. 2001. *Genetics, paleontology, and macroevolution*. Second edition. Cambridge University Press, Cambridge, 617 pp.
MAYR, E. 1969. The biological meaning of species. *Biological Journal of the Linnean Society*, **1**, 311–320.
McKINNEY, M. 1985. Distinguishing patterns of evolution from patterns of deposition. *Journal of Paleontology*, **59**, 561–567.
METZGER, R. A. 1994. Multielement reconstructions of *Palmatolepis* and *Polygnathus* (Upper Devonian, Famennian) from the Canning Basin, Australia, and Bactrian Mountains, Nevada. *Journal of Paleontology*, **68**, 617–647.
MURPHY, M. A. and SPRINGER, K. B. 1989. Morphometric study of the platform elements of *Amydrotaxis praejohnsoni* n. sp. (Lower Devonian, conodonts, Nevada). *Journal of Paleontology*, **63**, 349–355.
O'KEEFE, F. R. and SANDER, M. 1999. Paleontological paradigm and inferences of phylogenetic pattern: a case study. *Paleobiology*, **25**, 518–533.

OLEMPSKA, E. 1989. Gradual evolutionary transformations of ontogeny in an Ordovician ostracod lineage. *Lethaia*, **22**, 159–167.

PANCHEN, A. L. 1982. The use of parsimony in testing phylogenetic hypotheses. *Zoological Journal of the Linnean Society*, **74**, 305–328.

PATTERSON, C. 1981. Significance of fossils in determining evolutionary relationships. *Annual Reviews of Ecology and Systematics*, **12**, 195–223.

PEARSON, P. N. 1996. Cladogenetic, extinction and survivorship patterns from a lineage phylogeny: the Paleogene planktonic foraminifera. *Micropaleontology*, **42**, 179–188.

REIF, W.-E. 1983. Hilgendorf's (1863) dissertation on the Steinheim planorbids (Gastropoda: Miocene): the development of phylogenetic research program for paleontology. *Paläontologische Zeitschrift*, **57**, 7–20.

ROSE, K. D. and BOWN, T. M. 1984. Gradual phyletic evolution on the generic level in Early Eocene omomyid primates. *Nature*, **309**, 250–252.

SCHAEFFER, B., HECHT, M. K. and ELDREDGE, N. 1972. Phylogeny and paleontology. *Evolutionary Biology*, **6**, 31–46.

SCHÜLKE, I. 1999. Conodont multielement reconstructions from the early Famennian (Late Devonian) of the Montagne Noire (southern France). *Geologica et Palaeontologica, Sonderband*, **3**, 1–123.

SHELDON, P. R. 1990. Shaking up evolutionary patterns. *Nature*, **345**, 772.

—— 1996. Plus ça change – a model for stasis and evolution in different environments. *Palaeogeography, Palaeoclimatology, Palaeoecology*, **127**, 209–227.

SIMPSON, G. G. 1976. The compleat paleontologist? *Annual Review of Earth and Planetary Science*, **4**, 1–13.

SMITH, A. B. 1994. *Systematics and the fossil record: documenting evolutionary patterns*. Blackwell Scientific, Oxford, 223 pp.

SPRINGER, K. B. and MURPHY, M. A. 1994. Punctuated stasis and collateral evolution in the Devonian lineage of *Monograptus hercynicus*. *Lethaia*, **27**, 119–128.

STEWART, C. B. 1993. The powers and pitfalls of parsimony. *Nature*, **361**, 603–607.

SWEET, W. C. 1988. *The Conodonta: morphology, taxonomy, paleoecology, and evolutionary history of a long-extinct animal phylum*. Clarendon Press, Oxford, 212 pp.

VAN VALEN, L. M. 1989. The poverty of cladism. *Evolutionary Theory*, **9**, 109–110.

VERMEIJ, G. J. 1999. A serious matter with character-taxon matrices. *Paleobiology*, **25**, 431–433.

WAAGEN, W. 1869. Die Formernreihe des *Ammonites subradiatus*; Versuch einer paläontologische Monographie. *Benecke's Paläontologische Beiträge*, **2**, 181–256.

WÄGELE, J. W. 1994. Review of methodological problems of 'computer cladistics' exemplified with a case study on isopod phylogeny (Crustacea: Isopoda). *Zeitschrift für Zoologische Systematik und Evolutionsforschung*, **32**, 81–107.

WAGNER, P. J. 1998. A likelihood approach for evaluating estimates of phylogenetic relationships among fossil taxa. *Paleobiology*, **24**, 430–449.

WHITE, M. J. D. 1968. Models of speciation. *Science*, **159**, 1065–1070.

[Special Papers in Palaeontology, 73, 2005, pp. 185–218]

CLADOGRAMS, PHYLOGENIES AND THE VERACITY OF THE CONODONT FOSSIL RECORD

by LINDA M. WICKSTRÖM* *and* PHILIP C. J. DONOGHUE†

*School of Geography, Earth and Environmental Sciences, University of Birmingham, Edgbaston, Birmingham B15 2TT, UK; present address: Swedish Geological Survey (SGV), Box 670, Uppsala 751 25, Sweden; e-mail: linda.wickstrom@sgu.se

†Department of Earth Sciences, University of Bristol, Wills Memorial Building, Queens Road, Bristol BS8 1RJ, UK; e-mail: phil.donoghue@bristol.ac.uk

Abstract: Traditionally, conodont intrarelationships have been reconstructed following the evolutionary palaeontology paradigm, which lacks a formal methodology, renders hypotheses unrepeatable and takes little account of the imperfect nature of the fossil record. Cladistics provides a prescriptive approach to phylogeny reconstruction and, it is argued, a rigorous method of character analysis that is not incompatible with the aims of evolutionary palaeontology. To demonstrate this, we use the Silurian family Kockelellidae as an example of how cladistics can be used to reconstruct relative relationships, and how such hypotheses can be converted to phylogenies. We follow traditional, cladistic approaches to assessing the completeness of the fossil record and find that failure to conduct this within a milieu of absolute, rather than merely relative, relationships leads to spurious inferences of gaps in the fossil record. This appears to be a problem that is widespread in theory, but peculiar to species in practice, and parallels the observation that cladograms of fossil species tend to exhibit poorer correlation to stratigraphy than do cladograms of fossil higher taxa. We conclude that cladistics provides the only appropriate framework within which to conduct character analysis; phylogenies can be developed from cladograms but very often this additional inferential step is entirely superfluous to the aims of evolutionary studies.

Key words: phylogeny, stratigraphy, completeness, fossil record, conodont.

IN COMMON with other fossil groups of biostratigraphic utility, attempts at reconstructing conodont phylogeny have traditionally adopted a bottom-up approach, through the reconstruction of lineages from stratigraphically ordered collections by phenetic means or, more usually, based on the subjective perception of shape changes. Justification for the progressive linkage of such lineages becomes increasingly vague and subjective as potential relatives are identified on the basis of similarities and differences, where both the taxa and the characters are considered only from within a pool of candidates circumscribed by stratigraphy. This approach, often referred to as 'stratophenetics' (*non sensu* Gingerich 1979), is multifarious, hardly phenetic and is indistinguishable from the evolutionary palaeontology approach of, for example, Simpson (1974). However, all practitioners are united in the perspective that where the fossil record is sufficiently complete, stratigraphic data should be used *a priori* to constrain morphological comparisons in attempting to uncover phylogeny.

This approach is not without its problems, chief among which are concerns over how we determine whether the fossil record is sufficiently complete to justify applying the principles of evolutionary palaeontology. No guidelines are provided for a threshold before which the approach is redundant and, of even greater concern, no guidance is given on how to determine completeness. In practice, it appears that biostratigraphic utility provides confidence in the completeness of the fossil record as a record of the evolutionary history of a group. There is no doubt that the conodont fossil record is of great biostratigraphic value. After all, conodonts are the zone fossils of choice throughout much of the Palaeozoic and early Mesozoic, providing the basis of zonation schemes so refined that in certain intervals they surpass even the correlative power of famed biostratigraphic tools such as the graptolites and trilobites. Thus it has been assumed, both implicitly and explicitly, that the conodont fossil record is sufficiently complete that stratigraphic data can be used in reconstructing phylogeny. However, biostratigraphic utility is not enough in itself. It requires as a minimum only the qualities of repeatability and reproducibility in the chronological distribution of taxa (Sweet and Donoghue 2001) and this does not necessarily equate to 'completeness' in the fossil record of a taxon, or taxa. Even as a maximum we may expect that biostratigraphic utility equates to completeness in the fossil record of a taxon, but this may not be the same as achieving completeness for all taxa, or even representation of all taxa, as well as a record of morphological intermediates, assuming that such populations or individuals existed. These are prerequisites of attempts to reconstruct absolute evolutionary relationships such as ancestry and descent, which are the base currency of phylogenies steeped in the biostratigraphic tradition.

There are good reasons to doubt the completeness of the conodont fossil record. For instance, in graphical correlation, if biostratigraphic sections were complete, it would be expected that first and last appearance data would plot precisely on the line of correlation, but this expectation is never met (Cooper and Lindholm 1990, based on a graptolite database, is probably as good an approximation as is realizable). It is possible that a complete record may be achieved by piecing together a number of incomplete sections, but this assumes that a complete record is available to be sampled. This is an unrealistic assumption because of systematic biases in facies preservation in the rock record, and the impact that this has on the fossil record of organisms whose distribution is controlled by extrinsic variables (ecology, facies; Holland 1995, 2000; Smith 2001). The effects of such biases on the meaning of the conodont fossil record are potentially strongly limiting and yet their presence is only just beginning to be recognized (Donoghue *et al.* 2003; Barrick and Männik 2005; Lehnert *et al.* 2005).

The problem with the evolutionary palaeontology approach does not, however, begin and end with completeness of the record. In particular, the approach of confining consideration of relationships to within a time interval is problematic for a number of reasons: (1) the time interval is entirely arbitrary and never more than implicit; (2) because it fails to assess the relative importance of morphological similarities and/or differences between taxa within a fully historical context, which is the only appropriate means of determining the phylogenetic significance of characters; (3) because potential relatives are expected only to be found within a given time frame, it is an implicit assumption that gaps in the fossil record must be less than the span of this time frame; thus, this approach provides no scope for identifying bias and degrees of completeness in the fossil record because it has already been assumed that gaps in the record are negligible. Although techniques providing an aphylogenetic *a priori* assessment of fossil record completeness have been in development for many years (Paul 1982; Strauss and Sadler 1989; Marshall 1990, 1994, 1997; Weiss and Marshall 1999; Tavaré *et al.* 2002), they have not yet been employed to provide constraint in such approaches.

Above all else, the evolutionary palaeontology approach lacks a formal methodology and, as such, it is impossible to differentiate shared derived from shared primitive characters. Also there is no means of objectively choosing between competing hypotheses of relationships on the basis of morphology, nor of differentiating between morphological and stratigraphic data when attempting to judge the relative merits of competing hypotheses. Ultimately, because of the lack of clarity over how morphological characters are weighed against one another and stratigraphy, and because of the general failure to specify the stratigraphic context within which relatives can be identified, the 'analyses' are effectively unrepeatable and unreproducible. These are both qualities that are basic expectations (though not necessarily requisites) of scientific hypotheses. This is not to say that a hypothesis that arises from such an analysis, the phylogeny itself, is unscientific; all phylogenetic hypotheses are testable and are therefore scientific.

Many attempts have been made to formalize the evolutionary palaeontology approach (see Schoch 1986 for an excellent summary) but they remain vague. The most objective formalization is to be found in stratophenetics *sensu stricto* (Gingerich 1979) and a number of phylogenetic lineages of conodonts have been constructed following these established, or similar, procedures (e.g. Barnett 1971, 1972; Dzik and Trammer 1980; Murphy and Cebecioglu 1986, 1987; Dzik 1991, 1994, 1997, 1999, 2002, 2005). As seductively persuasive as many of these lineages are, the hypotheses of ancestor–descendant (direct) relationships between 'populations' or species do not preclude the possible existence of unrecognized, unsampled or, indeed, unsampleable intermediate taxa. Moreover, the potential for discovery of new taxa is inversely proportional to sampling density. As such, the recognized ancestor–descendant pairings can vary in meaning from direct relations to the pairing of taxa that are only very remotely related, and can be considered ancestors and descendants of one another in only the very vaguest sense (cf. Engelmann and Wiley 1977; Schoch 1986; Nelson 1989). Another limitation of stratophenetic studies is that they are usually limited to individual rock successions where the appearance and disappearance of taxa can represent migration, rather than evolutionary events. Finally, the characters considered in stratophenetic studies are generally limited to one, or just a few, and are chosen over other potential variates on a subjective and often unstated basis. These problems notwithstanding, stratophenetically derived lineages of conodonts are rare and have been limited to the species level and, more usually, below. Such evidence is wanting for most species, and our current understanding of the interrelationships of conodonts at the generic level and above is replete with uncertainty (see Sweet 1988, for the very best current assessment).

Dzik (1991, 2005) has attempted to formalize a methodology for evolutionary palaeontology in 'chronophyletics', but this still lacks any formal criteria for reconciling between competing data and datasets, and considers the significance of subjectively chosen characters within unspecified and arbitrarily circumscribed time frames, rather than within a fully historical context. The chief advantage of this methodology, as set out, is that it makes falsifiable predictions, such as morphological intermediates, through retrodiction. However, this is a quality of all phylogenetic hypotheses, rather than of a specific methodology, and is actually better suited to hypotheses that can be correlated directly to the morphological data on which they are based, in contrast to chronophyletics.

Cladistics

In contradistinction, cladistics provides a formal methodology for the treatment of characters in phylogenetic reconstruction. In eschewing stratigraphic data *a priori*, it provides a means of assessing the phylogenetic significance of characters in a context that is both atemporal and, at the same time, fully historical. In so doing, cladistic hypotheses and stratigraphic data provide a means of reciprocal testing *a posteriori*. For a variety of reasons, it is more appropriate to reconstruct relative, rather than direct relationships; these are just as testable and more likely to survive testing – thus, evolutionary studies of character evolution based on such hypotheses are less likely to be revised fundamentally with a new discovery. The data and character definitions on which cladistic hypotheses are based are (generally!) also presented for scrutiny, facilitating tests of both repeatability and reproducibility, in addition to tests of the hypotheses based on the data.

However, the cladistic approach is not beyond criticism. For instance, it can be and has been argued that:

1. *Characters and character states are artefacts of analysis, not properties of organisms.* This is a moot point, but even so, it is a limitation of all approaches to phylogenetic reconstruction, not just cladistics.
2. *Convergence (homoplasy) is so rife that it is not possible to reconstruct relationships independently of stratigraphic data.* It is unlikely that convergence is ever so prevalent that it affects all aspects of morphology. Indeed, if homoplasy were so widespread, it would be a wonder that fossils are of any use in biostratigraphy. Cladistics provides a means of discriminating homology and homoplasy, assuming that homology is dominant over homoplasy, but this is difficult to test (it is possible to compare statistical support for phylogenetic solutions to real and random datasets, e.g. see Donoghue 2001, but the assumption remains). It is possible to partition cladistic analyses by temporal interval (Fortunato 1998) although this removes the global test of homology provided by a fully historical (temporally unconstrained) analysis (cf. Donoghue *et al.* 1989). Because both stratophenetic and evolutionary palaeontological analyses rely heavily on morphological data they are just as subject to convergence as cladistic analyses and so stratigraphic data do not provide a solution to this potential problem.
3. *Characters have different degrees of phylogenetic significance.* This is true, and character selection is the most severe form of character weighting. For those characters included, it is not possible to determine objectively an appropriate weighting scheme without reference to a phylogeny; successive weighting (*a posteriori* reweighting) of cladistic analyses currently provides such a means, and Bayesian approaches are likely to provide more sophisticated approaches in the near future. Subjective methods of *a priori* weighting have always been available.
4. *Cladistics requires and can only reconstruct cladogenic, or divergent, speciation events.* This is a misconception of the goal of cladistics, which is to resolve relative relationships (in the non-familial sense), not to replace absolute, or direct, relationships, and cladistic analyses should be considered an interim step of rigorous character analysis towards the ultimate goal of phylogeny reconstruction. The successive branching between taxa in a cladogram reflects only the nested degrees of relationship between taxa, not (necessarily) the evolutionary process or pattern that underlies their distinction. The process of converting a cladogram to a phylogeny is non-trivial, and will be explored below, but there is no impediment to the conversion of a cladogram to a phylogenetic tree with an architecture common to those derived from evolutionary palaeontology, stratophenetics or chronophyletics.
5. *Evolution does not (always) proceed parsimoniously.* Quite likely, but parsimony is a universal scientific principle, not a tool employed solely by cladists, and it is the same principle employed in evolutionary palaeontology, stratophenetic and chronophyletic analyses when attempting to discriminate between competing hypotheses. Nevertheless, this raises an important point with respect to cladistics, which is that aside from the shibboleth, the key to cladistics is that it forces practitioners of cladistic methodology to confront their assumptions and provide justification for any hypothesis that they wish to present that is more extravagant than the most parsimonious solution. Thus, it is possible to provide uneven weights to the characters, it is possible to choose a tree that is less parsimonious but in greater concordance with stratigraphy, and it is

even possible to recognize taxa as ancestors within cladograms, but *a priori* justification needs to be provided for these extra assumptions. Even if justification is not forthcoming, the explicit acceptance of specific assumptions requires that their impact is considered in contrast to the most parsimonious solution, and the analysis is rendered repeatable and reproducible.

Thus, cladistics provides the most appropriate framework within which to attempt to resolve the interrelationships and systematics of conodonts. But it should go further, by attempting to uncover conodont phylogeny, and by considering the absolute relationships of taxa. This approach can provide a more defensible, stable framework on which to base evolutionary studies that require absolute rather than merely relative relationships. It can also provide a means of assessing the completeness of the conodont fossil record to determine whether there is some justification for the *a priori* use of stratigraphic data in phylogeny reconstruction. Determination of the fidelity with which the conodont fossil record reflects the evolutionary history of the group can further provide a measure of which questions asked of it can actually be answered. On a more general level, such studies will also provide a measure of whether the fossil records of biostratigraphically important groups are more complete than those of groups in which stratigraphic occurrence is an unquestioned guide to phylogenetic relations.

No such studies of the conodont fossil record have yet been undertaken and in this contribution we aim to reconstruct the phylogeny and assess the veracity of the fossil record of a putative clade of taxa that has been used widely in biostratigraphy, including a number of taxa employed in global biozonation.

THE CONODONT FAMILY KOCKELELLIDAE KLAPPER, 1981

The Silurian conodont family Kockelellidae was chosen as the subject of this study because it is widely regarded as monophyletic, because there is an established scheme of interrelationships for contrast that has been developed over a number of years in the biostratigraphic tradition (Barrick and Klapper 1976; Serpagli and Corradini 1999), and because its fossil record has been intensively studied and utilized in biostratigraphy. Indeed, many of the member taxa contribute to the global zonation scheme for the Silurian.

At least two genera, *Kockelella* (17 species, three in open nomenclature, five subspecies) and *Ancoradella* (one species), have been assigned to the family Kockelellidae (e.g. Sweet 1988), although Fordham (1991) has argued for the inclusion of *Polygnathoides* (one species). Despite an absence of natural assemblages, multielement reconstructions are available for the vast majority of these taxa, produced on the basis of recurrent associations (Walliser 1964; Klapper and Philip 1971; Barrick and Klapper 1976). Because the morphology of the elements constituting the apparatuses is conservative both within and without the family, their positional homologies can be inferred with reference to taxa that are known from natural assemblages, such as *Ozarkodina* (Pollock 1969; Mashkova 1972; Nicoll 1985; Nicoll and Rexroad 1987) and, thence, to more distantly related taxa such as members of the outgroup (Purnell and Donoghue 1998; Purnell *et al.* 2000). Positional homologies are expressed following the notation scheme of Purnell *et al.* (2000).

PHYLOGENETIC ANALYSIS

Taxa included in the analysis

Outgroup. Taxa representative of each of the three orders of 'complex' conodonts have been included among the outgroup. Specific taxa are selected either because of the degree to which they represent particular clades, or because they have previously been implicated in the origin, ancestry or diversification of the Kockelellidae. For instance, Sweet (1988) has suggested that *Ozarkodina hassi* and *Plectodina tenuis* are ancestral to *Kockelella*; species of *Oulodus* have been included in the analysis because Klapper and Murphy (1974) and Barrick and Klapper (1976) have suggested similarity to the S and M elements in *Kockelella variabilis variabilis*, possibly belying close phylogenetic relations; *Ancoradella ploeckensis* and *Polygnathoides siluricus* have been included in the analysis as well as they are supposedly closely related to *Kockelella* (Link and Druce 1972; Sweet 1988; Fordham 1991; Serpagli and Corradini 1999).

Ancoradella ploeckensis Walliser, 1964 following the reconstruction of Männik and Malkowski (1998: P_1 = Pa, P_2 = Pb, S_0 = Sa, $S_{1/2}$ = Sb, $S_{3/4}$ = Sc, M = ?).

Oulodus rohneri Ethington and Furnish, 1959 following the reconstruction of Nowlan and Barnes (1981: P_1 = priniodiniform, P_2 = oulodiform, S_0 = trichonodelliform, $S_{1/2}$ = zygognathiform, $S_{3/4}$ = eoligonodiniform, M = cyrtoniodiform).

Oulodus ulrichi (Stone and Furnish, 1959) following the reconstruction of Nowlan and Barnes (1981: P_1 = priniodiniform, P_2 = oulodiform, S_0 = trichonodelliform, $S_{1/2}$ = zygognathiform, $S_{3/4}$ = eoligonodiniform, M = cyrtoniodiform).

Ozarkodina confluens (Branson and Mehl, 1933) following the reconstruction of Jeppsson (1974: P_1 = sp, P_2 = oz, S_0 = tr, $S_{1/2}$ = pl, $S_{3/4}$ = hi, M = ne).

Ozarkodina excavata excavata (Branson and Mehl, 1933) following the reconstruction of Jeppsson (1974: P_1 = sp, P_2 = oz, S_0 = tr, $S_{1/2}$ = pl, $S_{3/4}$ = hi, M = ne).
Ozarkodina hassi (Pollock, Rexroad and Nicoll, 1970) following the reconstruction of Armstrong (1990: P_1 = Pa, P_2 = Pb, S_0 = Sa, $S_{1/2}$ = Sb, $S_{3/4}$ = Sc, M = M).
Polygnathoides siluricus Branson and Mehl, 1933 following the reconstruction of Jeppsson (1983: P_1 = sp, P_2 = oz, $S_{1/2}$ = pl, $S_{3/4}$ = hi, M = ne).
Plectodina tenuis Branson and Mehl, 1933 following the reconstruction of Sweet (1979: P_1 = Pa, P_2 = Pb, S_0 = Sa, $S_{1/2}$ = Sb, $S_{3/4}$ = Sc, M = M).

Ingroup
Kockelella abrupta (Aldridge, 1972) only known element represents the P_1 position. *K. abrupta* has previously been synonymized with *K. manitoulinensis* (McCracken and Barnes 1981; Zhang and Barnes 2002; contra Armstrong 1990), but without a reconstructed multielement apparatus such a step is premature and this taxon has been considered as distinct for the purpose of this study.
Kockelella amsdeni Barrick and Klapper, 1976 following the multielement reconstruction of Barrick and Klapper (1976: P_1 = Pa, P_2 = Pb, S_0 = Sa, $S_{1/2}$ = Sb, $S_{3/4}$ = Sc, M = M).
Kockelella corpulenta (Viira, 1975) only known element represents the P_1 position. *K. corpulenta* has previously been synonymized with *Kockelella walliseri* (Jeppsson 1983, 1997; contra Kleffner 1994; Serpagli and Corradini 1999), but without a multielement reconstruction such a step is premature and this taxon has been considered as distinct for the purpose of this study.
Kockelella crassa (Walliser, 1964) following the reconstruction of Serpagli and Corradini (1999: P_1 = Pa, P_2 = Pb, S_0 = ?, $S_{1/2}$ = ?, $S_{3/4}$ = ?, M = ?). The remaining positions within the apparatus were considered by Serpagli and Corradini (1999) to have been filled by elements that are indistinguishable from their homologues in *Kockelella variabilis variabilis*. However, in the absence of documentary support, the morphology of these elements must be considered unknown.
Kockelella latidentata Bischoff, 1986 (P_1 = Pa, P_2 = ?, S_0 = ?, $S_{1/2}$ = ?, $S_{3/4}$ = ?, M = ?). The validity of this taxon is uncertain due to the very close similarity of the Pa element to the Pa of *Kockelella ranuliformis* (L. Jeppsson, pers. comm. 2001, 2002). The elements assigned Pb and Sb by Bischoff (1986) may more appropriately be assigned to species of *Oulodus* and *Ozarkodina*, respectively.
Kockelella maenniki Serpagli and Corradini, 1999 following the original reconstruction (P_1 = Pa, P_2 = ?, S_0 = ?, $S_{1/2}$ = ?, $S_{3/4}$ = ?, M = M). The remaining positions within the apparatus were considered by Serpagli and Corradini (1999) to have been filled by elements that are indistinguishable from their homologues in *Kockelella variabilis variabilis*. However, in the absence of documentary support, the morphology of these elements must be considered unknown.
?*Kockelella manitoulinensis* (Pollock, Rexroad and Nicoll, 1970) following the reconstruction of Zhang and Barnes (2002: P_1 = Pa, P_2 = Pb, S_0 = Sa, $S_{1/2}$ = Sb, $S_{3/4}$ = Sc, M = M). Zhang and Barnes (2002) queried the assignment of this taxon to *Kockelella* because the element assigned to the P_1 position is typical neither of *Kockelella* nor *Ozarkodina*.
Kockelella ortus ortus (Walliser, 1964) following the reconstruction of Jeppsson (1997: P_1 = Pa, P_2 = Pb, S_0 = ?, $S_{1/2}$ = ?, $S_{3/4}$ = ?, M = M).
Kockelella ortus absidata Barrick and Klapper, 1976 following the reconstruction by Klapper *et al.* (1981: P_1 = Pa, P_2 = Pb, S_0 = ?, $S_{1/2}$ = Sb, $S_{3/4}$ = Sc, M = M). The M element of Rexroad *et al.* (1978) corresponds well with that found in association with *K. o. absidata* from both the type collection and collections from Dingle (Aldridge 1980) and Gotland (Jeppsson, unpublished collections), but it does not correspond to the M element of *Kockelella variabilis variabilis* (contra Barrick and Klapper 1976; Serpagli and Corradini 1999), even though the two morphotypes have, hitherto, been assigned to the same form species (*Neoprioniodus multiformis*).
Kockelella ortus sardoa Serpagli and Corradini, 1999 following the original reconstruction (P_1 = Pa, P_2 = ?, S_0 = ?, $S_{1/2}$ = ?, $S_{3/4}$ = ?, M = M). The remaining positions within the apparatus were considered by Serpagli and Corradini (1999) to have been filled by elements that are indistinguishable from their homologues in *Kockelella variabilis variabilis*. However, in the absence of documentary support, the morphology of these elements must be considered unknown.
Kockelella patula Walliser, 1964 following the reconstruction by Barrick and Klapper (1976: P_1 = Pa, P_2 = Pb, S_0 = Sa, $S_{1/2}$ = Sb, $S_{3/4}$ = Sc, M = M). The multielement reconstruction of Walliser (1972) is an unreliable synonymy list of form taxa lacking citation to specific specimens.
Kockelella ranuliformis (Walliser, 1964) following the reconstruction by Barrick and Klapper (1976: P_1 = Pa, P_2 = Pb, S_0 = Sa, $S_{1/2}$ = Sb, $S_{3/4}$ = Sc, M = M).
Kockelella stauros Barrick and Klapper, 1976 following the original multielement reconstruction (P_1 = Pa, P_2 = Pb, S_0 = Sa, $S_{1/2}$ = Sb, $S_{3/4}$ = Sc, M = M).
Kockelella variabilis ichnusae Serpagli and Corradini, 1999 (P_1 = Pa, P_2 = ?, S_0 = ?, $S_{1/2}$ = ?, $S_{3/4}$ = ?, M = M). The remaining positions within the apparatus were considered by Serpagli and Corradini (1999) to have been filled by elements that are indistinguishable from their homologues in *Kockelella variabilis variabilis*. However, in the absence of documentary support, the morphology of these elements must be considered unknown.
Kockelella variabilis variabilis Walliser, 1957 following the reconstruction of Aldridge (1980: P_1 = Pa, P_2 = Pb, S_0 = Sa, $S_{1/2}$ = Sb, $S_{3/4}$ = Sc, M = M). This apparatus was also reconstructed by Walliser (1964) as 'Conodonten apparat G'.
Kockelella walliseri (Helfrich, 1975) following the reconstructions of Bischoff (1986: P_1 = Pa, P_2 = Pb, S_0 = Sa, $S_{1/2}$ = Sb, $S_{3/4}$ = Sc, M = M) and Kleffner (1994: P_1 = Pa, P_2 = Pb, S_0 = Sa, $S_{1/2}$ = Sb, $S_{3/4}$ = Sc, M = ?).

Taxa excluded from the analysis

Three taxa assigned to *Kockelella* in open nomenclature (*Kockelella* n. sp. A Klapper and Murphy, 1974, *Kockelella* sp. A Barrick, 1983 and *Kockelella* sp. A Bischoff, 1986) are known from Pa (= P_1) elements alone. They are excluded from the analysis because they are considered to fall within the range of variation of established taxa.

Kockelella suglobovi Mashkova, 1979 exhibits only a superficial similarity to members of the Kockelellidae and may be more appropriately assigned to *Pterospathodus*, which is only remotely related to *Kockelella* (Aldridge and Smith 1993) and has a distinct apparatus composition (Männik 1998). Thus, *K. suglobovi* has been excluded from the analysis.

Characters and coding

The data matrix (Table 1) is composed of 49 discrete unordered morphological characters coded using a contingent coding strategy. This particular coding strategy was adopted because it is more biologically meaningful, theoretically defensible and practically valid than other commonly adopted coding strategies such as strict binary and inappropriate multistate coding (Hawkins *et al.* 1997; Lee and Bryant 1999; Strong and Lipscomb 1999; Forey and Kitching 2000; Hawkins 2000; Jenner 2002). The coding strategy is only operationally significant when a character is composed of more than two character states, some of which are not logically equivalent, such as absence of process versus presence in the form of a platform versus present in the form of a blade. Clearly, two of these states are different forms of 'process present', and the failure to reflect this in the design of the character leads to a loss of phylogenetic information and may lead to spurious phylogenetic hypotheses that require biologically incongruent character state transformations. There are many coding solutions to this problem, but contingent coding splits the character into two, one character pertaining to the presence or absence of a process, and a second character concerned with whether the process, if present, is manifest as a platform or a blade; if no process is present, the character is inapplicable, and the character state score will be the same as for missing data, '?'. Although this violates the cladistic axiom that all characters should be logically and biologically independent, it is an essentially theoretical problem, the significance of which is vastly outweighed by the advantages that the coding strategy bring to analyses in terms of facilitating only biologically valid character transformations, maximizing character congruence, and providing the most exacting test of homology.

Character coding is based on published and unpublished conodont collections from Oklahoma (USA), Gotland (Sweden), Dingle (Ireland), Britain and Sardinia (Italy), and from taxonomic descriptions and figures in the published literature. The usefulness of published taxonomic descriptions is limited, however, because of the tendency to emphasize taxonomic differences rather than similarities (possible homologous characters). The element notation scheme of Purnell *et al.* (2000) has been adopted herein to convey positional homologies between element morphotypes. Due to the morphological similarities of elements occupying S_1,S_2 and S_3,S_4 positions within the ingroup these have been coded as $S_{1/2}$ and $S_{3/4}$, respectively.

A note on homology

Homology is the backbone of any cladistic analysis. Positional homologies are well understood among the major groups of 'complex' conodonts (prioniodontids, prioniodinids and ozarkodinids; prioniodontids *sensu* Donoghue *et al.* 2000) and have previously been thoroughly discussed (e.g. Purnell 1993; Aldridge *et al.* 1995; Purnell and Donoghue 1998; Purnell *et al.* 2000). The homology of individual processes, however, has rarely been discussed (e.g. Donoghue 2001) and by convention processes are recognized and described according to their disposition relative to a landmark rather than in an attempt explicitly (or even implicitly) to identify homology with another process in an element in an homologous position in another taxon. Thus, traditional concepts of process homology are largely phenetic and it is necessary to tease apart the homologies that underlie this phenetic classification scheme using ancilliary data sources, such as patterns of ontogenetic development, to distinguish structures that are merely 'similar' from those that are 'the same' in a historical sense. For instance, we can consider homology between the elements occupying the P_1 positions in *Oulodus* and *Kockelella variabilis*. Both taxa possess two primary 'lateral' processes (primary in the conventional sense that the processes join the cusp), but are these processes homologous? The 'lateral' processes of the *Oulodus* P_1 element are manifest at every ontogenetic stage, while in the P_1 element of *K. variabilis*, the 'anterior' and 'posterior' processes begin to develop first; 'lateral' processes are not present in juveniles, and begin to develop much later, as evidenced by ontogenetic suites of specimens (e.g. Serpagli and Corradini 1999). On ontogenetic grounds, therefore, it appears more likely that the 'lateral' processes of *Oulodus* P_1 elements are homologous to the 'anterior' and 'posterior' processes of *K. variabilis*, and that the primary 'lateral' processes of *K. variabilis* are neomorphisms. Thus, we have devised a scheme for identifying homology of post-first generation processes (Text-fig. 1) that serves to obviate false statements of homology between the 'lateral' processes of *Ancoradella*, *Oulodus*, *Plectodina* and *Kockelella*, which we consider on ontogenetic grounds to be not the same, but merely similar.

Whether the first generation processes of these taxa should be considered homologous is best reflected upon after the result of a phylogenetic analysis. However,

TABLE 1. Data matrix.

	1	2	3	4	5	6	7	8	9	10	11	12	13	14	15	16	17	18	19	20	21	22	23	24	25	26	27	28	29	30	31	32	33	34	35	36	37	38	39	40	41	42	43	44	45	46	47	48	49
Ko. abrupta	1	1	0	0	?	?	?	?	0	0	0	0	1	1	?	1	0	1	1	0	0	0	1	?	0	0	?	?	?	?	?	?	?	?	?	?	?	?	?	?	?	?	?	?	?	?	?	?	?
Ko. o. absidata	1	1	0	0	?	?	?	?	0	0	1	1	0	1	?	0	0	1	1	0	0	0	0	?	0	0	1	1	0	0	0	0	0	1	?	?	?	0	0	?	0	1	0	1	1	0	?	?	?
Ko. o. sardoa	1	1	0	0	?	?	?	?	0	0	1	1	0	1	?	0	0	0	1	0	0	0	0	?	0	0	?	?	?	?	?	?	?	?	?	?	?	1	1	1	0	1	1	1	0	1	?	?	?
Ko. amsdeni	1	1	1	1	0	0	0	0	0	0	0	0	1	1	1	1	0	1	0	0	1	0	1	0	0	0	1	1	0	0	0	1	?	1	1	1	1	1	1	0	0	1	1	1	1	0	1	0	0
Ko. corpulenta	1	1	1	0	0	0	?	?	0	?	0	0	1	1	?	1	0	1	?	?	1	0	1	0	0	0	?	?	?	?	?	?	?	?	?	?	?	?	?	?	?	?	?	?	?	?	?	?	?
Ko. crassa	1	1	1	1	1	1	1	1	0	0	1	0	1	1	0	1	0	1	1	1	0	1	1	1	1	1	1	1	0	0	0	0	0	0	1	1	1	?	?	?	?	?	?	?	?	?	?	?	?
Ko. latidentata	1	1	0	0	?	?	?	?	0	0	0	0	1	1	?	1	0	1	0	0	1	0	1	0	0	0	?	?	?	?	?	?	?	?	?	?	?	?	?	?	?	?	?	?	?	?	?	?	?
Ko. manitoulensis	1	1	0	0	?	?	?	?	0	1	1	1	1	1	?	0	0	1	1	0	1	0	1	?	0	0	1	1	0	0	0	0	0	0	?	?	?	0	1	0	0	1	1	1	1	0	1	0	0
Ko. maenniki	1	1	1	1	0	0	0	0	0	?	1	0	1	1	0	0	0	1	1	0	0	1	1	0	1	1	?	?	?	?	?	?	?	?	?	?	?	1	1	?	?	1	0	?	0	0	?	?	?
Ko. o. ortus	1	1	0	0	?	?	?	?	0	?	1	1	0	1	?	0	0	1	?	?	0	0	?	?	0	0	1	1	0	0	0	0	0	0	?	?	?	0	0	?	0	1	1	1	0	0	?	?	?
Ko. patula	1	1	1	1	1	1	0	0	0	?	0	0	1	1	0	1	0	1	0	1	0	1	1	1	1	0	1	1	0	0	0	1	0	1	1	1	1	1	1	0	1	1	0	1	0	0	0	1	1
Ko. ranuliformis	1	1	0	0	?	?	?	?	0	0	0	0	1	1	?	1	0	1	0	0	1	0	1	0	0	0	1	1	0	0	0	1	?	1	0	0	1	1	1	0	0	1	1	1	1	0	1	0	0
Ko. stauros	1	1	1	1	0	0	0	0	0	0	0	0	1	1	0	1	0	1	0	1	0	1	1	1	1	0	1	1	0	0	?	1	?	0	?	?	1	1	1	0	1	1	0	0	1	0	1	0	0
Ko. v. ichnusae	1	1	1	1	1	1	0	0	0	?	1	0	1	1	1	1	0	1	1	0	0	1	1	0	1	1	?	?	?	?	?	?	?	?	?	?	?	?	?	?	?	?	?	?	?	?	?	?	?
Ko. v. variabilis	1	1	1	1	1	1	1	1	0	0	1	0	1	1	1	1	0	1	1	0	0	1	1	0	0	1	1	1	0	0	0	1	0	0	1	1	1	1	1	0	1	1	0	1	0	0	1	0	0
Ko. walliseri	1	1	1	0	0	0	?	?	0	0	0	0	1	1	0	1	0	1	0	1	1	0	1	0	0	0	1	1	0	0	0	1	?	0	?	?	1	1	1	0	?	?	?	?	0	0	1	0	0
Anc. ploeckensis	1	1	1	1	1	1	0	0	1	?	0	0	1	1	0	0	0	?	?	?	?	?	?	?	?	?	1	1	0	0	1	0	1	1	?	?	?	?	?	?	?	?	?	?	?	?	0	1	1
Ou. rhoneri	0	0	0	0	?	?	?	?	0	?	?	1	0	0	?	?	?	?	?	?	0	0	0	?	0	0	0	0	1	1	?	?	?	?	?	0	0	0	0	?	0	1	0	0	0	1	1	0	0
Ou. ulrichi	0	0	0	0	?	?	?	?	0	?	?	1	0	0	?	?	?	?	?	?	0	0	0	?	0	0	0	0	1	1	?	?	?	?	?	0	0	1	1	0	0	1	0	0	0	1	1	0	0
Oz.confluens	1	1	0	0	?	?	?	?	0	0	1	1	0	0	?	1	0	0	0	0	0	0	0	?	1	0	1	1	0	0	1	0	1	0	0	0	0	1	1	1	1	0	?	0	0	0	0	1	1
Oz. e. excavata	1	1	0	0	?	?	?	?	0	1	1	1	0	1	?	0	1	0	?	?	0	0	0	?	0	0	1	1	0	0	0	1	?	0	1	0	0	1	1	1	0	0	?	1	0	0	0	1	1
Oz. hassi	1	1	0	0	?	?	?	?	0	1	1	1	0	1	?	0	0	1	1	0	0	0	0	?	0	0	1	1	0	0	1	0	1	0	0	0	0	1	1	1	?	0	?	?	0	0	0	?	?
Pl. tenuis	1	1	0	0	?	?	?	?	0	1	1	1	0	0	?	0	1	0	0	1	0	0	0	?	0	0	0	0	1	1	?	?	?	?	?	1	0	0	0	?	0	0	?	0	0	0	1	0	?
Pol. siluricus	1	1	0	0	?	?	?	?	1	?	0	1	0	0	?	0	1	?	?	?	?	?	?	?	?	0	1	1	0	0	0	0	0	0	?	0	0	1	1	0	1	1	1	1	0	0	?	1	1

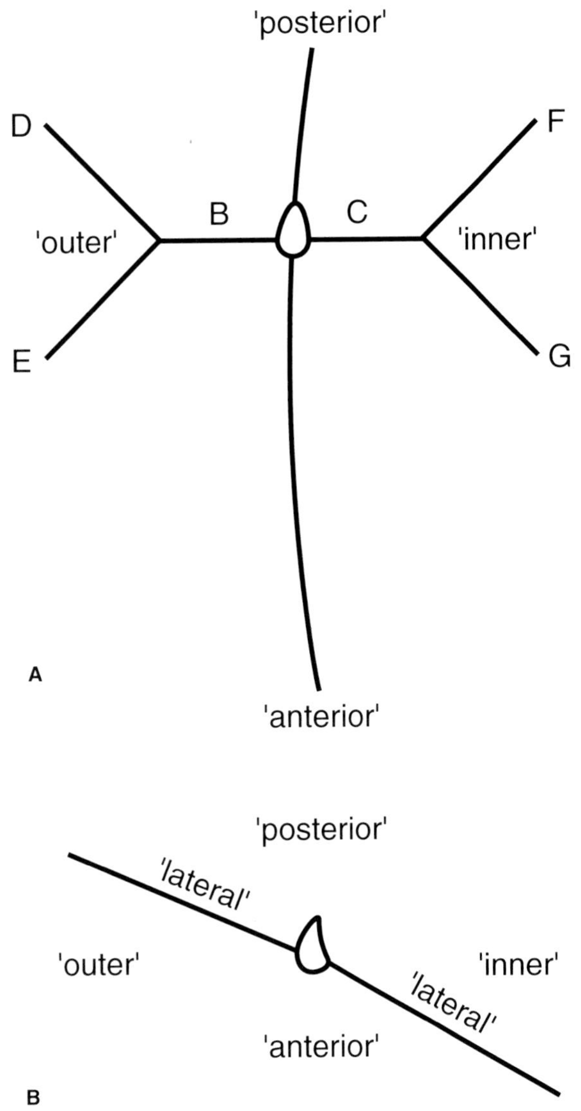

TEXT-FIG. 1. Scheme of process homology followed showing the relationship between conventional terminology and our revised terminology. A, the scheme for taxa with lateral processes, but lacking first generation lateral processes. B, the conventional scheme applied to taxa with first generation 'lateral' processes.

within the context of this analysis, to do so would be phylogenetically uninformative, as all of the taxa considered possess these putative homologies. It would be necessary to resort to an additional tier of homological statements considering homology of position of these processes, following the traditional sense of orientation relative to the cusp, which would result in codings common to presence/absence of 'anterior', 'posterior', 'lateral' first generation processes. Thus we have chosen to follow convention in identifying homology of first generation processes.

1. P_1 'anterior' process: absent (0), present (1). For identification of P_1 process homologies see Text-figure 3.
2. P_1 'posterior' process: absent (0), present (1). Contrary to the conventional interpretation, a 'posterior' process can be demonstrated to be present in the P_1 element of *Kockelella ranuliformis* on the basis that the tip of the basal cavity lies 'anterior' to the 'posterior'-most denticles.
3. P_1 process B: absent (0), present (1). See Text-figure 1.
4. P_1 process C: absent (0), present (1). See Text-figure 1.
5. P_1 process D: absent (0), present (1). See Text-figure 1.
6. P_1 process E: absent (0), present (1). See Text-figure 1.
7. P_1 process F: absent (0), present (1). See Text-figure 1.
8. P_1 process G: absent (0), present (1). See Text-figure 1.
9. P_1 element planate with a recessive basal margin: absent (0), present (1).
10. P_1 basal margin of the 'posterior' process is in the same plane as the basal margin of the 'anterior' process: absent (0), present (1). This character is coded inapplicable in taxa with a recessive basal margin as these taxa do not possess a comparable basal margin.
11. P_1 'posterior' process free: absent (0), present (1). The 'posterior' process has no basal flare. This character is coded inapplicable in taxa with a recessive basal margin as these taxa do not possess a basal cavity and, consequently, cannot have a basal flare.
12. P_1 cusp is larger than other denticles: absent (0), present (1).
13. P_1 cusp is erect in gerontic specimens: absent (0), present (1).
14. P_1 'posterior' process significantly proportionately shorter than the 'anterior' process: absent (0), present (1).
15. P_1 small ridge connects the cusp with the first denticle on B and/or C process(es): absent (0) present (1)
16. P_1 fan: absent (0), present (1). A fan is considered present when the 'anterior'-most denticles on the 'anterior' process are bigger than remaining denticles (Murphy and Valenzuela-Ríos 1999).
17. P_1 denticles on the 'anterior' process: erect (0), directed toward the cusp (1). The orientation of the denticles is considered relative to the basal margin of the process.
18. P_1 basal cavity extends under the 'posterior' process: absent (0), present (1).
19. P_1 basal cavity extends under the 'posterior' process as a shallow groove: absent (0), present (1).
20. P_1 basal cavity deepens under the 'posterior' process: absent (0), present (1).
21. P_1 basal cavity rounded in outline: absent (0), present (1).
22. P_1 basal cavity outline constrained by process disposition: absent (0), present (1).
23. P_1 ventral constriction of the basal cavity: absent (0), present (1).
24. P_1 basal cavity extends and deepens under the B and/or C processes: absent (0), present (1).
25. P_1 'anterior' margin of the platform is approximately perpendicular to the 'anterior-posterior' axis of the element: absent (0), present (1).

26. P_1 'outer lateral' margin of the basal flare is approximately planar and parallel/subparallel to the 'anterior-posterior' of the element: absent (0), present (1).
27. P_2 'anterior' process absent: (0), present (1).
28. P_2 'posterior' process: absent (0), present (1).
29. P_2 'inner lateral' process: absent (0), present (1). The 'lateral' process in *K. crassa* (but not present in the holotype of the form taxon) appears to be secondarily derived from a 'posterior' process and, thus, is not homologous to either of the 'lateral' processes in either *Oulodus* or *Plectodina*.
30. P_2 'outer lateral' process: absent (0), present (1).
31. P_2 denticles fused on the 'anterior' process: absent (0), present (1).
32. P_2 denticle spacing on the 'posterior' process: close (0), wide (1).
33. P_2 denticles are fused on the 'posterior' process: absent (0), present (1).
34. P_2 bar extends higher on the 'anterior' process than the bar on the 'posterior' process: absent (0), present (1).
35. P_2 basal cavity extends shallowly and broadly under the 'posterior' process: absent (0), present (1).
36. P_2 twisted basal cavity: absent (0), present (1).
37. P_2 'outer' basal cavity flare: absent (0), present (1).
38. M 'outer-lateral' process: absent (0), present (1).
39. M element: dolabrate (0), makellate (1).
40. M 'inner-lateral' process of makellate elements: straight (0), bowed/curved (1).
41. M small, constricted basal cavity: absent (0), present (1).
42. M 'inner' basal cavity flare: absent (0), present (1).
43. M 'posterior' basal cavity flare extends under the 'inner-lateral' process: absent (0), present (1).
44. M basal cavity extends broadly and shallowly under the 'inner-lateral' process: absent (0), present (1).
45. M spacing of denticles on the 'inner-lateral' process: close (0), wide (1).
46. M denticles are reclined on the 'inner-lateral' process: absent (0), present (1).
47. S_0 denticles are discretely spaced: absent (0), present (1).
48. $S_{1/2}$ spacing of denticles on the 'inner-lateral' process: wide (0), close (1).
49. $S_{3/4}$ spacing of denticles on the 'posterior' process: wide (0), close (1).

Search techniques

Cladistic analyses were undertaken using PAUP 4.0b10 (Swofford 2002) using 'branch-and-bound' and heuristic (random sequence addition, 100 replicates, ten trees retained at each step, steepest descent on) search options; character evolution was resolved using MacClade 4.05 (Maddison and Maddison 2002). *Plectodina tenuis* was designated as the root in all analyses. Although all characters were unordered and unweighted in primary analyses, these results were subjected to *a posteriori* reweighting in all instances. This methodology, which is equivalent to the successive weighting technique of Farris (1969), assigns a weight to each character based upon on how well they perform in the primary, unweighted analysis, such that a character exhibiting perfect/good fit to the tree(s) derived from the primary analysis will be assigned a high weighting, while a character exhibiting poor fit to the primary tree(s) will be assigned a low/no weight. The weighted dataset is then reanalysed and the process repeated until assigned weights cease to vary between analyses. The methodology achieves results that are both internally consistent and least affected by spurious hypotheses of homology (Platnick *et al.* 1991, 1996; Goloboff 1993). Furthermore, *a posteriori* reweighting provides a test of minimal trees achieved through consensus of poorly fitting characters (Platnick *et al.* 1991; Kitching *et al.* 1998).

Results, experimental analysis of the dataset, and discussion

The primary unweighted analysis of the dataset yielded 48 equally most-parsimonious trees (CI 0·4667; RI 0·7143; RCI 0·3333; 105 steps; strict consensus presented in Text-fig. 2A), which differ primarily in the interrelationships of *Kockelella ranuliformis* and *K. latidentata*, *K. amsdeni*, *K. corpulenta*, and *K. walliseri*, and a clade of four taxa: *K. crassa*, *K. maenniki*, *K. variabilis ichnusae* and *K. variabilis variabilis*. Successive weighting yielded nine equally most-parsimonious trees (CI 0·7091; RI 0·9003; RCI 0·6384; 33·23254 steps; strict consensus presented in Text-fig. 2B) that differ only in the interrelationships of two pairings, *K. ranuliformis* and *K. latidentata*, and *K. walliseri* and *K. corpulenta*; all nine tree topologies are found among the 48 most-parsimonious trees recovered from the primary, unweighted, analysis. The nine competing trees represent every possible solution to the interrelationships within each of these two pairings, all of which are supported by zero length branches. Almost all of the unresolved components of the consensus trees are attributable to poorly known taxa. *K. corpulenta*, *K. v. ichnusae*, *K. latidentata* and *K. abrupta* (the last not contributing to conflict) are known from just one among the eight paired and one unpaired element positions that comprise the apparatus of kockelellid conodonts. Thus, rather than character conflict, the basis of poor resolution lies in an absence of data. To constrain the influence of the rogue taxa, they were excluded from the dataset and the analysis repeated, yielding two equally most-parsimonious trees that differed only in the relative relationships of *K. walliseri* and *K. amsdeni*, which are resolved in reciprocal positions in the two trees (CI 0·4757; RI 0·6949; RCI 0·3306; 103 steps; Text-fig. 2C). Reweighting yields a single most-parsimonious tree (CI 0·7175; RI 0·8936; RCI

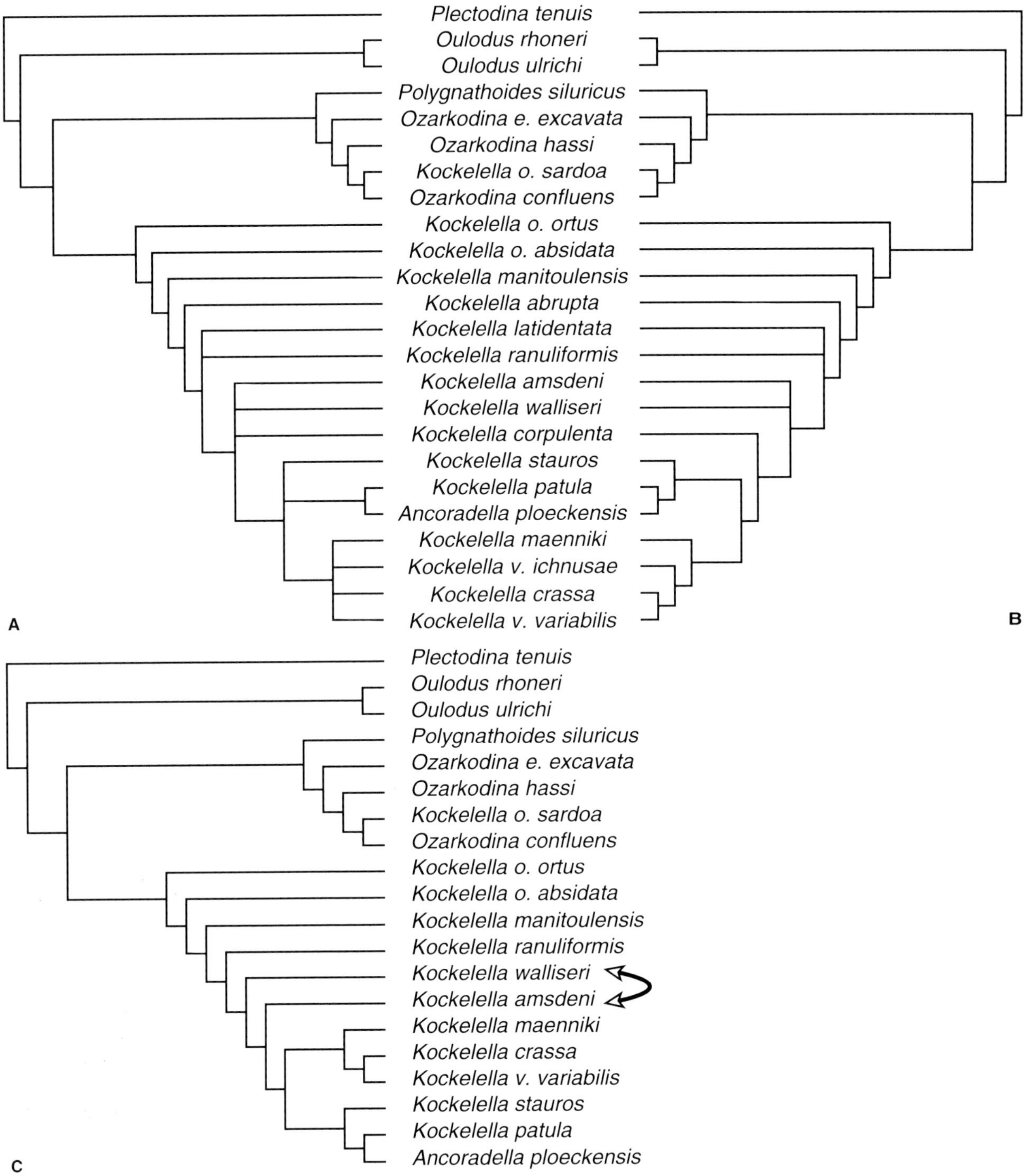

TEXT-FIG. 2. Trees derived from stratigrahically unconstrained analysis of the cladistic dataset. A, strict consensus of 48 trees derived from preliminary, unweighted analysis of the dataset. B, strict conseneus of nine trees derived from analysis of the dataset reweighted *a posteriori* according to the Rescaled Consistency Index (RCI) values for the characters in the preliminary unweighted analysis. C, tree derived from unweighted analysis of the dataset, excluding those taxa known only from P_1 elements; the two trees differed in the relative positions of *Kockelella walliseri* and *Kockelella amsdeni*; *a posteriori* reweighting resulted in a single tree in which *K. amsdeni* was resolved as the more derived of the two taxa.

0·6412; 32·75476 steps; Text-fig. 2C) in which *K. amsdeni* is resolved as more derived than *K. walliseri*, compatible with one of the trees obtained from reweighting of the pruned dataset, and all nine of the trees arrived at from reweighting of the complete dataset (and, in corollary, nine of the 48 trees obtained from analysis of the original unweighted complete dataset). The alternative tree topology (Text-fig. 2C) is not compatible with any of the nine trees obtained from reweighting of the complete dataset, but is compatible with 15 trees that are slightly longer (CI 0·6896; RI 0·8907; RCI 0·6142; 34·17004 steps).

The difference in the original 48 trees represents a conflation of the conflicting positions of *K. amsdeni*, *K. variabilis ichnusae*, and the two pairs *K. ranuliformis* and *K. latidentata*, and *K. walliseri* and *K. corpulenta*. The phylogenetic positions of *K. amsdeni* and *K.* vs. *ichnusae* are resolved through successive weighting, reducing the competing topologies to nine, representing all possible solutions to the two remaining unresolved taxon pairs. Resolution within the two taxon pairs cannot be obtained because none of the competing solutions is supported by data. Thus, the nine trees are identified as optimal for the purposes of the analysis.

The results confirm that the vast majority of taxa assigned hitherto to the family Kockelellidae, and the genus *Kockelella*, constitute a monophyletic group, and that the Kockelellidae are members of the Order Ozarkodinida. However, there are some exceptions. *Polygnathoides*, a putative member of the ingroup, is resolved as part of the outgroup, in a position which suggests that it is a plesiomorphic member of the Order Ozarkodinida and that its similarities to members of the Kockelellidae are little more than ozarkodinid symplesiomorphies. *Ancoradella* is resolved as a member of the ingroup, confirming earlier suggestions to this effect (Link and Druce 1972; Serpagli and Corradini 1999) and rendering *Kockelella* paraphyletic; we suggest that this monospecific genus should in future be considered a junior synonym of *Kockelella*, though this judgement should await a full multielement reconstruction of *Ancoradella ploeckensis*. *Kockelella ortus sardoa* is also resolved among the outgroup, and among species of *Ozarkodina* in particular; it is more appropriately assigned to *Ozarkodina*. *Kockelella ortus absidata* and *K. ortus ortus* are resolved as successive sister taxa to all other members of the Kockelellidae and can be included within the clade. However, it is possible that this hypothesis will not stand up to scrutiny using additional outgroup taxa, specifically including more species of *Ozarkodina*. *K. o. absidata* and *K. o. ortus* are retained within the concept of the Kockelellidae for the purposes of this study. Finally, we note that none of the putative ancestors of the Kockelellidae are resolved as close relatives.

The results of the analysis are also interesting in more general terms, with reference to the question of whether multielement taxonomy is necessary for phylogenetic reconstruction. Donoghue (2001) demonstrated that although contemporary conodont phylogenies are based almost exclusively on the characteristics of the P_1 element position, at least with regard to the Late Devonian palmatolepidids, the majority of phylogenetically informative characters are to be found in element positions other than the P_1. Of the 49 phylogenetic informative characters utilized in this analysis, only 26 are based on P_1 element position, a clear indication that although this element position is integral to phylogenetic reconstruction, alone it is not sufficient in this endeavour. The remaining 23 characters are concentrated on the P_2 position especially. Most of the synapomorphies of the Kockelellidae are present on the P_1 element, re-enforcing the importance of characteristics of this element position not only for taxonomy, but also in the systematics of Kockelellidae.

CLADOGRAM-STRATIGRAPHY CONGRUENCE: CLADISTIC ASSESSMENT OF FOSSIL RECORD COMPLETENESS

The fit of stratigraphy to a cladogram can be determined both qualitatively and quantitatively by calibrating the branches of a cladogram to the stratigraphic range of the taxa that the branches subtend ('X-trees' in the terminology of Eldredge 1979); successive sister taxa and sister groups then provide a series of corroborative tests of the early fossil record of one another. Because sister taxa derive from a common ancestor, the first appearance of both lineages in the fossil record should be synchronous. Thus, gaps in the early fossil record of taxa (so-called 'ghost lineages'; Norell 1992) can be inferred.

This can be assessed qualitatively, but a variety of metrics are also available to assess quantitatively the degree of congruence. The most widely employed of these are the Stratigraphic Consistency Index (SCI; Huelsenbeck 1994) and the Relative Completeness Index (RCI; Benton and Storrs 1994); more recently, the Gap Excess Ratio (GER) has been developed by Wills (1999). The SCI assesses the degree to which the hierarchical branching of a cladogram is congruent with the appearance of taxa in the fossil record (Huelsenbeck 1994).

$$\mathrm{SCI} = \text{number of stratigraphically consistent nodes} / \text{total number of nodes}$$

The Relative Completeness Index (RCI; Benton and Storrs 1994) goes beyond determining whether or not cladogram branching order is stratigraphically consistent and actually attempts to measure the overall level of inconsistency in a tree. This is achieved by quantifying the ghost range implied by the difference in age between the ages of the origin of branches subtending sister taxa, and dividing this value by the observed range length. The RCI is given by the complement of this value expressed as a percentage.

$$RCI = \left\{1 - \frac{\Sigma(\text{implied ghost range})}{\Sigma(\text{stratigraphic range})}\right\} \times 100\%$$

Thus, while the RCI is based on the identification and existence of ghost ranges and attempts to quantify degree of incompleteness, the SCI refers only to the known stratigraphic ranges of taxa and assesses only gross congruence between cladogram branching order and appearance of taxa in the fossil record.

The Gap Excess Ratio (GER; Wills 1999) combines aspects of both SCI and RCI into a metric that expresses the proportion of total implied ghost range necessitated by the constraints of a cladogram. The GER is calculated as follows:

$$GER = 1 - \frac{(MIG - G_{min})}{(G_{max} - G_{min})}$$

where MIG is the sum of the implied ghost ranges, G_{min} is the sum of minimum possible ghost ranges (i.e. the implied ghost ranges from a hypothesis of relationships based solely upon stratigraphic order) and G_{max} is the maximum possible sum of ghost ranges (obtained by summing the difference in origination times between the oldest taxon and all other taxa). GER values range between 0·00 and 1·00, where MIG is equal to G_{max} (worst possible) and G_{min} (best possible), respectively.

SCI, RCI and GER metrics were calculated using GHOST 2.4 (Wills 1999). The stratigraphic ranges of taxa are determined on a presence/absence basis with reference to a timescale divided into time bins of equal length, referred to below as temporal spacing. Because the taxa included in the analysis have been used widely in biostratigraphy their stratigraphic ranges are more finely and accurately resolved than is normally possible. Thus, the temporal spacing applied herein is considerably shorter than has been utilized hitherto in assessments of cladogram-stratigraphy congruence (e.g. Wills 2001) and is based on the graphic correlation composite standard developed for the Silurian by Kleffner (1995), revised after the time scale of Tucker and McKerrow (1995), where each time unit averages to approximately 300,000 years. The stratigraphic ranges of taxa are taken from Aldridge (1972), Barrick and Klapper (1976), Kleffner (1995), Jeppsson (1997), Serpagli and Corradini (1999) and Zhang and Barnes (2002). Stratigraphy and data files used in GHOST analyses are presented in the Appendix (section 1).

Results

The stratigraphy-calibrated consensus cladogram for *Kockelella* is presented in Text-figure 3. This reveals significant ghost lineages throughout the tree, and particularly along the branches subtending the taxa *Kockelella abrupta*, *K. ranuliformis*, *K. walliseri* and their respective sister groups, implying that there are very significant gaps in the fossil record of the Kockelellidae. Results of the quantitative assessments of cladogram-stratigraphy congruence for the nine optimal trees are presented in Table 2; these exhibit SCI values in the range 0·42–0·5 indicating that half, or slightly fewer, of the nodes in the cladogram are stratigraphically consistent; significance values of 1·5–16·5 indicate that some of these trees (2, 4, 5, 7, 8) exhibit significant ($P > 0·95$) stratigraphic congruence (relative to the permuted dataset) while the others exhibit a level of congruence that is certainly not significantly worse than can be achieved by randomly reassigning the data across respective trees. RCI values of − 49 to −59 for the various trees indicate that the fossil record of the Kockelellidae is more gap than record. GER values of 0·78–0·80 indicate that the sum of ghost lineages required by these trees is not close to the minimum possible given the tree topology, but they are nevertheless high. Results from the permutation test for the RCI and GER analyses indicate that all trees are significantly better than randomized data ($P > 0·95$).

The results of this now standard approach to assessing fossil record completeness are somewhat contradictory. The contrasting signal from the SCI and GER metrics could indicate a rich, but significantly incomplete fossil record, the depressed SCI arising from a number of taxa with origination dates that approximate one another. However, the RCI appears to indicate that the kockelellid fossil record is only a very poor reflection of its evolutionary history. In either instance, the low SCI values indicate that stratigraphic data should definitely not be used as a guide to phylogeny reconstruction. However, because fossil record completeness is not the only variable in cladogram-stratigraphy incongruence, it is worth considering alternative explanations before reaching a conclusion.

Causes of incongruence between cladograms and stratigraphic data

Benton *et al.* (1999) identified some major causes of incongruence between cladistic hypotheses and stratigraphic data: (1) cladogram quality; (2) fossil record quality; (3) sampling density; and (4) stratigraphic problems and taxonomic focus. These are explored below as potential explanations of cladogram-stratigraphy incongruence in the fossil record of *Kockelella*.

Cladogram quality. The veracity of any hypothesis of relationships can always be questioned. However, in compar-

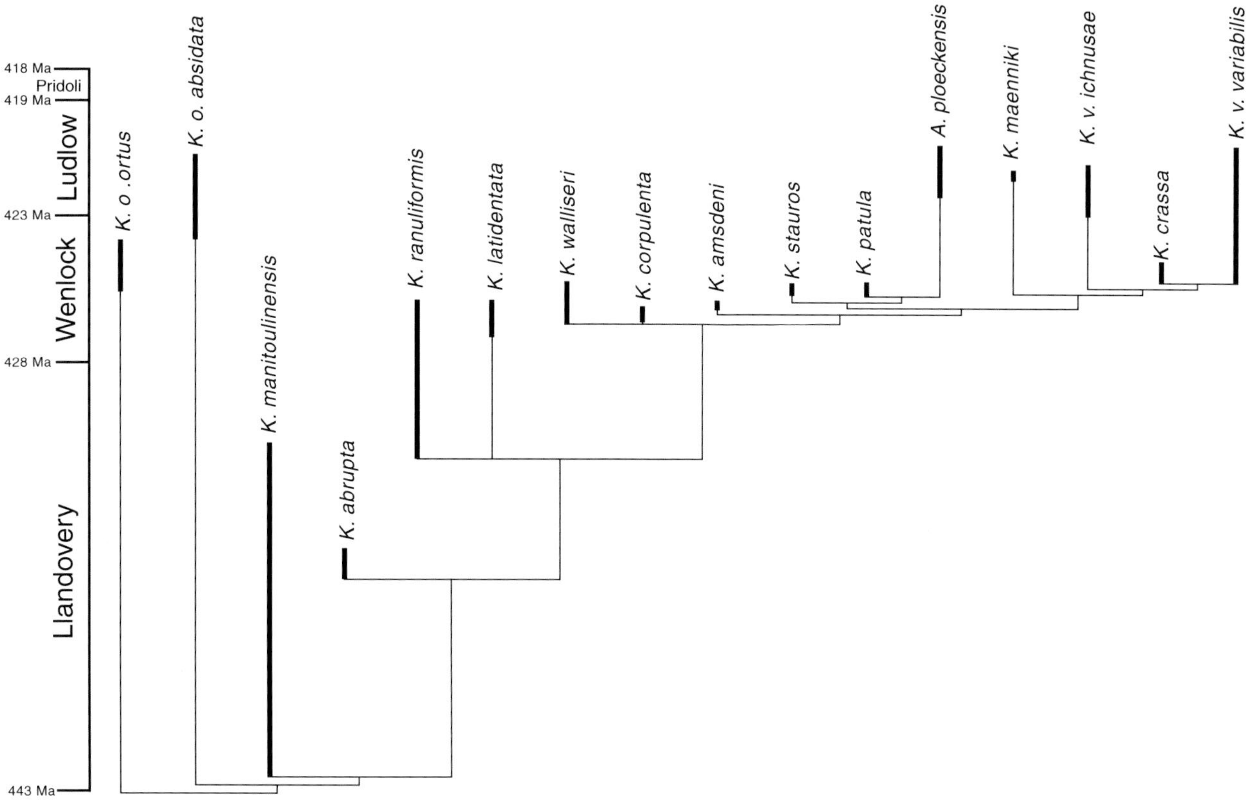

TEXT-FIG. 3. 'X'-tree for the Kockelellidae based upon the strict consensus of the nine optimal trees. Resolution of the polytomies has little implication for the extent of the total inferred ghost lineages.

ison with previously published hypotheses of relationships within *Kockelella* (Barrick and Klapper 1976; Serpagli and Corradini 1999) the hypothesis of relationships presented here provide the best explanation of the available morphological data.

Fossil record quality. All three cladogram-stratigraphy congruence metrics imply surprisingly low values for a group widely appreciated for its rich and well-documented fossil record. It is not possible to preclude the possibility that inaccurate dating of fossils is a contributing factor towards incongruence, given that the conodonts themselves invariably provide the dating, but this risk has been minimized by employing most of the stratigraphic range data from a graphical correlation composite standard (Kleffner 1995). While we have gone some way to control for, and exclude, the possibility of imperfect sampling of the fossil record, the possibility remains that there are systematic biases affecting the rock record itself, which particularly affect the fossil record of organisms whose distribution was facies controlled (Holland 1995), which appears to have been the case in *Kockelella* (Aldridge and Mabillard 1981; Armstrong 1990).

Sampling density. The nature of the relationship between cladogram-stratigraphy congruence metrics and sampling density (e.g. the number of taxa included in an analysis) remains problematic (Benton *et al.* 1999). However, analyses including all taxa, as here, must provide the most realistic assessment of fossil record quality. Tree balance is another known bias of these metrics such that the full range of SCI values is not open to fully balanced trees, and fully imbalanced (pectinate) trees will always show the minimum MIG and, thus, exhibit low RCI values (Sidall 1996, 1997; Wills 1999). All of our trees are extremely, but not completely, imbalanced and so it is possible that this is a contributing factor towards the low (negative) RCI scores.

Stratigraphic problems and taxonomic focus. Duration of taxa and temporal spacing of nodes affect both the SCI and the RCI (Hitchin and Benton 1997; Benton *et al.* 1999). If the nodes in the cladogram are dated too precisely, very low SCI and RCI values can be obtained, whereas if all nodes are found at the same temporal level, i.e. dated less precisely, misleadingly high scores can result (Benton *et al.* 1999). To test whether this applied to our analysis, the temporal spacing was increased and the

TABLE 2. Cladogram-congruence metric scores for the cladograms derived from stratigraphically constrained and unconstrained analysis. These metrics were calculated using GHOSTS 2.4 (Wills 1999).

Tree no.	Raw RCI	RCI & GER Sig.	Raw SCI	SCI Sig.	Raw GER
Unconstrained – fine time bins					
1	−59	2·7	0·428571	7·6	0·78389
2	−49	2·2	0·5	2·5	0·803536
3	−59	2·8	0·428571	16·5	0·78389
4	−59	2·8	0·428571	4·3	0·78389
5	−59	2·8	0·428571	4·3	0·78389
6	−59	2·8	0·428571	7·6	0·78389
7	−49	1·8	0·5	1·5	0·803536
8	−49	1·8	0·5	1·7	0·803536
9	−59	2·8	0·428571	8·2	0·78389
Unconstrained –coarse time bins					
1	−42·105263	2·8	0·428571	10·3	0·779923
2	−33·333333	2·1	0·5	4·2	0·799228
3	−42·105263	3	0·428571	22·6	0·779923
4	−42·105263	2·6	0·428571	6·3	0·779923
5	−42·105263	2·5	0·428571	6	0·779923
6	−42·105263	2·7	0·428571	10·6	0·779923
7	−33·333333	2·1	0·5	2·2	0·799228
8	−33·333333	2·1	0·5	2·1	0·799228
9	−42·105263	2·7	0·428571	11·2	0·779923
Constrained –fine time bins					
1	−18·75	1·4	0·6	0·7	0·846715
2	−17·857143	1·1	0·6	0·4	0·84854
3	−10·714286	0·8	0·6	0·6	0·863139
4	−18·75	1	0·6	0·7	0·846715
5	−18·75	1·1	0·6	0·5	0·846715
6	−9·821429	0·5	0·6	0·4	0·864964
7	−17·857143	0·8	0·6	0·6	0·84854
8	−10·714286	0·6	0·6	0·5	0·863139
9	−18·75	0·8	0·6	0·6	0·846715

cladogram-stratigraphy congruence metrics recalculated; the results are presented in Table 2. Less precisely dated nodes result in a modest increase of the RCI, and although SCI values are unchanged, fewer trees exhibit SCI values that are significantly better than random (2, 7, 8). Neither of these factors indicates that the temporal resolution of the analysis is introducing artefacts into the analysis. Another possible bias, identified by Benton and Storrs (1994), arises in the analysis of low-level taxonomic groups, such as species, where artificially low RCI values are achieved when range lengths approach the finest level of stratigraphic resolution. Given that the stratigraphic ranges of almost all the taxa included in our analysis extend through more than one time bin, this potential bias can be excluded.

Other sources of error: artefact. None of the potential biases outlined by Benton *et al.* (1999) obviously provides an adequate explanation for the mismatch between cladograms and stratigraphic data. An alternative interpretation is that some ghost lineages are artefacts of cladistic analysis. This can arise for two (or more) reasons. First, spurious ghost lineages can be inferred because one of the basic assumptions of cladistic analysis, the monophyly of component taxa, has not been met. Thus, the inclusion of paraphyletic taxa leads to the inference of a ghost lineage in their sister group of duration equal to their own known stratigraphic range. Secondly, spurious ghost lineages can also be inferred where cladograms, which are only intended as explanations of the distributions of synapomorphies, are reinterpreted as phylogenies without further qualification other than calibrating the branching events to time using stratigraphic range data from the fossil record (so-called X trees; Eldredge 1979). This enforces a model or phylogenesis limited to cladogenesis, which may not be entirely appropriate (Bretsky 1979, among many others); this can be particularly problematic when undertaking cladistic analyses at the species level (see also Wagner 1995, 1998; Foote 1996; Paul 1998, 2003). In such instances, indices based on the existence of such ghost lineages would not be an appropriate measure of the quality of the fossil record. Both factors would lead to disparity in the signal from RCI and GER metrics.

Before considering whether the analysis has been affected by such artefacts, the problem arises as to how to distinguish spurious ghost lineages from real gaps in the fossil record. It has been suggested that confidence intervals may be the most appropriate means of verifying ghost lineages. However, confidence intervals and the results of cladogram-stratigraphy congruence tests are not logically equable. Confidence intervals differ in that they do not incorporate phylogenetic assumptions in their formulation and assess the veracity of the fossil record on a taxon by taxon basis, providing inference of gaps in the range of known taxa. Meanwhile, cladogram-stratigraphy congruence tests assess the overall completeness of the fossil record, providing inference not only of gaps in the range of known taxa but also of the existence of unrecognized and unknown taxa, such as known taxa that have not, hitherto, been recognized as members of a particular in-group, or taxa that have yet to be discovered.

Thus, confidence intervals do not provide an appropriate test of ghost lineages inferred from cladogram-stratigraphy congruence tests. However, in contrasting the results of the two methods, it is possible to discriminate which component of an inferred ghost lineage can be attributed to one of these two types of fossil record incompleteness: (1) gaps in the fossil record of known taxa, determined directly from confidence intervals, and (2) gaps in the fossil record due to unrecognized or unknown taxa, determined by subtracting confidence intervals on range extensions from inferred ghost lineages.

CONFIDENCE INTERVALS

Confidence intervals estimate the 'true' stratigraphic range of a taxon within given a level of statistical confidence (e.g. 95%) and follow the common sense principle that the greater the number of stratigraphic/chronological levels at which a taxon has been recovered, the lower the probability that future discoveries will lie far beyond the currently known distribution. Strauss and Sadler (1989) provide the following formula for their calculation, expressed as a fraction of the known stratigraphic range:

$$(1 - \text{confidence level})^{-1/(\text{number of known fossiliferous horizons})} - 1.$$

The calculation of confidence intervals requires a number of assumptions and, where these assumptions are not met, there are a variety of techniques that can be implemented to overcome such limitations. At the simplest level, the calculation assumes constant fossil recovery potential, but techniques have been developed where recovery potential may vary, or where the assumption of continuous sampling is not achieved. Thus, it is possible to account for variable recovery potential that may result from, for example, biases in facies preservation arising from changes in relative sea level and affecting the distribution of organisms whose distribution is facies controlled (Holland 1995, 2000; Marshall 1997; Tavaré *et al.* 2002). It is also possible to account for strategies other than continuous sampling, which is particularly prevalent in micropalaeontology where sections are routinely sampled as a series of arbitrarily spaced discrete horizons (Weiss and Marshall 1999). Sampling strategies that are biased by the requirements of laboratory techniques, such as the sampling restricted to carbonates, also violate the constant recovery potential assumption and should be accounted for in confidence interval calculations. However, the assumptions required of such techniques are both non-trivial and overly simplistic, and can only be readily applied to stratigraphic ranges in local sections and are, thus, not compatible with our aim of assessing the degree to which a composite of global records reflects the evolutionary history of particular taxa. We follow the general method of confidence interval calculation with the caveat that some of the assumptions on which it is based may not be met and that our confidence intervals may be longer or shorter than would otherwise be anticipated.

A database of known stratigraphic occurrences of species of kockelellid on which confidence intervals were calculated, using the general method, is presented in the Appendix (section 2). Occurrences unsupported by illustrated examples of the taxonomic concepts followed were not included in the database. Stratigraphic occurrences in which the total number of fossiliferous horizons is unstated were recorded in the database as single occurrences.

Results and discussion

Confidence intervals on the stratigraphic ranges of in-group members of the Kockelellidae are presented in Text-figure 4 and the Appendix (section 2). These reveal that although the expected stratigraphic ranges of all taxa consistently exceed the observed stratigraphic range, in the vast majority of taxa these coincide approximately, thus providing confidence in the veracity of these taxa as zone fossils and their utility in biostratigraphy, which has been assumed previously only on the basis of repeatability and reproducibility. However, there are very significant exceptions to this overall observation. Most notably, the expected stratigraphic range of *Kockelella abrupta* exceeds the known observed stratigraphic ranges by millions of years, indicating that we can have

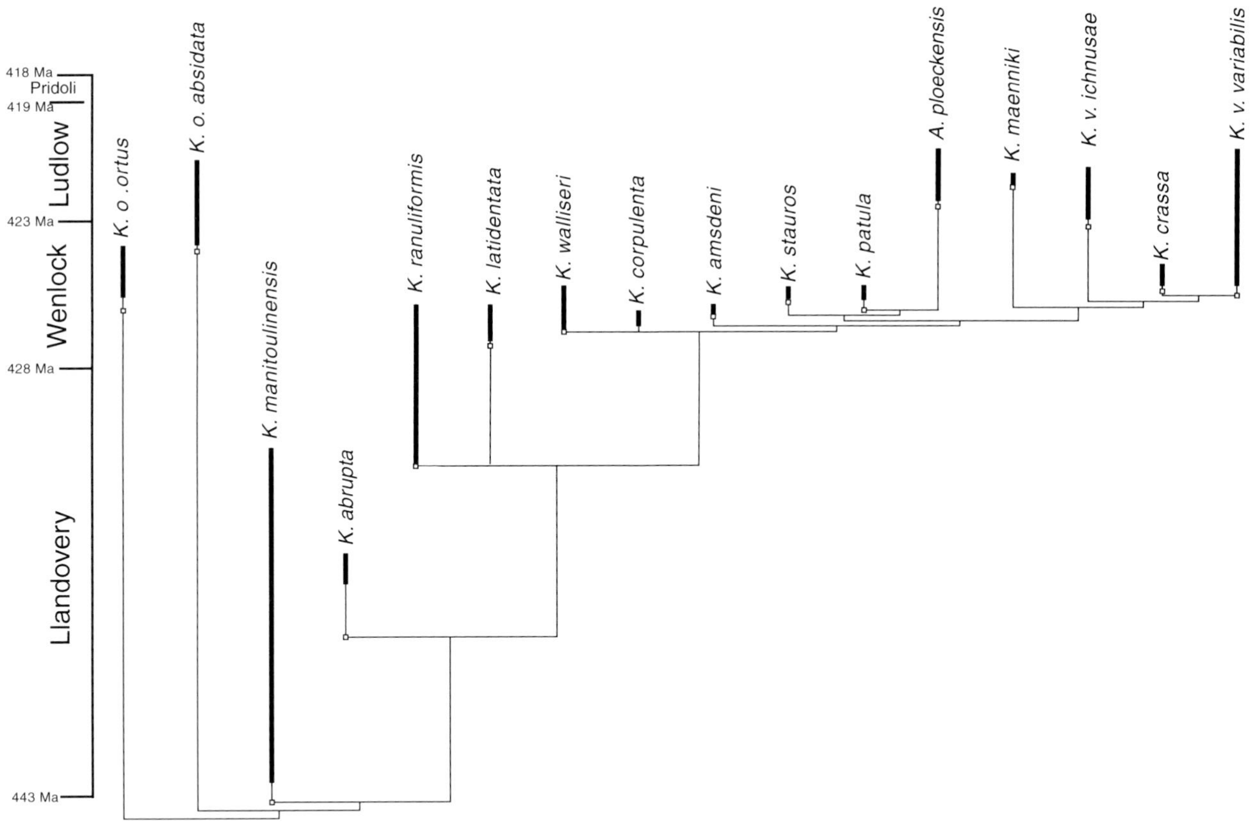

TEXT-FIG. 4. 'X'-tree for the Kockelellidae based upon the strict consensus of the nine optimal trees. Small open boxes represent 95 per cent confidence intervals on the time of first appearance of the component taxa.

little confidence that future discoveries of this taxon will lie within or close to the known limits on its stratigraphic distribution.

Neither cladogram-stratigraphy congruence nor confidence intervals indicate that the fossil record of *Kockelella* is complete, but the results of these analyses infer degrees of incompleteness that vary very widely. Confidence intervals on the stratigraphic ranges of individual taxa in all instances indicate some prolongation of the known record, although in almost all instances the confidence intervals are sufficiently short to be considered beyond stratigraphic resolution. Even in those instances where a significant (i.e. greater than 5% of the known range) range extension is inferred, these are very minor in comparison with the ghost lineages inferred from stratigraphically calibrated cladograms, which suggest that the fossil record for this group represents considerably less than half of its evolutionary history, as evidenced by negative RCI values. Given the disparity in magnitude of the inferred missing record it must be concluded that the vast majority of the sum of the ghost lineages cannot be attributed to gaps in the fossil records of known taxa. Thus, the vast majority of the total sum of inferred ghost lineages must be attributed to hitherto unrecognized and/or unknown and/or to an analytical artefact. It is potentially possible to evaluate, at least qualitatively, the probability that the remaining ghost lineages represent either unknown/unrecognized taxa or an analytical artefact with reference to the discovery curve for *Kockelella*.

DISCOVERY CURVES

Discovery curves are usually employed in ecological or palaeoecological studies as a means of determining the degree to which sampling of taxa is representative. As collecting of individuals begins, it is expected that the rate of discovery of new species will be very high, but as collecting continues the rate of new discoveries will slow progressively. Thus, in a cumulative plot of new discoveries against time, it is expected that the curve will be steep initially, reflecting the high probability of new discoveries through continued collecting in a poorly sampled record, eventually shallowing asymptotically and reflecting the reduced probability of new discoveries as the record is progressively more completely sampled. The utility of discovery curves has long been appreciated outside of palaeoecology and they have proven instructive in assessing the degree to which existing sampling of a group is representative of its fossil record (Paul 1982; though not necessarily its evolutionary history as the fossil record itself can be affected by systematic bias, e.g. Holland 1995).

Text-figure 5A is a cumulative plot of *Kockelella* species descriptions against their date of publication (compiled on the basis of the data presented in the Appendix, section 2). The steep curve suggests that the diversity of *Kockelella* has been only very partially sampled and that continued collecting will be rewarded with the discovery of many more species. However, this plot is potentially misleading because it incorporates new taxa that were known from earlier discoveries (synonyms), and previously assigned to form taxa; these clearly do not represent discoveries of new taxa as a result of continued collecting effort, but are merely the recognition of additional taxa among the existing database. The revised plot (Text-fig. 5B), which dates the discovery of new taxa to the earliest description of an assigned synonym, suggests that sampling of the record has matured. This plot can be filtered further for taxa that have been erected as a result of splitting known taxa hitherto assigned to *Kockelella*, as these new taxa cannot be considered 'genuinely new'. The flat profile of the revised curve (Text-fig. 5C) further suggests that the sampling of *Kockelella* is mature and that continued effort will not prove profitable.

Finally, to exclude further the possibility of artefacts in the discovery curve, it is worth considering whether the flattening of the filtered discovery curves, reflecting the continued lack of new discoveries, is an artefact of diminishing collector effort, rather than from diminishing returns of continued effort maintained at approximately the same level. This possibility can be tested by plotting collector effort, represented by number of new publications describing new records of *Kockelella*, against time. The resulting plot (Text-fig. 5D; data from the Appendix, section 2) reveals that collector effort has been episodic over time but that, most significantly, there is no evidence of diminishing collector effort co-ordinate with diminishing discoveries of new taxa.

With all artefacts and potential artefacts controlled for, we may conclude from the shape of the discovery curves that knowledge of the fossil record of *Kockelella* is now mature. Thus, it is unlikely that the vast majority of the ghost lineages (that portion remaining after subtraction of confidence intervals) are valid. Thus, it is pertinent to consider sources of analytical artefacts, rather than fossil record completeness, as an explanation.

ANALYTICAL ARTEFACTS IN GHOST LINEAGE INFERENCE: CLADOGRAMS VERSUS PHYLOGENIES

At least one further possibility remains, that incongruence between cladograms and stratigraphic data arises because cladograms are abstractions of phylogenetic trees and spurious ghost lineages are inferred because of the artefact of considering only relative relationships, to the exclusion of absolute relations. The problem, or lack thereof, in considering direct, or ancestor–descendant, relationships in phylogenetic analysis is an old one (e.g. Engelmann and Wiley 1977 versus Szalay 1977; Eldredge 1979 versus Bretsky 1979). In general, absolute relationships are superfluous, such as in attempts to trace character evolution, where it is more appropriate to follow the evolution of characters through the inferred conditions of hypothetical common ancestors of successive sister groups. This practice neither denies the contribution of potential ancestors (conventionally identified as taxa lacking autapomorphies, e.g. Engelmann and Wiley 1977; Smith 1994), nor is it handicapped by potentially spurious identifications of ancestors. Absolute relationships are also of little concern when attempting to uncover the interrelationships, or when assessing the completeness of the fossil record of higher taxa, which do not have ancestors (Engelmann and Wiley 1977).

However, in attempting to assess fossil record completeness at the species level and/or in attempting to constrain the time of origin of a group, a consideration of absolute relationships is a necessity. This is because, unlike higher taxa, populations and, perhaps, species can be considered to have ancestors (Engelmann and Wiley 1977; but see Nelson 1989). Thus, it is not possible to assume monophyly of taxa at the species level, which is one of the basic requirements of both cladistic analysis and ghost lineage inference. Analyses of the completeness of the fossil record at species level are likely therefore to be subject to the same kind of artefact introduced by paraphyletic higher taxa, and the controls routinely employed to test for paraphyly in higher taxa are not readily applicable to species. Although species cannot strictly be considered ancestral because they arise from differentiation among populations rather than from other species (Wiley 1981), it is a reality that many, particularly fossil, species will include ancestral populations. Such taxa will be found among metataxa, i.e. plesiomorphic grades of organisms united by convention, rather than biological integrity, which cannot be further subdivided on the basis of available evidence (but which may be further resolved on the basis of additional data). Metataxa are recognized operationally on the basis of an absence of autapomorphies, and potentially ancestral groups can be identified among metataxa on the basis of stratigraphic evidence (Smith 1994). Such consideration of potentially ancestral groups, regardless of whether any are identified, has the effect of producing what Eldredge (1979) termed an 'A' tree, which is equivalent to a traditional phylogenetic tree. More to the point, it has the effect of

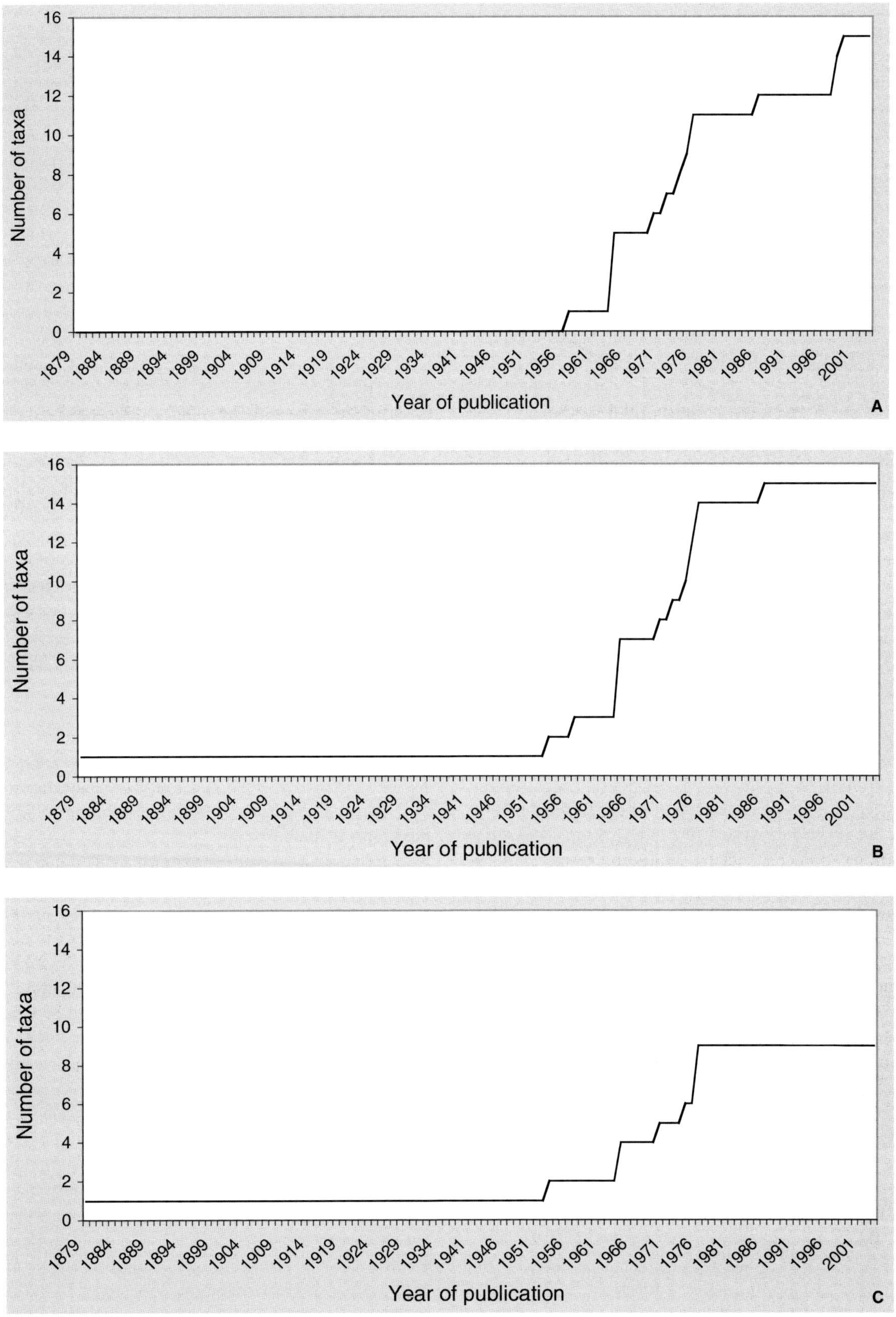
Number of taxa
0 2 4 6 8 10 12 14 16
1879 1884 1889 1894 1899 1904 1909 1914 1919 1924 1929 1934 1941 1946 1951 1956 1961 1966 1971 1976 1981 1986 1991 1996 2001
Year of publication
A
Number of taxa
0 2 4 6 8 10 12 14 16
1879 1884 1889 1894 1899 1904 1909 1914 1919 1924 1929 1934 1941 1946 1951 1956 1961 1966 1971 1976 1981 1986 1991 1996 2001
Year of publication
B
Number of taxa
0 2 4 6 8 10 12 14 16
1879 1884 1889 1894 1899 1904 1909 1914 1919 1924 1929 1934 1941 1946 1951 1956 1961 1966 1971 1976 1981 1986 1991 1996 2001
Year of publication
C

TEXT-FIG. 5

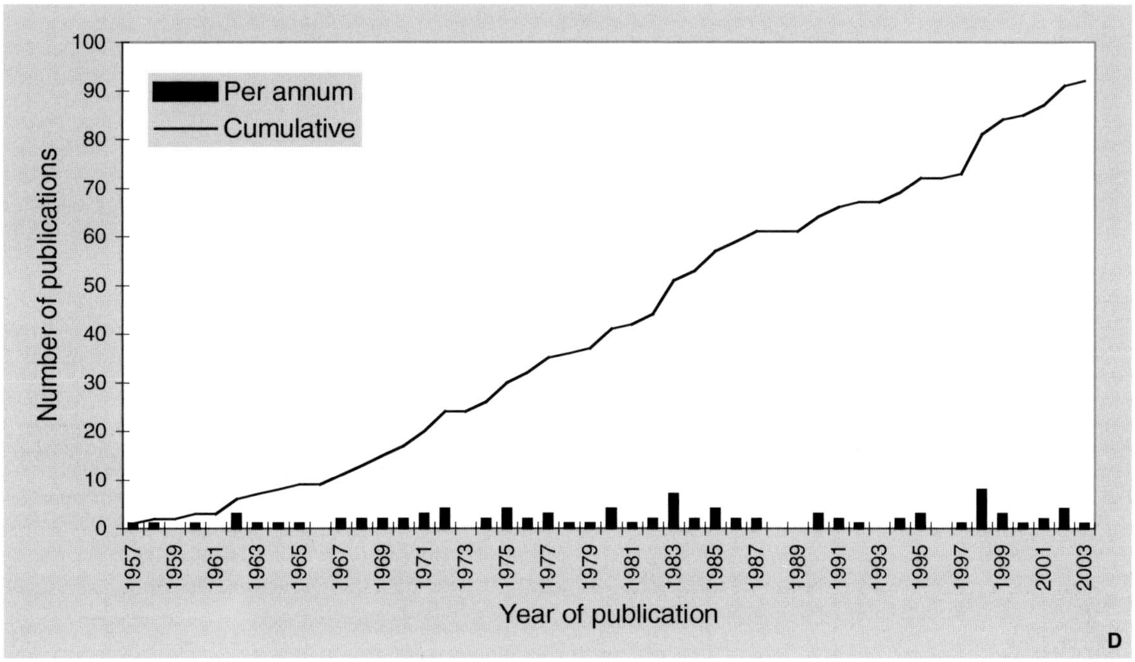

TEXT-FIG. 5. A–C, discovery curves for the Kockelellidae; D, research effort over time for the Kockelellidae. A, cumulative plot of the discovery of new species and subspecies of kockelellid. B, cumulative plot of the discovery of new species and subspecies of kockelellid, corrected such that the date of discovery of new taxa is referred to the date of description of the earliest synonym. C, cumulative plot of the discovery of new species and subspecies of kockelellid, corrected such that the date of discovery of new taxa is referred to the date of description of the earliest synonym, and such that only genuinely new discoveries are considered (i.e. by excluding 'new' taxa identified by massaging the concept of existing taxa). D, research effort, measured according to the number of publications describing 'new' collections of kockelellids, plotted per annum and as a cumulative total.

excluding ghost lineage artefacts from assessment of fossil record completeness.

Returning to *Kockelella*, it is possible to identify a number of metataxa by examining the character change tree for terminal branches lacking apomorphies (autapomorphies) (Text-fig. 6); these are *K. abrupta*, *K. corpulenta*, *K. latidentata*, *K. manitoulinensis*, *K. ortus ortus*, *K. patula* and *K. variabilis ichnusae*. Among these species, the stratigraphic range of *K. v. ichusae* is incompatible with its identification as including groupings potentially ancestral to its sister group. *K. latidentata* and *K. corpulenta* are problematic in that they not only operationally lack autapomorphies but also all possible cladistic solutions to their relationship to *K. ranuliformis* and *K. walliseri*, respectively, are supported by zero-length branches. However, their morphology plus stratigraphic range are compatible only with a sister-taxon relationship within their respective taxon pairings, or as more derived than their respective taxon pairs and ancestral to their sister-groups. Given that *K. latidentata* can only be distinguished from *K. ranuliformis*, and *K. corpulenta* from *K. walliseri* on the basis of gerontic specimens (i.e. juveniles of *K. latidentata* and *K. corpulenta* fall within the taxonomic concept of *K. ranuliformis* and *K. walliseri*, respectively), and that the operationally apparent autapomorphies of *K. ranuliformis* and *K. walliseri* pertain to unknown characteristics of *K. latidentata* and *K. corpulenta*, the solution most parsimonious with stratigraphic data (equally parsimonious with morphological data) is to consider *K. ranuliformis* as ancestral to *K. latidentata* and *K. walliseri* as ancestral to *K. corpulenta*. The resulting 'A'-tree (Text-fig. 7) demonstrates the effect upon inferred ghost ranges of considering metataxa; the total sum of implied gap in the record is significantly diminished through the abolition of a ghost lineage subtending the sister groups to *K. manitoulinensis*, *K. abrupta* and *K. patula*.

Remaining ghost lineages and the addition of ad hoc *assumptions*

Significant ghost lineages remain that are not corroborated by confidence intervals. In particular, these include the ghost lineage subtending the sister-group to *Kockelella ranuliformis* and *Kockelella latidentata*, which is imposed exclusively by the stratigraphic range of *K. ranuliformis*, interpreted previously as the direct ancestor of many members of its sister-group. These hypotheses of ancestry

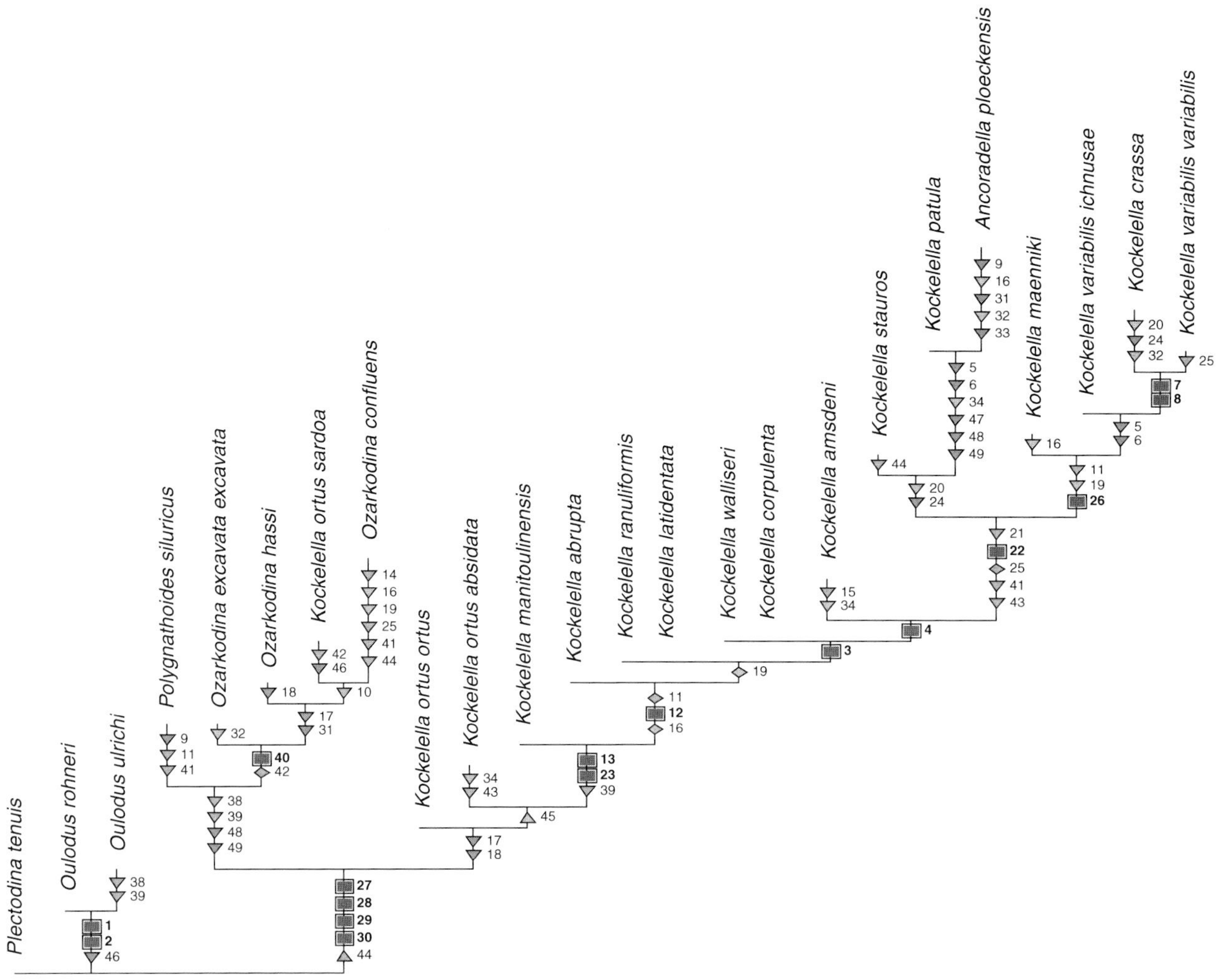

TEXT-FIG. 6. Character changes for the consensus of the nine optimal trees derived from stratigraphically unconstrained analysis. Rectangles represent synapomorphies, triangles represent homoplastic character changes (upward facing triangles reflect character changes that are reversed above, downward facing triangles represent character changes that are reversed outside the clade/taxon in question), diamonds represent homoplastic character changes above and outside; in all instances the number adjacent to the character symbol represents the number in the character descriptions.

are precluded on the identification of a single, homoplastic, autapomorphy of *K. ranuliformis*, the presence of a deeply excavated basal cavity extending beneath the 'posterior' process of the P_1 element (character 20).

Kockelella variabilis ichnusae has previously been identified as a descendant of *K. variabilis variabilis* (Serpagli and Corradini 1999) and within such a framework, a plausible case could also be made for *Kockelella maenniki*; such hypotheses of relationship would erase significant, uncorroborated, ghost lineages, but appear to be precluded by the absence of processes F and G from the P_1 elements of *K. maenniki* and *K. v. ichnusae*. These characters are acquired during the latest stages of ontogeny of the P_1 elements of *K. v. variabilis* (and its sister taxon, *Kockelella crassa*), after passing through a sequence common to the P_1 elements of *K. maenniki* and *K. v. ichnusae*. Thus, it is possible that the absence of these characters could be the result of secondary loss through paedomorphosis.

Either or all of these additional assumptions could be accepted in reconstructing the phylogeny of *Kockelella* and in attempting to assess the completeness of its fossil record. This would serve to bring the phylogenetic scheme into still closer accord with stratigraphic data than was indicated by the original X tree, at the cost of a few additional steps in the parsimony argument. After all, we acknowledge that evolution may proceed less frugally than strict parsimony. However, ghost lineages inferred elsewhere in the phylogeny of *Kockelella* cannot be as easily explained away and if we are to accept the existence of such gaps, it is perhaps more advisable and scientific to challenge our desire to tidy away such aberrations and to challenge their existence, by increasing the sampling den-

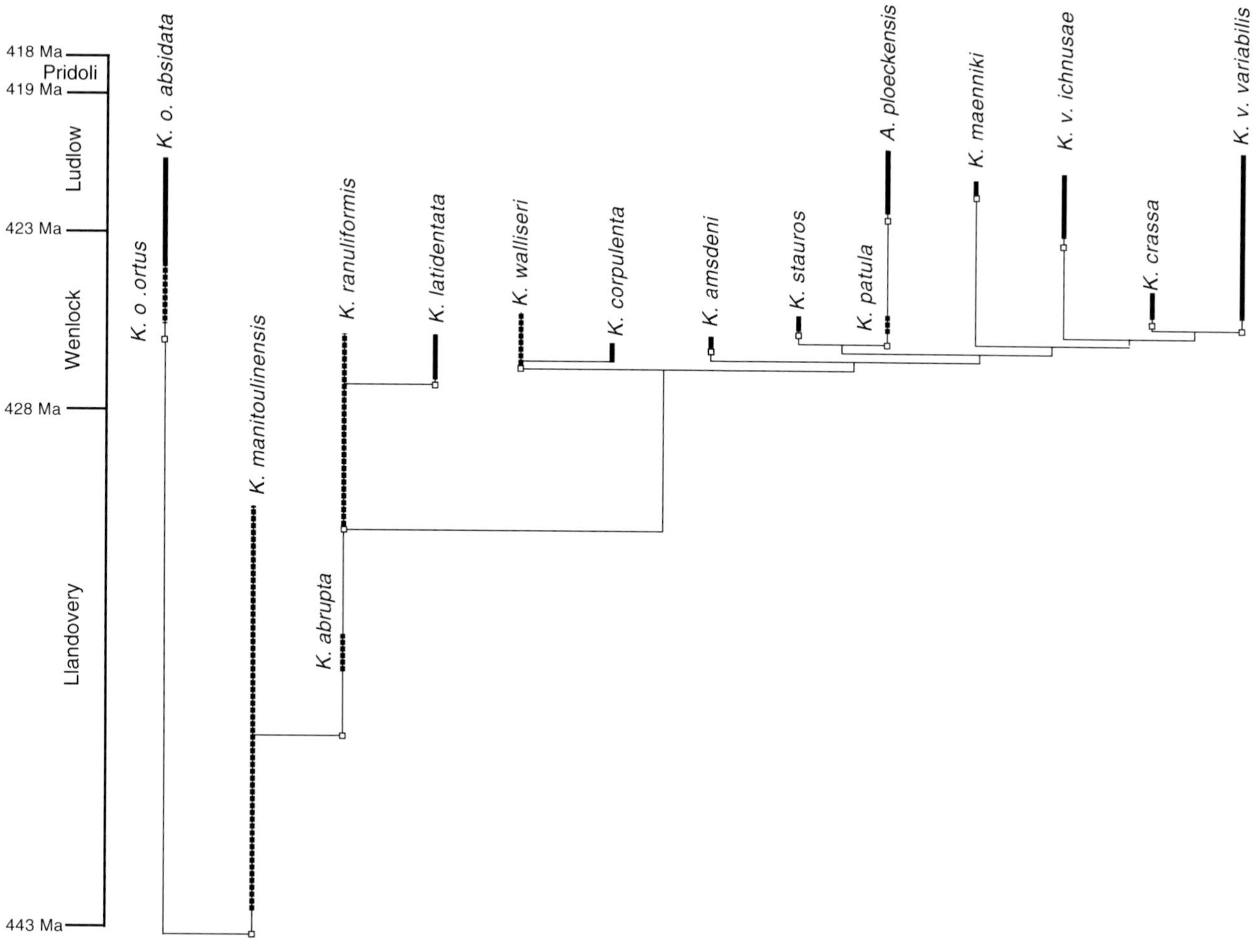

TEXT-FIG. 7. 'A'-tree for the Kockelellidae. Metataxa are denoted by coarse dashed lines; thin solid lines and boxes represent confidence intervals on the first appearance of a taxon.

sity of the fossil record, by challenging the cladistic hypothesis with full multielement reconstructions for those taxa whose apparatus is currently known only in part, and through more refined taxonomic analysis of identified metataxa.

Comparison with established phylogenies of the Kockelellidae

It is pertinent to consider the results of the phylogenetic analysis with respect to phylogenies derived hitherto following the practice of evolutionary palaeontology. Two attempts have been made to devise a phylogeny for *Kockelella*, first by Barrick and Klapper (1976), then subsequently developed more fully by Serpagli and Corradini (1999); because the latter encompasses the former we shall confine comparison to the Serpagli and Corradini (1999) phylogeny (Text-fig. 8).

The first point to note is that even the phylogeny devised by evolutionary palaeontological principles indicates that the fossil record of the Kockelellidae is incomplete. This is particularly obvious in the case of *Kockelella crassa* and its proposed descendants, and *Kockelella patula* and its proposed ancestor and descendant. These inferred gaps notwithstanding, the Serpagli and Corradini (1999) phylogeny requires far fewer gaps in the known fossil record of the Kockelellidae than do either the X- or the A-trees presented herein.

Greater stratigraphic congruence, but at what price? To constrain the impact of increased stratigraphic congruence upon morphological evolution, we devised a cladogram compatible with the Serpagli and Corradini (1999) phylogeny (Text-fig. 9A) and implemented it as a constraint on the analysis of the cladistic dataset, with the aim of finding the shortest compatible trees. In doing so we excluded *K. crassa* from the constraint tree because Serpagli and Corradini (1999) were com-

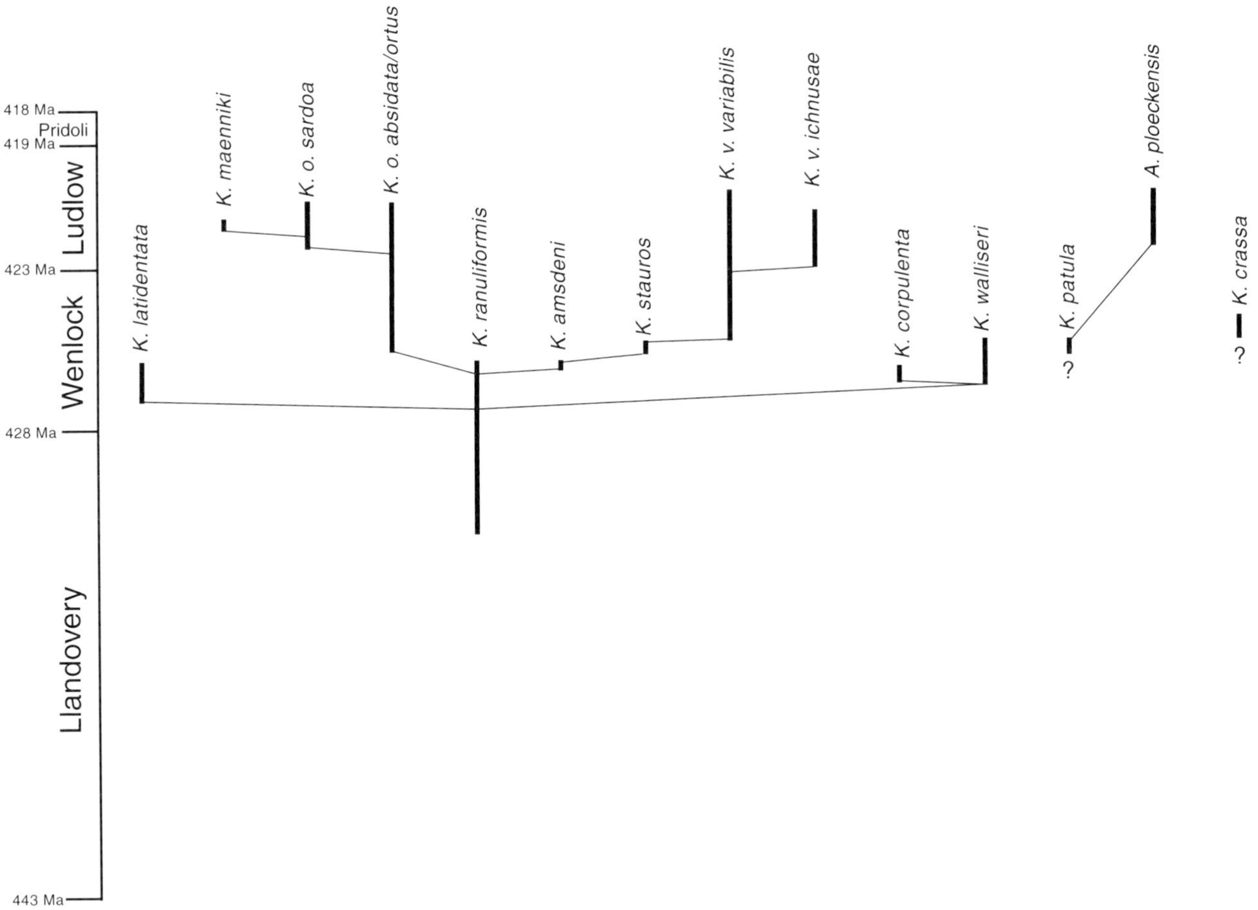

TEXT-FIG. 8. Phylogeny for the Kockelellidae according to Serpagli and Corradini (1999), stratigraphic ranges of certain taxa corrected and plotted according to a revised timescale.

pletely unable to resolve its relationships to any other formal member of the Kockelellidae. Thus, during the subsequent constraint analysis, resolution of the interrelationhips of *K. crassa* is unconstrained; this diminishes the severity of the morphological-congruence test of the Serpagli and Corradini (1999) phylogeny. *Kockelella patula* and *Ancoradella ploeckensis* were placed at the base of the constraint tree to reflect the view that they represent a 'different stock' to the remainder of the Kockelellidae (Serpagli and Corradini 1999, p. 277). Constrained analysis of the dataset produced nine equally most-parsimonious trees at 135 steps (CIe 0·3630; RI 0·5612; RC 0·2037; the strict consensus is presented in Text-fig. 9B). This is 27 per cent longer than the most-parsimonious trees derived from the unconstrained analysis; it is also a good deal less parsimonious than the *ad hoc* solutions to greater stratigraphic congruence discussed above. These trees do perform marginally better in all stratigraphic congruence metrics (all trees achieving significant congruence in all metrics; Table 2; GHOST file presented in Appendix, section 1), although this is not altogether surprising given that they were constructed in large part on the basis of stratigraphic data.

To put the disparity in tree length into perspective, we can compare differences in the levels of homoplasy in morphological evolution as implied by the constrained and unconstrained most-parsimonious solutions to the levels expected of a dataset containing the same number of taxa. Based on an empirical study, Sanderson and Donoghue (1989) resolved that there is a positive relationship between the number of taxa included in a cladistic dataset and the consistency index, which is the reciprocal of inferred levels of homoplasy. The expected consistency index can be calculated as follows:

$$\mathrm{CI} = 0\cdot 90 - 0\cdot 022\,(\text{number of taxa}) + 0\cdot 000213\,(\text{number of taxa})^2.$$

For the cladistic dataset presented herein the expected consistency index value is 0·494688. The value of the consistency index for the most-parsimonious trees derived

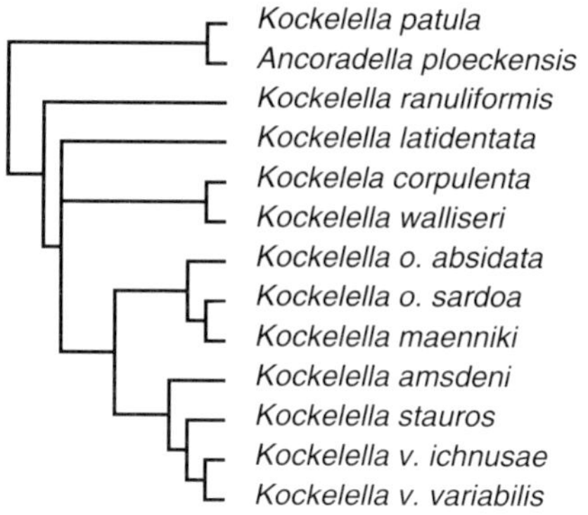

A

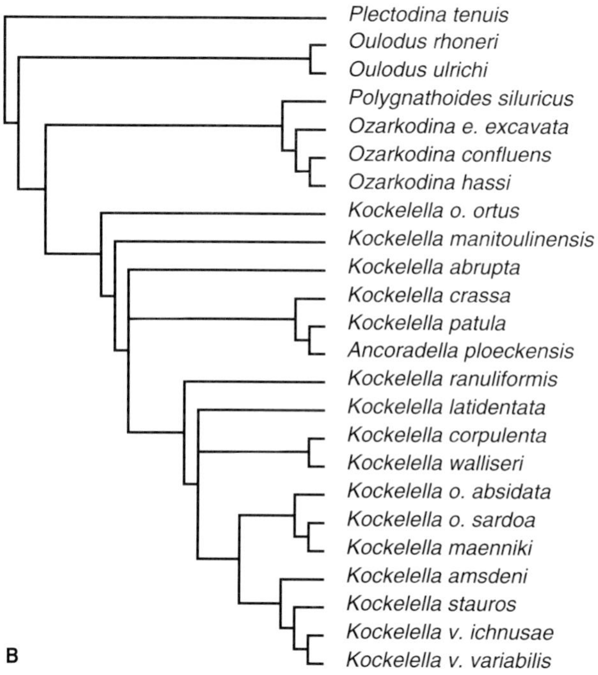

B

TEXT-FIG. 9. A, constraint tree compatible with the Serpagli and Corradini (1999) kockelellid phylogeny. B, strict consensus of the nine optimal trees compatible with the constraint tree and according to the same data matrix utilized in the stratigraphically unconstrained analyses.

from the unconstrained analysis is slightly less than this expected value (CI 0·4667), but it falls well within the range of variation seen in the dataset upon which Sanderson and Donoghue's regression line is derived. In contrast, the consistency index value for the most-parsimonious trees from the constrained analysis (CI 0·3630) is considerably lower than the expected value, falling outside the envelope delimited by the Sanderson and Donoghue (1989) dataset.

This all appears to bear out a perception among evolutionary palaeontologists that homoplasy is rife, particularly in groups utilized in biostratigraphy, and a major limitation on phylogenetic inference (e.g. Lazarus and Prothero 1984). However, levels of inferred homoplasy are contingent upon a particular hypothesis of relationships and it is just as likely that the prevalence of homoplasy is not so great, that many inferred homoplasies are misidentified homologies, precluded from interpretation as such by a misplaced reliance upon the *a priori* use of stratigraphic constraint in phylogenetic reconstruction.

Is the fossil record of conodonts any more or less complete than any other fossil group?

At the outset of this analysis we observed that the conodont fossil record is generally considered sufficiently complete for stratigraphic data to be not just useful but vital guides to reconstructing evolutionary relationships. In the absence of any other supporting evidence, this implicit assumption must be based in biostratigraphic experience; indeed it is an assumption common to all fossil groups that are exploited for their biostratigraphic utility. The fossil record of groups that lack biostratigraphic utility, such as most vertebrate clades, is generally considered with considerable circumspection and the drive for an atemporal approach to phylogenetic reconstruction has been most forcibly argued for by specialists in these fields. But, using the same atemporal approaches to phylogenetic reconstruction, are the fossil records of groups that are biostratigraphically useful, such as conodonts, any better than those that are not? Of course, our analysis represents just one small clade that may hardly be representative of conodonts let alone all such groups exploited for their biostratigraphical utility. Nevertheless, it is interesting to compare the stratigraphic congruence metrics derived from cladistic analysis of this entire clade to the average results derived from the fossil record. These results compare favourably to an average of stratigraphic congruence analyses, based on 1000 cladograms derived from a full range of fossil groups (Benton *et al.* 2000); the SCI of our unconstrained analysis is slightly lower than the average (0·42–0·5 vs. 0·55), while the GER of our unconstrained analysis is significantly higher than the average (0·78–0·8 vs. 0·56). The RCI of our unconstrained analysis deviates significantly from the fossil record average (−49 to −59 vs. +31); the reason for this lies, almost certainly, with the number of metataxa included in our analysis, which introduce ghost lineage artefacts, and also contribute to a depressed SCI, as discussed above. The disparity in results is based, more than anything, in the disparity in the taxonomic scale of our analysis versus those included in the study by Benton *et al.* (2000). Although, metataxa are not the preserve of fossil species (Smith 1994), they have been more readily identified among higher taxa, and revised. Thus, analyses of higher taxa, and of distantly related lower taxa, are much less likely to be prone to the kinds of analytical artefacts discussed herein.

CONCLUDING DISCUSSION

We hope to have shown that the fossil record of the conodont clade *Kockelella* is incomplete. This has been demonstrated through the use of phylogenetic (cladogram-stratigraphy congruence) and aphylogenetic (confidence interval) approaches to assessing the completeness of the fossil record. Given the close temporal proximity in the origination of many taxa included in these analyses, both approaches indicate that stratigraphic data are a poor guide to relationships. Nevertheless, the two approaches reach conclusions concerning the extent to which the fossil record of the Kockelellidae is incomplete, which differ by as much as an order of magnitude. It is possible that this discrepancy arises because aphylogenetic methods assess only the quality of the record of a known taxon, while phylogenetic methods assess this and also allow for the existence of unknown taxa and for known taxa of unrecognized significance. However, we have resolved that much of the discrepancy arises from the use of raw cladograms, calibrated to time and stratigraphy (so-called 'X-trees'), but without further qualification, in assessing the completeness of the fossil record. This is a problem because many fossil taxa are metataxa (Smith 1994) that, if left unrecognized and unconsidered, can lead to spurious inference of missing fossil record. The consideration of metataxa as paraphyletic assemblages of hitherto unrecognized component monophyletic taxa and, potentially, ancestors, leads to the conversion of X-trees to A-trees, which are equivalent to phylogenies in the general sense. Confidence intervals can then be used to determine the likely time of origination of descendants from among metataxa, therein controlling for incompleteness in the record.

By implication, clade-stratigraphy congruence metrics are handicapped if they fail to entertain the presence of metataxa. Indeed, in light of the potential artefact introduced by metataxa, the utility of such metrics is reduced to assessing the relative merits of competing cladograms, as the implied ghost lineages of X-trees will vastly exceed that implied by A-trees. However, although metataxa are not the preserve of species (Smith 1994), metataxa of higher taxonomic status are, in practice, more readily identifiable because their paraphyletic nature is implicit in their definition and, as such, they are usually removed or remedied *a priori* to preclude the introduction of the artefact in cladistic analyses. Thus it is unlikely that most assessments of fossil record quality that entertain higher taxa, or consider only distantly related species as proxies for higher taxa, are subject to this kind artefact. This equates well with the observation that species-level cladograms of fossil taxa exhibit significantly poorer correlation to stratigraphic data than do cladograms of higher taxa (Benton and Storrs 1994). Comprehensive species-level analyses of fossil clades are relatively uncommon, at present, but those that have been undertaken with the aim of assessing fossil record quality should be re-evaluated in this light. This is particularly so in instances where a high inferred fossil record quality is relied upon for accessory theories, such as the fossil record of the synapsid–diapsid divergence and its widespread use in molecular clock calibration.

Finally, we hope to have shown that the commonplace criticisms of cladistics are misplaced, conflate theory with practice, and character analysis with phylogeny reconstruction. Cladistic analysis is not incompatible with recognition of ancestor–descendant lineages, it is just that for most practical purposes such hypotheses of relationships are a superfluous inferential step. Clearly then, cladistic analysis does not prescribe cladogenesis as the sole mechanism of evolutionary change; in fact it does not prescribe any pattern of evolutionary change as it is not the objective of cladistic analysis to reveal anything concerning evolutionary processes, although these may be inferred through the construction of phylogenies – which are abstractions of cladograms. Furthermore, cladistic analyses do not preclude the use of stratigraphic data in phylogeny reconstruction; whether stratigraphic data are employed or eschewed in the process of phylogenetic reconstruction is quite a distinct matter. Only cladistics provides a framework within which the consequences of increased stratigraphic congruence upon diminished comparative anatomy can be readily determined and weighed. Above all else, cladistics is not (necessarily) incompatible with the evolutionary palaeontological approach to phylogeny reconstruction, in any of its many guises. Rather, it provides the prescriptive method of character analysis that evolutionary palaeontology traditionally lacks.

Acknowledgements. We express our gratitude to the following: Peter Roopnarine (California Academy of Sciences, USA) and Dick Aldridge (Leicester, USA), who provided constructively critical and insightful reviews of the manuscript, and the latter also for access to published and unpublished collections from the Silurian of Ireland, Bohemia and the Welsh Borderlands; Jim Barrick (Texas Tech, USA) and Julie Golden (Iowa, USA), who provided access to the collections on which Barrick and Klapper (1976) is based; Lennart Jeppsson (Lund, Sweden), who provided access to unpublished collections from the Silurian of Gotland; Matthew Wills (Bath, UK), who provided assistance with GHOST analyses. This work was funded through a University of Birmingham Studentship (to LMW) and NERC Research Fellowship GT5/99/ES/5 (to PCJD).

REFERENCES

ALDRIDGE, R. J. 1972. Llandovery conodonts from the Welsh borderland. *Bulletin of the British Museum (Natural History) Geology*, **22**, 125–231.

—— 1980. Notes on some Silurian conodonts from Ireland. *Irish Journal of Earth Science*, **3**, 127–132.

—— 1985. Conodonts of the Silurian System from the British Isles. 68–92. *In* AUSTIN, R. L. and HIGGINS, A. C. (eds). *A stratigraphical index of conodonts*. Ellis Horwood Ltd, Chichester, 263 pp.

—— and MABILLARD, J. E. 1981. Local variations in the distribution of Silurian conodonts: an example from the *amorphognathoides* interval of the Welsh Basin. 10–17. *In* NEALE, J. and BRASIER, M. (eds). *Microfossils from Recent and fossil shelf seas*. Ellis Horwood Ltd, Chichester, 296 pp.

—— and MOHAMED. 1982. Conodont biostratigraphy of the Early Silurian of the Oslo region. 109–120. *In* WORSLEY, D. (ed.). *Subcommisson on Silurian Stratigraphy Field Meeting, Oslo Region.*

—— and SMITH, M. P. 1993. Conodonta. 563–572. *In* BENTON, M. J. (ed.). *The fossil record 2*. Chapman & Hall, London, 845 pp.

—— PURNELL, M. A., GABBOTT, S. E. and THERON, J. N. 1995. The apparatus architecture and function of *Promissum pulchrum* Kovács-Endrödy (Conodonta, Upper Ordovician), and the prioniodontid plan. *Philosophical Transactions of the Royal Society of London, Series B*, **347**, 275–291.

ARMSTRONG, H. A. 1990. Conodonts from the Upper Ordovician – Lower Silurian carbonate platform of North Greenland. *Grønlands Geologiske Undersøgelse, Bulletin*, **159**, 1–151.

BARCA, S., FERRETTI, A., MASSA, P. and SERPAGLI, E. 1992. The Hercynian arburese tectonic unit of SW Sardinia: new stratigraphic and structural data. *Rivista Italiana di Paleontologia e Stratigrafia*, **98**, 119–136.

BARNETT, S. G. 1971. Biometric determination of the evolution of *Spathognathodus remscheidensis*: a method for precise intrabasinal time correlations in the northern Appalachians. *Journal of Paleontology*, **45**, 274–300.

—— 1972. The evolution of *Spathognathodus remscheidensis* in New York, New Jersey, Nevada, and Czechoslovakia. *Journal of Paleontology*, **46**, 900–917.

BARRICK, J. E. 1983. Wenlockian (Silurian) conodont biostratigraphy, biofacies, and carbonate lithofacies, Wayne Formation, central Tennessee. *Journal of Paleontology*, **57**, 208–239.

—— and KLAPPER, G. 1976. Multielement Silurian (late Llandoverian–Wenlockian) conodonts of the Clarita Formation, Arbuckle Mountains, Oklahoma, and phylogeny of *Kockelella*. *Geologica et Palaeontologica*, **10**, 59–100.

—— and MÄNNIK, P. 2005. Silurian conodont biostratigraphy and palaeobiology in stratigraphic sequences. 103–116. *Special Papers in Palaeontology*, **73**, 218 pp.

BENFRIKA, M. 1999. Some upper Silurian – middle Devonian conodonts from the northern part of Western Meseta of Morocco: systematic and palaeogeographical relationships. *Bollettino della Società Paleontologica Italiana*, **37**, 311–319.

BENTON, M. J. and STORRS, G. W. 1994. Testing the quality of the fossil record: paleontological knowledge is improving. *Geology*, **22**, 111–114.

—— HITCHIN, R. and WILLS, M. A. 1999. Assessing congruence between cladistic and stratigraphic data. *Systematic Biology*, **48**, 581–596.

—— WILLS, M. A. and HITCHIN, R. 2000. Quality of the fossil record through time. *Nature*, **403**, 534–537.

BISCHOFF, G. C. O. 1986. Early and Middle Silurian conodonts from midwestern New South Wales. *Courier Forschungsinstitut Senckenberg*, **89**, 1–337.

BRANSON, E. B. and MEHL, M. G. 1933. Conodont studies, number 1. *University of Missouri Studies*, **8**, 1–72.

BRETSKY, S. S. 1979. Recognition of ancestor–descendant relationships in invertebrate paleontology. 113–163. *In* ELDREDGE, N. and CRACRAFT, J. (eds). *Phylogenetic analysis and paleontology*. Columbia University Press, New York, 233 pp.

COCKLE, P. 1999. Conodont data in relation to time, space and environmental relationships in the Silurian (Late Llandovery–Ludlow) succession at Boree Creek (New South Wales, Australia). *Abhandlugen der Geologischen Bundesanstalt*, **54**, 107–133.

COOPER, R. A. and LINDHOLM, K. 1990. A precise worldwide correlation of early Ordovician graptolite sequences. *Geological Magazine*, **127**, 497–525.

CORRADINI, C., FERRETTI, A., SERPAGLI, E. and BARCA, S. 1998*a*. Stop 1.3 The Ludlow–Pridoli section Genna Ciuerciu, west of Silius. *Geologia*, **60**, 112–118.

—— —— —— 1998*b*. Stop 3.3 Wenlock and Pridoli conodonts from Argiola, east of Domusnovas. *Geologia*, **60**, 194–198.

DONOGHUE, M. J., DOYLE, J., GAUTHIER, J., KLUGE, A. and ROWE, T. 1989. The importance of fossils in phylogeny reconstruction. *Annual Review of Ecology and Systematics*, **20**, 431–460.

DONOGHUE, P. C. J. 2001. Conodonts meet cladistics: recovering relationships and assessing the completeness of the conodont fossil record. *Palaeontology*, **44**, 65–93.

—— FOREY, P. L. and ALDRIDGE, R. J. 2000. Conodont affinity and chordate phylogeny. *Biological Reviews*, **75**, 191–251.

—— SMITH, M. P. and SANSOM, I. J. 2003. The origin and early evolution of chordates: molecular clocks and the fossil record. 190–223. *In* DONOGHUE, P. C. J. and SMITH, M. P. (eds). *Telling the evolutionary time: molecular clocks and the fossil record*. CRC Press, London, 296 pp.

DRYGANT, D. M. 1969. Konodonty restevskogo, kitaigorodskogo i miskshingskogo gorizontov silura Podolii. *Palaeontologicheskii Sbornik*, **6**, 49–55.

—— 1971. Konodontova zona *Spathognathodus crispus* i vik skal's'koho horyzontu (Sylur Volyno Podillia). *Akademiya Nauk Ukrainskoi RSR, Dopovidi, Seriya B, Geologiya, Khimiya ta Biologiya*, **33**, 780–783.

—— 1984. *Correlation and the conodonts of Silurian and Devonian deposits of the Volyn-Podolia*. Izdatelstvo Nauk Dumka, Kiev, 210 pp.

DZIK, J. 1991. Features of the fossil record of evolution. *Acta Palaeontologica Polonica*, **36**, 91–113.

—— 1994. Conodonts of the Mójcza Limestone. *Palaeontologia Polonica*, **53**, 43–128.

—— 1997. Emergence and succession of Carboniferous conodont and ammonoid communities in the Polish part of the Variscan Sea. *Acta Palaeontologica Polonica*, **42**, 57–170.

—— 1999. Relationship between rates of speciation and phyletic evolution: stratophenetic data on pelagic conodont chordates and benthic ostracods. *Geobios*, **32**, 205–221.

—— 2002. Emergence and collapse of the Frasnian conodont and ammonoid communities in the Holy Cross Mountains. *Acta Palaeontologica Polonica*, **47**, 565–650.

—— 2005. The chronophyletic approach: stratophenetics facing an incomplete fossil record. 159–183. *Special Papers in Palaeontology*, **73**, 218 pp.

—— and TRAMMER, J. 1980. Gradual evolution of conodontophorids in the Polish Triassic. *Acta Palaeontologica Polonica*, **25**, 55–89.

ELDREDGE, N. 1979. Cladism and common sense. 165–198. *In* ELDREDGE, N. and CRACRAFT, J. (eds). *Phylogenetic analysis and paleontology*. Columbia University Press, New York, 233 pp.

ENGELMANN, G. F. and WILEY, E. O. 1977. The place of ancestor–descendant relationships in phylogeny reconstruction. *Systematic Zoology*, **26**, 1–11.

ETHINGTON, R. L. and FURNISH, W. M. 1959. Ordovician conodonts from northern Manitoba. *Journal of Paleontology*, **33**, 540–546.

—— and FURNISH, W. M. 1962. Silurian and Devonian conodonts from the Spanish Sahara. *Journal of Paleontology*, **36**, 1253–1290.

FÅHRAEUS, L. E. 1969. Conodont zones in the Ludlovian of Gotland and correlation with Great Britain. *Sveriges Geologiska Undersökning*, **63**, 1–33.

FARRIS, J. S. 1969. A successive approximations approach to character weighting. *Systematic Zoology*, **18**, 374–385.

FERRETTI, A., CORRADINI, C. and SERPAGLI, E. 1998. Stop 2.2 Wenlock–Ludlow conodonts from Perd'e Fogu (Fluminimaggiore). *Geologia*, **60**, 156–167.

FLAJS, G. 1967. Conodontenstratigraphische Untersuchungen im Raum von Eisenerz, Nordliche Grauwackenzone. *Mitteilungen der Geologischen Gesellschaft in Wien*, **59**, 157–212.

—— and SCHÖNLAUB, H. P. 1976. Die biostratigraphie Gliederung des Altpaläozoikums am Polster bei Eisenerz (Nödliche Grauwackenzone, Österreich). *Verhandlungen der Geologischen Bundesanstalt Österreich*, **2**, 257–303.

FOOTE, M. 1996. On the probability of ancestors in the fossil record. *Paleobiology*, **22**, 141–151.

FORDHAM, B. G. 1991. A literature-based phylogeny and classification of Silurian conodonts. *Palaeontographica Abteilung A*, **217**, 1–136.

FOREY, P. L. and KITCHING, I. J. 2000. Experiments in coding multistate characters. 54–80. *In* SCOTLAND, R. W. and PENNINGTON, R. T. (eds). *Homology and systematics: coding characters for phylogenetic analysis*. Taylor & Francis, London, 232 pp.

FORTUNATO, H. 1998. Reconciling observed patterns of temporal occurrence with cladistic hypotheses of phylogenetic relationship. *American Malacological Bulletin*, **14**, 191–200.

GINGERICH, P. 1979. Paleontology, phylogeny, and classification: an example from the mammalian fossil record. *Systematic Zoology*, **28**, 451–464.

GOLOBOFF, P. A. 1993. Estimating character weights during tree search. *Cladistics*, **9**, 83–91.

HAWKINS, J. A. 2000. A survey of homology assessment: different botanists perceive and define characters in different ways. 22–53. *In* SCOTLAND, R. W. and PENNINGTON, R. T. (eds). *Homology and systematics: coding characters for phylogenetic analysis*. Taylor & Francis, London, 232 pp.

—— HUGHES, C. E. and SCOTLAND, R. W. 1997. Primary homology assessment, characters and character states. *Cladistics*, **13**, 275–283.

HELFRICH, C. T. 1975. Silurian conodonts from Wills Mountain Anticline, Virginia, West Virginia, and Maryland. *Geological Society of America, Special Paper*, **161**, 1–198.

—— 1980. Late Llandovery–Early Wenlock conodonts from the upper part of the Rose Hill and basal part of the Mifflintown formations, Virginia, West Virginia and Maryland. *Journal of Paleontology*, **54**, 557–569.

HITCHIN, R. and BENTON, M. J. 1997. Stratigraphic indices and tree balance. *Systematic Biology*, **46**, 563–569.

HOLLAND, S. M. 1995. The stratigraphic distribution of fossils. *Paleobiology*, **21**, 92–109.

—— 2000. The quality of the fossil record: a sequence stratigraphic perspective. *Paleobiology*, **26** (Supplement), 148–168.

HUELSENBECK, J. P. 1994. Comparing the stratigraphic record to estimates of phylogeny. *Paleobiology*, **20**, 470–483.

IGO, H. and KOIKE, T. 1967. Ordovician and Silurian conodonts from the Langkawi Islands, Malaya. *Contributions to the Geology and Paleontology of Southeastern Asia*, **3**, 1–29.

—— —— 1968. Ordovician and Silurian conodonts from the Langkawi Islands, Malaya. *Contributions to the Geology and Paleontology of Southeastern Asia*, **4**, 1–21.

JENNER, R. A. 2002. Boolean logic and character state identity: pitfalls of character coding in metazoan cladistics. *Contributions to Zoology*, **71**, 67–91.

JEPPSSON, L. 1974. Aspects of late Silurian conodonts. *Fossils and Strata*, **6**, 1–54.

—— 1979. Conodonts. *Sveriges Geologiska Undersökning*, **C 762**, 225–248.

—— 1983. Silurian conodont faunas from Gotland. *Fossils and Strata*, **15**, 121–144.

—— 1997. A new latest Telychian, Sheinwoodian and Early Homerian (Early Silurian) standard conodont zonation. *Transactions of the Royal Society of Edinburgh (Earth Sciences)*, **88**, 91–114.

—— and ALDRIDGE, R. J. 2000. Ludlow (late Silurian) oceanic episodes and events. *Journal of the Geological Society, London*, **157**, 1137–1148.

—— and CALNER, M. 2003. The Silurian Mulde Event and a scenario for secundo–secundo events. *Transactions of the Royal Society of Edinburgh (Earth Sciences)*, **93**, 135–154.

—— —— and DORNING, K. J. 1995. Wenlock (Silurian) oceanic episodes and events. *Journal of the Geological Society, London*, **152**, 487–498.

—— VIIRA, V. and MÄNNIK, P. 1994. Silurian conodont-based correlations between Gotland (Sweden) and Saaremaa (Estonia). *Geological Magazine*, **131**, 201–218.

JOHNSON, M. E., RONG, J. Y., WANG, C. Y. and WANG, P. 2001. Continental island from the Upper Silurian (Ludfordian Stage) of Inner Mongolia: implications for eustasy and paleogeography. *Geology*, **29**, 955–958.

KITCHING, I. J., FOREY, P. L., HUMPHRIES, C. J. and WILLIAMS, D. M. 1998. *Cladistics: the theory and practice of parsimony analysis*. Oxford University Press, Oxford, 228 pp.

KLAPPER, G. and MURPHY, M. A. 1974. Silurian–lower Devonian conodont sequence in the Roberts Mountains Formation of central Nevada. *University of California Publications in Geological Sciences*, **111**, 1–62.

—— and PHILIP, G. M. 1971. Devonian conodont apparatuses and their vicarious skeletal elements. *Lethaia*, **4**, 429–452.

—— LINDSTRÖM, M., SWEET, W. C. and ZIEGLER, W. 1981. *Catalogue of conodonts*, Volume iv. E. Schweizerbart'sche Verlagsbuchhandlung, Berlin, 445 pp.

KLEFFNER, M. A. 1985. Conodont biostratigraphy of the Stray 'Clinton' and 'Packer Shell' (Silurian, Ohio subsurface) and its bearing on correlation. 219–230. *In* GRAY, J., MASLOWSKI, A., McCULLOUGH, W. and SHAFER, W. E. (eds). *The New Clinton Collection – 1985*. The Ohio Geological Society, Columbus.

KLEFFNER, M. A. 1987. Conodonts of the Estill Shale and Bisher Formation (Silurian, southern Ohio): biostratigraphy and distribution. *Ohio Journal of Science*, **87**, 78–89.

—— 1990. Wenlockian (Silurian) conodont biostratigraphy, depositional environments, and depositional history along the eastern flank of the Cincinnati Arch in Southern Ontario. *Journal of Paleontology*, **64**, 319–328.

—— 1991. Conodont biostratigraphy of the upper part of the Clinton Group and the Lockport Group (Silurian) in the Niagara Gorge region, New York and Ontario. *Journal of Paleontology*, **65**, 500–511.

—— 1994. Conodont biostratigraphy and depositional history of strata comprising the Niagaran sequence (Silurian) in the northern part of the Cincinnati Arch region, west–central Ohio, and evolution of *Kockelella walliseri* (Helfrich). *Journal of Paleontology*, **68**, 141–153.

—— 1995. A conodont- and graptolite-based Silurian chronostratigraphy. 159–176. *In* MANN, K. O. and LANE, H. R. (eds). *Graphic correlation*. SEPM Special Publication, **53**, 263 pp.

KOCKEL, F. 1958. Conodonten aus dem Paläozoikum von Malaga (Spanien). *Neues Jahrbuch für Geologie und Paläontologie, Monatschefte*, **1958**, 255–263.

KRIZ, J., JAEGER, H., PARIS, F. and SCHÖNLAUB, H. P. 1986. Pridoli – the fourth subdivision of the Silurian. *Jahrbuch der Geologischen Bundesanstalt*, **129**, 291–360.

LANE, H. R. and ORMISTON, A. R. 1979. Siluro-Devonian biostratigraphy of the Salmontrout River area, east-central Alaska. *Geologica et Palaeontologica*, **13**, 39–96.

LAZARUS, D. B. and PROTHERO, D. R. 1984. The role of stratigraphic and morphologic data in phylogeny. *Journal of Paleontology*, **58**, 163–172.

LEE, D. C. and BRYANT, H. N. 1999. A reconsideration of the coding of inapplicable characters: assumptions and problems. *Cladistics*, **15**, 373–378.

LEHNERT, O., MILLER, J. F., LESLIE, S. A., REPETSKI, J. E. and ETHINGTON, R. L. 2005. Cambro-Ordovician sea-level fluctuations and sequence boundaries: the missing record and the evolution of new taxa, 117–134. *Special Papers in Palaeontology*, **73**, 218 pp.

LIEBE, R. M. and REXROAD, C. B. 1977. Conodonts from Alexandrian and early Niagaran rocks in the Joliet, Illinois, area. *Journal of Paleontology*, **51**, 844–857.

LIN BAO-YU and QIU HONG-RONG 1983. The Silurian System of Xizang (Tibet). *Contributions to the Geology of Qinghai–Xizang Plateau*, **8**, 25–36.

LINK, A. G. and DRUCE, E. C. 1972. Ludlovian and Gedinnian conodont stratigraphy of the Yass Basin, New South Wales. *Bureau of Mineral Resources, Bulletin*, **134**, 1–136.

LOYDELL, D. K., KALJO, D. and MÄNNIK, P. 1998. Integrated biostratigraphy of the lower Silurian of the Ohesaare core, Saaremaa, Estonia. *Geological Magazine*, **135**, 769–783.

—— MULLINS, G. L., MÄNNIK, P., MIKULIC, D. G. and KLUESSENDORF, J. 2002. Biostratigraphical dating of the Thornton Fossil Konservat-Lagerstätte, Silurian, Illinois, USA. *Geological Journal*, **37**, 269–278.

MABILLARD, J. E. and ALDRIDGE, R. J. 1983. Conodonts from the Coralliferous Group (Silurian) of Marloes Bay, south-west Dyfed, Wales. *Geologica et Palaeontologica*, **17**, 29–43.

MADDISON, W. and MADDISON, D. R. 2002. MacClade, version 4.05: analysis of phylogeny and character evolution. Program and Documentation. Sinauer Associates, Inc., Sunderland, MA, 404 pp.

MÄNNIK, P. 1983. Silurian conodonts from Severnaya Zemlya. *Fossils and Strata*, **15**, 111–119.

—— 1998. Evolution and taxonomy of the Silurian conodont *Pterospathodus*. *Palaeontology*, **41**, 1001–1050.

—— 2002. Conodonts in the Silurian of Severnaya Zemlya and Sedov archipelagos (Russia), with special reference to the genus *Ozarkodina* Branson and Mehl, 1933. *Geodiversitas*, **24**, 77–97.

—— and MALKOWSKI, K. 1998. Silurian conodonts from the Goldap Core, Poland. Proceedings of the Sixth European Conodont Symposium (ECOS VI). *Palaeontologica Polonica*, **58**, 141–151.

MARSHALL, C. R. 1990. Confidence intervals on stratigraphic ranges. *Paleobiology*, **16**, 1–10.

—— 1994. Confidence-intervals on stratigraphic ranges – partial relaxation of the assumption of randomly distributed fossil horizons. *Paleobiology*, **20**, 459–469.

—— 1997. Confidence intervals on stratigraphic ranges with non-random distribution of fossil horizons. *Paleobiology*, **23**, 165–173.

MASHKOVA, T. V. 1972. *Ozarkodina steinhornensis* (Ziegler) apparatus, its conodonts and biozone. *Geologica et Palaeontologica*, **1**, 81–90.

—— 1977. New conodonts of the *amorphognathoides* Zone from the Lower Silurian of Podolia. *Paleontological Journal*, **11**, 513–517.

—— 1979. New Silurian conodonts from central Siberia. *Paleontological Journal*, **13**, 215–221.

McCRACKEN, A. D. 1991. Taxonomy and biostratigraphy of Llandovery (Silurian) conodonts in the Canadian Cordillera, northern Yukon Territory. *Geological Survey of Canada, Bulletin*, **417**, 65–95.

—— and BARNES, C. R. 1981. Part 2: conodont biostratigraphy and paleoecology of the Ellis Bay Formation, Anticosti Island, Quebec, with special reference to late Ordovician – early Silurian chronostratigraphy and the systematic boundary. *Geological Survey of Canada, Bulletin*, **329**, 50–134.

MILLER, R. H. 1972. Silurian conodonts from the Llano Region, Texas. *Journal of Paleontology*, **46**, 556–564.

MURPHY, M. A. and CEBECIOGLU, M. K. 1986. Statistical study of *Ozarkodina excavata* (Branson and Mehl) and *O. tuma* Murphy and Matti (Lower Devonian, *delta* Zone conodonts, Nevada). *Journal of Paleontology*, **60**, 865–869.

—— —— 1987. Morphometric study of the genus *Ancyrodelloides* (Lower Devonian, conodonts), central Nevada. *Journal of Paleontology*, **61**, 583–594.

—— and VALENZUELA-RÍOS, J. I. 1999. *Lanea* new genus, lineage of Early Devonian conodonts. *Bollettino della Società Paleontologica Italiana*, **37**, 321–334.

NELSON, G. 1989. Species and taxa: systematics and evolution. 60–81. *In* OTTO, D. and ENDLER, J. A. (eds). *Speciation and its consequences*. Sinauer, Sunderland, MA, 679 pp.

NI YU-NAN, CHEN TING-EN, CAI CHONG-YANG, LI GUO-HUA, DUAN YAN-XUE and WANG JU-DE 1982. The Silurian rocks of western Yunnan. *Acta Palaeontologica Sinica*, **21**, 119–132.

NICOLL, R. S. 1985. Multielement composition of the conodont species *Polygnathus xylus xylus* Stauffer, 1940 and *Ozarkodina brevis* (Bischoff and Ziegler, 1957) from the Upper Devonian of the Canning Basin, Western Australia. *Bureau of Mineral Resources, Journal of Australian Geology and Geophysics*, **9**, 133–147.

—— and REXROAD, C. B. 1968. Stratigraphy and conodont paleontology of the Salamonie Dolomite and Lee Creek Member of the Brassfield Limestone (Silurian) in southeastern Indiana and adjacent Kentucky. *Indiana Geological Survey, Bulletin*, **40**, 1–73.

—— —— 1987. Re-examination of Silurian conodont clusters from northern Indiana. 49–61. *In* ALDRIDGE, R. J. (ed.). *Palaeobiology of conodonts*. Ellis Horwood, Chichester, 180 pp.

NORELL, M. A. 1992. Taxic origin and temporal diversity: the effect of phylogeny. *In* NOVACEK, M. J. and WHEELER, Q. D. (eds). *Extinction and phylogeny*. Columbia University Press, New York, 253 pp.

NORFORD, B. S. and ORCHARD, M. J. 1985. Early Silurian age of rocks hosting lead–zinc mineralization at Howards Pass, Yukon Territory and District of Mackenzie, local biostratigraphy of Road River and Earn Goup. *Geological Survey of Canada, Paper*, **83-11**, 1–35.

—— NOWLAN, G. S., HAIDL, F. M. and BEZYS, R. K. 1998. The Ordovician-Silurian Boundary interval in Saskatchewan and Manitoba. 27–45. *In* CHRISTOPHER, J. E., GILBOY, C. F., PATERSON, D. F. and BNED, S. L. (eds). *Eighth International Williston Basin Symposium*. Saskatchewan Geological Survey, Special Publication, **13**.

NOWLAN, G. S. and BARNES, C. R. 1981. Part 1: Late Ordovician conodonts from the Vaureal Formation, Anticosti Island, Quebec. *Geological Survey of Canada, Bulletin*, **329**, 1–49.

OVER, D. J. and CHATTERTON, B. D. E. 1987. Silurian conodonts from the southern Mackenzie Mountains, North-West Territories, Canada. *Geologica et Palaeontologica*, **21**, 1–49.

PAUL, C. R. C. 1982. The adequacy of the fossil record. 75–117. *In* JOYSEY, K. A. and FRIDAY, A. E. (eds). *Problems of phylogenetic reconstruction*. Academic Press, London, 442 pp.

—— 1998. Adequacy, completeness and the fossil record. 1–22. *In* DONOVAN, S. K. and PAUL, C. R. C. (eds). *The Adequacy of the fossil record*. John Wiley and Sons, Chichester, 312 pp.

—— 2003. Ghost ranges. *In* DONOGHUE, P. C. J. and SMITH, M. P. (eds). *Telling the evolutionary time: molecular clocks and the fossil record*. CRC Press, London, 296 pp.

PLATNICK, N. I., CODDINGTON, J. A., FORSTER, R. R. and GRISWOLD, C. E. 1991. Spinneret morphology and the phylogeny of haplogyne spiders (Araneae, Araneomorphae). *American Museum Novitates*, **3016**, 1–73.

—— HUMPHRIES, C. J., NELSON, G. and WILLIAMS, D. M. 1996. Is Farris Optimization perfect?: three-taxon statements and multiple branching. *Cladistics*, **12**, 243–252.

POLLOCK, C. A. 1969. Fused Silurian conodont clusters from Indiana. *Journal of Paleontology*, **43**, 929–935.

—— REXROAD, C. B. and NICOLL, R. S. 1970. Lower Silurian conodonts from northern Michigan and Ontario. *Journal of Paleontology*, **44**, 743–764.

PURNELL, M. A. 1993. The *Kladognathus* apparatus (Conodonta, Carboniferous): homologies with ozarkodinids and the prioniodinid Bauplan. *Journal of Paleontology*, **67**, 875–882.

—— and DONOGHUE, P. C. J. 1998. Skeletal architecture, homologies and taphonomy of ozarkodinid conodonts. *Palaeontology*, **41**, 57–102.

—— —— and ALDRIDGE, R. J. 2000. Orientation and anatomical notation in conodonts. *Journal of Paleontology*, **74**, 113–122.

QUI HONG-RONG 1984. Xicang Gushengdai he Sandieji di yaxingshi Dongwuqun. 109–131. *In* LI, G.-C. and MERCIER, J. L. (eds). *Sino-French cooperative investigation in the Himalayas*. Geological Publishing House, Beijing.

REICHSTEIN, M. 1962. Conodonten und Graptolithen aus einem Kalk-Mergel-Geschiebe des Unter Ludlow. *Geologie*, **11**, 538–547.

REMACK-PETITOT, M.-L. 1960. Contribution a l'étude des conodontes du Sahara (Bassins de Fort-Polignac, d'Adrar Reganne et du Jebel Bechar) comparaison avec les Pyrenees et la Montagne Noir. *Bulletin de la Société Geologique de France*, **99**, 471–479.

REXROAD, C. B. and CRAIG, W. W. 1971. Restudy of conodonts from the Bainbridge Formation (Silurian) at Lithium, Missouri. *Journal of Paleontology*, **45**, 684–703.

—— NOLAND, A. V. and POLLOCK, C. A. 1978. Conodonts from the Louisville Limestone and the Wabash Formation (Silurian) in Clark County, Indiana and Jefferson County, Kentucky. *Department of Natural Resources, Kentucky, Geological Survey, Special Report*, **16**, 1–15.

RHODES, F. H. T. and NEWALL, G. 1963. Occurence of *Kockelella variabilis* Walliser in the Aymestry Limestone of Shropshire. *Nature*, **199**, 166–167.

SANDERSON, M. J. and DONOGHUE, M. J. 1989. Patterns of variation in levels of homoplasy. *Evolution*, **43**, 1781–1795.

SANZ-LÓPEZ, J., GIL-PEÑA, I. and RODRÍGUEZ-CAÑERO, R. 2002. Conodont content and stratigraphy of the Llessui Formation from the south central Pyrenees. 391–401. *In* GARCÍA-LÓPEZ, S. and BASTIDA, F. (eds). *Palaeozoic conodonts from northern Spain*. Instituto Geológico y Minero de España, Serie Cuadernos del Museo Geominero, **1**.

SARMIENTO, G. N., PIÇARRA, J. M., REBELO, J. A., ROBARDET, M., GUTIÉRREZ-MARCO, J. C., STORCH, P. and RÁBANO, I. 1998. Le Silurien du Synclinorium de Moncorvo (NE du Portugal): biostratigraphie et importance paléogéographique. *Geobios*, **32**, 749–767.

SAVAGE, N. M. 1985. Silurian (Llandovery–Wenlock) conodonts from the base of the Heceta Limestone, southeastern Alaska. *Canadian Journal of Earth Sciences*, **22**, 711–727.

SCHOCH, R. M. 1986. *Phylogeny reconstruction in paleontology*. Van Nostrand Reinhold, New York, 353 pp.

SCHÖNLAUB, H. P. 1971. Zur problematik der Conodontenchronologie an der Wende Ordoviz. Silur mit besonderer Berucksichtigung der Verhaltnisse im Llandovery. *Geologica et Palaeontologica*, **5**, 35–57.

—— 1975. Conodonten aus dem Llandovery der Westkarawanken (Österreich). *Verhandlungen der Geologischen Bundesanstalt Österreich*, **2**, 45–65.

—— (ed.) 1980. Second European Conodont Symposium – ECOS II. *Abhandlungen Geologischen Bundesanstalt*, **35**, 1–213.

—— and ZEZULA, G. 1975. Silur-Conodonten aus einer Phyllonitzone im Muralpen-Kristallin Lungau-Salzburg. *Verhandlungen der Geologischen Bundesanstalt Österreich*, **4**, 235–269.

SERPAGLI, E. 1970. Uppermost Wenlockian–Upper Ludlovian (Silurian) conodonts from Western Sardinia. *Bollettino della Società Paleontologica Italiana*, **9**, 76–96.

—— and CORRADINI, C. 1998. New taxa of *Kockelella* (Conodonta) from Late Wenlock–Ludlow (Silurian) of Sardinia. *Geologia*, **60**, 79–83.

—— —— 1999. Taxonomy and evolution of *Kockelella* (Conodonta) from the Silurian of Sardinia (Italy). *Bollettino della Società Paleontologica Italiana*, **37**, 275–298.

—— —— and FERRETTI, A. 1998. Conodonts from a Ludlow–Pridoli section near the Silius village. *Geologia*, **60**, 104–111.

SIDALL, M. E. 1996. Stratigraphic consistency and the shape of things. *Systematic Biology*, **45**, 111–115.

—— 1997. Stratigraphic indices in the balance: a reply to Hitchin and Benton. *Systematic Biology*, **46**, 569–573.

SIMPSON, A. J. 1995. Silurian conodont biostratigraphy in Australia: a review and critique. *Courier Forschungsinstitut Senckenberg*, **182**, 325–345.

—— and TALENT, J. A. 1995. Silurian conodonts from the headwaters of the Indi (upper Murray) and Buchan rivers, southeastern Australia, and their implications. *Courier Forschungsinstitut Senckenberg*, **182**, 79–215.

SIMPSON, G. G. 1974. The compleat paleontologist? *Annual Review of Earth and Planetary Science*, **4**, 1–13.

SMITH, A. B. 1994. *Systematics and the fossil record: documenting evolutionary patterns*. Blackwell Scientific, Oxford, 223 pp.

—— 2001. Large-scale heterogeneity of the fossil record: implications for Phanerozoic biodiversity studies. *Philosophical Transactions of the Royal Society of London, Series B*, **356**, 351–367.

STONE, G. L. and FURNISH, W. M. 1959. Bighorn conodonts from Wyoming. *Journal of Paleontology*, **33**, 211–228.

STRAUSS, D. and SADLER, P. M. 1989. Classical confidence-intervals and Bayesian probability estimates for ends of local taxon ranges. *Mathematical Geology*, **21**, 411–421.

STRONG, E. E. and LIPSCOMB, D. 1999. Character coding and inapplicable data. *Cladistics*, **15**, 363–371.

SWEET, W. C. 1979. Late Ordovician conodonts and biostratigraphy of the Western Midcontinent Province. *Brigham Young University Geology Studies*, **26**, 45–86.

—— 1988. *The Conodonta: morphology, taxonomy, paleoecology, and evolutionary history of a long-extinct animal phylum*. Clarendon Press, Oxford, 212 pp.

—— and DONOGHUE, P. C. J. 2001. Conodonts: past, present and future. *Journal of Paleontology*, **75**, 1174–1184.

SWOFFORD, D. L. 2002. PAUP* Phylogenetic Analysis Using Parsimony. Smithsonian Institution, Washington, DC.

SZALAY, F. S. 1977. Ancestors, descendants, sister groups and testing of phylogenetic hypotheses. *Systematic Zoology*, **26**, 12–18.

TAVARÉ, S., MARSHALL, C. R., WILL, O., SOLIGO, C. and MARTIN, R. D. 2002. Using the fossil record to estimate the age of the last common ancestor of extant primates. *Nature*, **416**, 726–729.

TUCKER, R. D. and McKERROW, W. S. 1995. Early Paleozoic chronology – a review in the light of new U–Pb zircon ages from Newfoundland and Britain. *Canadian Journal of Earth Sciences*, **32**, 369–379.

UYENO, T. T. 1977. Summary of conodont biostratigraphy of the Read Bay Formation in its type sections and adjacent areas, eastern Cornwallis Island, District of Franklin. *Geological Survey of Canada, Paper*, **77–1**, 211–216.

—— 1980. Stratigraphy of conodonts of Upper Silurian and Lower Devonian rocks in the environs of the Boothia Uplift, Canadian Arctic Archipelago. Part 2: Systematic study of conodonts. *Geological Survey of Canada, Bulletin*, **292**, 39–75.

—— 1990. Biostratigraphy and conodont faunas of Upper Ordovician through Middle Devonian rocks, eastern Arctic Archipelago. *Geological Survey of Canada, Bulletin*, **401**, 1–211.

—— and BARNES, C. R. 1983. Conodonts of the Jupiter and Chicotte formations (Lower Silurian), Anticosti Island, Quebec. *Geological Survey of Canada, Bulletin*, **355**, 1–49.

VAN DEN BOOGAARD, M. 1965. Two conodont faunas from the Paleozoic of the Betic of Malaga near Vélez Rubio, S. E. Spain. *Proceedings of the Koninklijke Nederlandse Akademie Van Wetenschappen*, **68**, 33–37.

VIIRA, V. 1975. A new species of *Spathognathodus* from the Jaani Stage of the East Baltic. *Proceedings of the Estonian Academy of Sciences: Geology*, 233–236.

WAGNER, P. J. 1995. Stratigraphic tests of cladistic hypotheses. *Paleobiology*, **21**, 153–178.

—— 1998. Phylogenetic analysis and the quality of the fossil record. 165–187. *In* DONOVAN, S. K. and PAUL, C. R. C. (eds). *The adequacy of the fossil record*. John Wiley and Sons Ltd, Chichester, 312 pp.

WALLISER, O. H. 1957. Conodonten aus dem oberen Gotlandium Deutschlands und der Karnischen Alpen. *Notizblatt des Hessischen Landesamtes für Bodenforschung zu Wiesbaden*, **85**, 28–52.

—— 1962. Conodontenchronologie des Silurs (=Gotlandiums) und des tieferen Devons mit besonderer Berücksichtigung der Formations-grenze. 281–287. *In* ERBEN, H. K. (ed.). *Symposiums – Band der 2 Internat. Arbeitstagung uber die Silur/Devon-Grenze und die stratigraphie von Silur und Devon, Bonn Bruxelles. E.* Schweizerbart'sche Verlagsbuchhandlung, Stuttgart.

—— 1964. Conodonten des Silurs. *Abhandlungen des Hessischen Landesamtes für Bodenforschung, Wiesbaden*, **41**, 1–106.

—— 1972. Conodont apparatuses in the Silurian. *Geologica et Palaeontologica*, **SB1**, 75–79.

WALMSLEY, V. G., ALDRIDGE, R. J. and AUSTIN, R. L. 1974. Brachiopod and conodont faunas from the Silurian and Devonian of Bohemia. *Geologica et Palaeontologica*, **8**, 39–47.

WANG CHEN-YUAN and ZIEGLER, W. 1983. Conodonten aus Tibet. *Neues Jahrbuch für Geologie und Paläontologie, Monatschefte*, **1983**, 69–79.

WEISS, R. E. and MARSHALL, C. R. 1999. The uncertainty in the true endpoint of a fossil's stratigraphic range when stratigraphic sections are sampled discretely. *Mathematical Geology*, **31**, 435–543.

WILEY, E. O. 1981. *Phylogenetics: the theory and practice of phylogenetic systematics*. John Wiley, New York, 439 pp.

WILLS, M. A. 1999. Congruence between phylogeny and stratigraphy: randomization tests and the gap excess ratio. *Systematic Biology*, **48**, 559–580.

—— 2001. How good is the fossil record of arthropods? An assessment using the stratigraphic congruence of cladograms. *Geological Journal*, **36**, 187–201.

ZHANG SHUNXIN and BARNES, C. R. 2002. A new Llandovery (Early Silurian) conodont biozonation and conodonts from the Becscie, Merrimack, and Gun River formations, Anticosti Island, Quebec. *Journal of Paleontology, Memoir*, **57**, 1–46.

APPENDIX

1. Files used in the GHOST analyses

Optimal trees derived from stratigraphically unconstrained analysis of the cladistic dataset:

16

1. ((((B(((H((((KL)M)N)((GO)I)))F)E)(DC)))A)Q)P)
2. ((((B(((H((((KL)M)N)((GO)I)))(FE))D)C))A)Q)P)
3. ((((B(((H((((KL)M)N)((GO)I)))(FE))(DC)))A)Q)P)
4. ((((B((((H((((KL)M)N)((GO)I)))E)F)C)D))A)Q)P)
5. ((((B((((H((((KL)M)N)((GO)I)))F)E)C)D))A)Q)P)
6. ((((B(((H((((KL)M)N)((GO)I)))E)F)(DC)))A)Q)P)
7. ((((B((((H((((KL)M)N)((GO)I)))F)E)D)C))A)Q)P)
8. ((((B((((H((((KL)M)N)((GO)I)))E)F)D)C))A)Q)P)
9. ((((B(((H((((KL)M)N)((GO)I)))(FE))C)D))A)Q)P)

Optimal trees derived from analysis of the cladistic dataset while implementing a constraint tree compatible with the phylogeny of Serpagli and Corradini (1999):

17

1. (P(A((K(GO))(B(C((D(FE))((Q(JN))(H(I(ML)))))))))
2. (P(A((K(GO))(B(C(D((FE)((Q(JN))(H(I(ML))))))))))
3. (P(A(B((K(GO))(C((D(FE))((Q(JN))(H(I(ML)))))))))
4. (P(A((B(K(GO)))(C((D(FE))((Q(JN))(H(I(ML))))))))
5. (P(A((K(GO))(B(C((FE)(D((Q(JN))(H(I(ML))))))))))
6. (P(A(B((K(GO))(C(D((FE)((Q(JN))(H(I(ML))))))))))
7. (P(A((B(K(GO)))(C(D((FE)((Q(JN))(H(I(ML)))))))))
8. (P(A(B((K(GO))(C((FE)(D((Q(JN))(H(I(ML))))))))))
9. (P(A((B(K(GO)))(C((FE)(D((Q(JN))(H(I(ML)))))))))

Stratigraphic ranges formatted in accordance with the relevant Ghost 2.4 stratigraphy files:

Taxon code	Taxon	Start (fine)	End (fine)	Start (coarse)	End (coarse)
A	*Kockelella manitoulinensis*	bbb	fff	aba	dad
B	*Kockelella abrupta*	dcc	eae	cbc	ccb
C	*Kockelella ranuliformis*	eed	gag	dad	eee
D	*Kockelella latidentata*	fef	gag	dcc	eee
E	*Kockelella walliseri*	ffe	gga	dcc	eae
F	*Kockelella corpulenta*	ffe	ggg	dcc	eee
G	*Kockelella patula*	ffe	ggg	dcc	eee
H	*Kockelella amsdeni*	ggg	gag	eee	eee
I	*Kockelella stauros*	gag	gga	eee	eae
J	*Kockelella ortus sardoa*	gga	hhh	–	–
K	*Kockelella crassa*	gga	ggb	eae	eea
L	*Kockelella variabilis variabilis*	gga	hhh	eae	eec
M	*Kockelella variabilis ichnusae*	ggd	hhh	eeb	eec
N	*Kockelella maenniki*	ggf	gfg	ece	eec
O	*Acoradella ploeckensis*	ggd	hah	eeb	eec
P	*Kockelella ortus ortus*	gga	gcg	eae	eea
Q	*Kockelella ortus absidata*	ggc	hhh	ebe	eec

Stratigraphy files

Fine		Coarse	
61		29	
aaa	61	aaa	29
aba	60	aba	28
aab	59	aab	27
bbb	58	bbb	26
bab	57	bab	25
bba	56	bba	24
ccc	55	ccc	23
cac	54	cac	22
cca	53	cca	21
cbc	52	cbc	20
ccb	51	ccb	19
cab	50	cab	18
cba	49	cba	17
ddd	48	ddd	16
dad	47	dad	15
dda	46	dda	14
dbd	45	dbd	13
dbb	44	dbb	12
dcd	43	dcd	11
dcc	42	dcc	10
eee	41	eee	9
eae	40	eae	8
eea	39	eea	7
ebe	38	ebe	6
eeb	37	eeb	5
ece	36	ece	4
eec	35	eec	3
ede	34	ede	2
eed	33	eed	1
fff	32		

Stratigraphy files (continued)

Fine	
faf	31
ffa	30
fbf	29
ffb	28
fcf	27
ffc	26
fdf	25
ffd	24
fef	23
ffe	22
ggg	21
gag	20
gga	19
gbg	18
ggb	17
gcg	16
ggc	15
gdg	14
ggd	13
geg	12
gge	11
gfg	10
ggf	9
hhh	8
hah	7
hha	6
hbh	5
hhb	4
hch	3
hhc	2
hdh	1

2. *Dataset used in the compilation of confidence intervals, discovery curves and plot of research effort*

Many of the publications are not scored as recording new occurrences of kockelellids because the taxonomic assignments to species level and below were not verifiable (no figures). These publications were nevertheless included in the research effort plot.

Author(s)	Year	Number of fossiliferous horizons per taxon															
		Anc. ploeckensis	*Ko. abrupta*	*Ko. amsdeni*	*Ko. corpulenta*	*Ko. crassa*	*Ko. latidentata*	*Ko. maenniki*	*Ko. manitoulinensisss*	*Ko. ortus absidata*	*Ko. ortus ortus*	*Ko. patula*	*Ko. ranuliformis*	*Ko. stauros*	*Ko. variabilis ichnusae*	*Ko. variabilis variabilis*	*Ko. walliseri*
Walliser	1957															5	
Kockel	1958																
Remack-Petitot	1960																
Walliser	1962																
Reichstein	1962																
Ethington and Furnish	1962																
Rhodes and Newall	1963																
Walliser	1964	7				1					1	3	3			35	
van den Boogaard	1965																
Igo and Koike	1967																
Flajs	1967																
Nicoll and Rexroad	1968												6				
Igo and Koike	1968																
Drygant	1969											1	1				
Fåhreus	1969	1															
Serpagli	1970					2										4	
Pollock *et al.*	1970								1								
Drygant	1971																
Schönlaub	1971																
Rexroad and Craig	1971															1	
Link and Druce	1972	3				2					3					3	
Miller	1972																
Aldridge	1972		4										7				
Walliser	1972																
Walmsley *et al.*	1974												1				
Klapper and Murphy	1974	5				1		1									
Schönlaub	1975												5				
Schönlaub and Zezula	1975																
Helfrich	1975																8
Viira	1975				1												
Barrick and Klapper	1976			7		1				13		2	30	8		2	4
Flajs and Schönlaub	1976							1									
Mashkova	1977												1				

(continued)

Author(s)	Year	Number of fossiliferous horizons per taxon														
		Anc. ploeck-ensis	*Ko. abrupta*	*Ko. amsdeni*	*Ko. corpu-lenta*	*Ko. crassa*	*Ko. lati-dentata*	*Ko. maenniki*	*Ko. manitou-linensisss*	*Ko. ortus absidata*	*Ko. ortus ortus*	*Ko. patula*	*Ko. ranuli-formis*	*Ko. stauros*	*Ko. variabilis ichnusae*	*Ko. variabilis variabilis*	*Ko. walliseri*
Uyeno	1977																
Liebe and Rexroad	1977								1				1				
Rexroad *et al.*	1978									2						5	
Lane and Ormiston	1979															1	
Helfrich	1980																1
Aldridge	1980															1	
Schönlaub	1980									1							
Uyeno	1980	2															
McCracken and Barnes	1981								11								
Ni *et al.*	1982																
Aldridge and Mohamed	1982																
Lin and Qiu	1983	1											8	1			
Wang and Ziegler	1983									1						1	
Männik	1983																
Barrick	1983			1						3							6
Uyeno and Barnes	1983												1				
Mabillard and Aldridge	1983												10				
Jeppsson	1983															1	1
Qui	1984																
Drygant	1984																
Kleffner	1985																
Norford and Orchard	1985	1															
Aldridge	1985																
Savage	1985												6				
Kriz *et al.*	1986	1															
Bischoff	1986			8			24						28				16
Kleffner	1987																
Over and Chatterton	1987										3		7				1
Armstrong	1990								32								
Uyeno	1990	1															1
Kleffner	1990												3				6
McCracken	1991	1															
Kleffner	1991																
Barca *et al.*	1992	1															
Jeppsson *et al.*	1994	1															

(continued)

Author(s)	Year	*Anc. ploeckensis*	*Ko. abrupta*	*Ko. amsdeni*	*Ko. corpulenta*	*Ko. crassa*	*Ko. latidentata*	*Ko. maenniki*	*Ko. manitoulinensisss*	*Ko. ortus absidata*	*Ko. ortus ortus*	*Ko. patula*	*Ko. ranuliformis*	*Ko. stauros*	*Ko. variabilis ichnusae*	*Ko. variabilis variabilis*	*Ko. walliseri*
		Number of fossiliferous horizons per taxon															
Kleffner	1994			1									10	1			6
Jeppsson *et al.*	1995																
Simpson and Talent	1995	4															
Simpson	1995												6			17	
Jeppsson	1997									1	1	1	1				
Sarmiento *et al.*	1998																
Männik and Malkowski	1998										1				1	1	1
Serpagli and Corradini	1998					1				24			1	5	25	26	
Loydell *et al.*	1998																
Norford *et al.*	1998								1								
Serpagli *et al.*	1998																
Corradini *et al.* (a)	1998																
Corradini *et al.* (b)	1998																
Ferretti *et al.*	1998																
Benfrika	1999													1			
Serpagli and Corradini	1999																
Cockle	1999						1	16					6				
Jeppsson and Aldridge	2000																
Johnson *et al.*	2001	1															
Zhang and Barnes	2002								4								
Sanz-Lopez *et al.*	2002																
Männik	2002																
Loydell *et al.*	2002																
Jeppsson and Calner	2003																
Total number of horizons		30	4	17	1	9	25	18	50	45	9	7	142	16	26	103	51
Confidence interval (%)		10·8	171	20·6	–	45·4	13·2	19·2	6·3	7·0	45·4	64·8	2·1	22·1	12·7	3·0	6·2